Beurling Generalized Numbers

Mathematical
Surveys
and
Monographs

Volume 213

Beurling Generalized Numbers

Harold G. Diamond
Wen-Bin Zhang (Cheung Man Ping)

American Mathematical Society
Providence, Rhode Island

2010 *Mathematics Subject Classification*. Primary 11N80.

For additional information and updates on this book, visit
www.ams.org/bookpages/surv-213

Library of Congress Cataloging-in-Publication Data

Names: Diamond, Harold G., 1940–. Zhang, Wen-Bin (Cheung, Man Ping), 1940– .
Title: Beurling generalized numbers / Harold G. Diamond, Wen-Bin Zhang (Cheung Man Ping).
Description: Providence, Rhode Island : American Mathematical Society, [2016] | Series: Mathematical surveys and monographs ; volume 213 | Includes bibliographical references and index.
Identifiers: LCCN 2016022110 | ISBN 9781470430450 (alk. paper)
Subjects: LCSH: Numbers, Prime. | Numbers, Real. | Riemann hypothesis. | AMS: Number theory – Multiplicative number theory – Generalized primes and integers. msc
Classification: LCC QA246 .D5292 2016 | DDC 512/.2–dc23 LC record available at https://lccn.loc.gov/2016022110

∞ The paper used in this book is acid-free and falls within the guidelines established to ensure permanence and durability.
Visit the AMS home page at `http://www.ams.org/`

10 9 8 7 6 5 4 3 2 1 21 20 19 18 17 16

Dedicated to our wives, Nancy Diamond and Kun-Ming Luo Zhang,
and to the memory of our friend and colleague, Paul T. Bateman

Contents

Preface

Generalized numbers are a multiplicative structure introduced by A. Beurling [**Be37**] in 1937 to investigate the degree to which prime number theory is independent of the additive properties of the natural numbers.

Beyond their own interest, the results and techniques of this theory apply to several other systems having the character of prime numbers and integers. Indeed, such ideas occurred already in a 1903 paper of E. Landau [**La03**] proving the prime number theorem for ideals of algebraic number fields. We shall introduce and use continuous (!) analogues of generalized (briefly: g-) numbers. As another application, these distributions provide an attractive path to the theories of Dickman and Buchstab for integers whose prime factors lie only in restricted ranges.

A central question that we shall examine is the following: if a sequence of g-integers is generated by a sequence of g-primes, and if one of the collections is "reasonably near" its classical counterpart, does the other collection also have this property? This monograph does not examine all facets of g-number theory; some interesting topics that are largely ignored include probabilistic theory, oscillatory counting functions, and collections of primes and integers that are unusually dense or sparse. We hope that the accompanying list of references will help interested readers to explore these topics further.

Our intended audience is readers having some familiarity with mathematical analysis and analytic number theory, particularly an analytic proof of the prime number theorem. Background material that we assume can be found in such books as those of Apostol [**Ap76**], Bateman-Diamond [**BD04**], Chandrasekharan [**Ch68**], [**Ch70**], Davenport [**Da00**], Ingham [**In32**], Montgomery-Vaughan [**MV07**], or Tenenbaum [**Te95**]. Specialized results will be developed as needed.

Many examples are provided to illustrate how various hypotheses affect the behavior of g-number systems. They are important! But readers put off by details are encouraged to at least note the point of each example.

This work contains published and new work of the authors. Also, we have benefited from the contributions of many others, and it is our pleasant duty to thank them here. These include Beurling, who originated the study and established its first important result; P. J. Cohen, who introduced the first author to this subject; P. T. Bateman and H. L. Montgomery, with whom we have worked in this area; and J.-P. Kahane, who established one of our main results. Also, we thank A. J. Hildebrand for his mathematical and TEXnical advice.

The authors request that readers advise us of errors or obscurities they find. Our email address is `beurlingbook@illinois.edu`. We maintain corrections and comments at `www.math.illinois.edu/~hdiamond/hgdwbz/corrigenda.pdf`.

Urbana and Chicago IL
May, 2016

CHAPTER 1

Overview

What can you say if you drop all *the hypotheses?*

SUMMARY. Some questions we shall consider, main functions, and examples of generalized number systems.

1.1. Some questions about primes

The positive integers can be generated from the number 1 by the operation of addition, or alternatively, from the prime numbers by multiplication. Generalized (henceforth g-) numbers were introduced by Arne Beurling to examine prime number theory in a purely multiplicative setting. This theory also provides a framework for studying several related topics of interest. As a brief preview of some areas we shall visit, we pose a few questions:

(1) How sensitive is the prime number theorem (PNT) to the additive structure of the integers?
(2) How important is this additive structure in the quest to establish the Riemann hypothesis?
(3) What is the exponential of a measure? Is the counting measure of integers representable as the exponential of something interesting?
(4) Is there a continuous analogue of prime number theory?
(5) Why has it been so hard to improve upon the 1899 PNT error estimate of Ch. J. de la Vallée Poussin?

For a glimpse at the scenery to be encountered enroute, here are some brief, incomplete, and perhaps tantalizing answers.

(1) Suppose we use the rational primes exceeding 2 to generate g-integers. This set, the odd numbers, does not admit addition. However, it has just one fewer prime than in the classical case, and so the primes here obey the PNT. Thus we may suspect—and shall show—that the PNT can be established without use of the additive structure of integers. What *is* needed is that the counting function of the g-integers be "suitably close" to that of the rational integers.

(2) There exist g-number systems whose integer counting function $N(x)$ is, like that of the rational integers, very close to x but have neither an additive structure nor a "zeta functional equation," and for which the analogue of the Riemann hypothesis does not hold. Thus, a proof of the Riemann hypothesis will require more than just the multiplicative structure of integers and smallness of $N(x) - x$.

(3) With multiplicative convolution as the product operation on measures, the exponential of a measure can be defined by its power series. In these terms,

the counting measure dN of integers can be represented as the exponential of the weighted prime and prime power counting measure $d\Pi$.

(4) If we replace the counting functions of primes and integers by suitable continuous functions and recognize Chebyshev's elementary prime number identity as the key relation connecting primes and integers, then we can develop an elegant and useful parallel version of prime number theory.

(5) The only improvements ever made in the PNT error estimate of de la Vallée Poussin have been achieved by exploiting the additive as well as the multiplicative structure of integers. We shall show by a construction that for g-numbers (where additive structure is missing) the dlVP error term is in fact optimal.

1.2. The cast

As a replacement for the natural primes, we introduce a sequence of real numbers $\mathcal{P} = \{p_i\}$ satisfying

$$1 < p_1 \leq p_2 \leq p_3 \leq \dots, \quad p_i \to \infty.$$

We call $\mathcal{P}$ a generalized prime number system or, briefly, a g-prime system.

It is not important that the elements of $\mathcal{P}$ be distinct – e.g. prime ideals may have equal norms, and we shall meet further examples in which g-primes and g-integers occur with multiplicity. However, the smallest g-prime must exceed 1, for we do not admit 1 as a prime and any g-prime in the interval (0, 1) would generate a system of g-integers having an infinite number of elements in (0, 1). Also, the sequence of primes should be unbounded. Failure of either of the last two conditions would yield a system that little resembles the natural numbers.

Using notation that parallels that of classical prime number theory, we introduce the g-prime counting functions

$$\begin{aligned}
\pi_{\mathcal{P}}(x) &:= \#\{i \geq 1\colon\, p_i \leq x\},\\
\Pi_{\mathcal{P}}(x) &:= \pi_{\mathcal{P}}(x) + \frac{1}{2}\pi_{\mathcal{P}}(x^{1/2}) + \frac{1}{3}\pi_{\mathcal{P}}(x^{1/3}) + \dots,\\
\psi_{\mathcal{P}}(x) &:= \sum_{{p_i}^{\alpha_i} \leq x} \log p_i = \int_1^x \log u \, d\Pi_{\mathcal{P}}(u).
\end{aligned}$$

When the context is clear, we shall generally suppress the $\mathcal{P}$ subscripts.

The function Π is given above by a formally infinite series. Here we show that it is well-defined and, importantly, close to the prime counting function.

LEMMA 1.1. *For each $x > 1$, the formally infinite series for Π is terminating, and if we assume that $\pi(x) \ll x$, then $\Pi(x) - \pi(x) \ll x^{1/2}$.*

PROOF. We have $x^{1/n} < p_1$ for $n > \log x/\log p_1$. For such n, $\pi(x^{1/n}) = 0$, so the series for $\Pi(x)$ is in fact a finite sum. The claimed inequality holds because

$$0 \leq \Pi(x) - \pi(x) \leq \pi(x^{1/2}) + \frac{\log x}{\log p_1}\,\pi(x^{1/3}) \ll x^{1/2}. \qquad \square$$

Associated with $\mathcal{P}$ we have the semigroup of g-integers it generates,

$$\mathcal{N} = \mathcal{N}_{\mathcal{P}}\colon\; 1 = n_1 < n_2 \leq n_3 \leq n_4 \leq \dots.$$

In view of possible nonunique representation of integers, we give a formal definition of $\mathcal{N}$ in terms of exponents in the product representation. Let $\mathcal{E}$ denote the family of sequences of nonnegative integers $\boldsymbol{\nu} = \{\nu_1, \nu_2, \dots\}$ with $\nu_i = 0$ from some point onward. The family $\mathcal{E}$ is countable, and to each element $\boldsymbol{\nu}_i = \{\nu_{i1}, \nu_{i2}, \dots, 0, 0, \dots\}$ associate a g-integer by the rule

$$n_i \longleftrightarrow \prod_{j=1}^{\infty} p_j^{\nu_{ij}}.$$

Given a positive number x, at most a finite number of g-integers n_i satisfy

$$n_i = \prod_{j=1}^{\infty} p_j^{\nu_{ij}} \leq x. \tag{1.1}$$

We define a counting function $N(\cdot) = N_{\mathcal{P}}(\cdot)$ on $[1, \infty)$ by setting $N(x)$ to be the number of g-integers for which (1.1) holds.

The central theme of this work is the relation between $\mathcal{P}$ and $\mathcal{N}$, generally expressed as a hypothesis upon one of $N_{\mathcal{P}}$ or $\Pi_{\mathcal{P}}$, and a conclusion expressed in terms of the other function.

For the rest of this chapter, we suppose that

$$\limsup_{x\to\infty} N(x)/x < \infty, \tag{1.2}$$

which we describe as an *O-density condition.* This bound is a reasonable condition for g-integer systems; moreover, as we shall show in Chapter 4, it is not a real restriction for our work. In case $\lim_{x\to\infty} N(x)/x$ exists and equals δ, we say that *$\mathcal{N}$ has a density* and δ is its value.

Continuing with our list of functions, we have the *g-zeta function*, the analogue of the Riemann zeta function, defined on a half plane in $\mathbb{C}$ by

$$\zeta(s) = \zeta_{\mathcal{N}}(s) := \sum_i n_i^{-s} = \int_{1-}^{\infty} u^{-s} dN(u). \tag{1.3}$$

The sum is called a generalized Dirichlet series and the integral is called the *Mellin transform* of dN, with each convergent for $\Re s > 1$. (See §2.3 for the general definition of Mellin transform. Also, the "1−" in the last integral denotes inclusion of $dN(\{1\})$; integrals having lower limit 1 omit this contribution.)

The g-zeta function has an Euler product representation

$$\prod_i \left(1 - p_i^{-s}\right)^{-1}, \quad \Re s > 1, \tag{1.4}$$

which is established in the same way as the classical formula. Also, in this region,

$$\log \zeta(s) = -\sum_i \log\left(1 - p_i^{-s}\right) = \sum_i \sum_{\nu \geq 1} p_i^{-\nu s}/\nu = \int_1^{\infty} u^{-s}\, d\Pi(u) \tag{1.5}$$

and

$$-\frac{\zeta'(s)}{\zeta(s)} = -\frac{d}{ds}\int_1^{\infty} u^{-s}\, d\Pi(u) = \int_1^{\infty} u^{-s} \log u\, d\Pi(u) = \int_1^{\infty} u^{-s} d\psi(u)\,. \tag{1.6}$$

1.3. Examples

EXAMPLE 1.2. ODD NUMBERS (again). Let

$$\mathcal{P}_1 = \ 3,\ 5,\ 7,\ 11,\ \ldots,$$

so that $\mathcal{N}_1$ is the odd numbers. By inspection, $N_1(x) = \lfloor (x+1)/2 \rfloor$, whence $\mathcal{N}_1$ has density $1/2$. A moral to be drawn from this example is that the density of integers is sensitive to small changes in g-primes (whereas, as we shall see, the truth of the PNT is not). The zeta function of $\mathcal{N}_1$ can be expressed in terms of the Riemann zeta function as

$$\zeta_1(s) = \sum_{n\geq 1,\ n\,\mathrm{odd}} n^{-s} = \prod_{p\geq 3} (1-p^{-s})^{-1} = (1-2^{-s})\,\zeta(s), \quad \Re s > 1,$$

with the product extending over all natural primes from 3 onward. We found the last expression by multiplying and dividing the product by $(1-2^{-s})$. Note that the residue of $\zeta_1(s)$ at $s=1$ is $1/2$, which is the density of $\mathcal{N}_1$. *If a g-number system has a density*, then it equals the residue of the associated zeta function at $s=1$. We shall return to this point and qualify use of the word "residue" in Chapter 7.

EXAMPLE 1.3. A REPEATED PRIME. Let

$$\mathcal{P}_2 = 3,\ 3,\ 4,\ 5,\ 7,\ 11,\ 13,\ \ldots .$$

Here the (natural) prime 2 is suppressed, 3 is introduced as a g-prime (in addition to its usual occurrence), the number 4 also is taken as a g-prime, and all the natural primes from 5 onward are g-primes. The sequence $\mathcal{N}_2$ of g-integers begins

$$1,\ 3,\ 3,\ 4,\ 5,\ 7,\ 9,\ 9,\ 9,\ 11,\ 12,\ 12,\ 13,\ 15,\ 15,\ 16,\ 17,\ 19,\ \ldots .$$

To find the counting function N_2 we determine the associated zeta function, first as an Euler product

$$\zeta_2(s) = \left(1-3^{-s}\right)^{-2}\left(1-4^{-s}\right)^{-1}\prod_{p\geq 5}\left(1-p^{-s}\right)^{-1}, \quad \Re s > 1.$$

A small manipulation gives $\zeta_2(s)$ in terms of the Riemann zeta function:

$$\zeta_2(s) = \left(1+2^{-s}\right)^{-1}\left(1-3^{-s}\right)^{-1}\zeta(s)\,. \tag{1.7}$$

Expanding each of the factors in (1.7), we find that

$$\begin{aligned}\zeta_2(s) &= (1-2^{-s}+2^{-2s}\mp\ldots)(1+3^{-s}+3^{-2s}+\ldots)\\ &\qquad\qquad\times(1+2^{-s}+3^{-s}+4^{-s}+\ldots)\\ &= \sum_{j\geq 0}\sum_{k\geq 0}\sum_{n\geq 1}(-1)^j(2^j3^kn)^{-s}.\end{aligned}$$

(The rearrangement is valid, because the series is absolutely convergent for $\Re s > 1$.) Since $\zeta_2(s)$ also has the Mellin transform representation

$$\zeta_2(s) = \int_{1-}^{\infty} u^{-s}\,dN_2(u),$$

we have

$$N_2(x) = \sum_{\substack{j,k\geq 0,\,n\geq 1\\ 2^j3^kn\leq x}} (-1)^j = \sum_{j,k\geq 0}(-1)^j\left\lfloor\frac{x}{2^j3^k}\right\rfloor.$$

The last sum has $O(\log^2 x)$ terms, and approximating $\lfloor t \rfloor$ by t, we find that $\mathcal{N}_2$ has a density and its value (cf. §7.1) is

$$\delta_2 = \sum_{j,k \geq 0} \frac{(-1)^j}{2^j 3^k} = \frac{1}{1+1/2} \frac{1}{1-1/3} = 1.$$

This example provides a g-number system that consists entirely of natural numbers, has density 1, and yet is different from the natural numbers.

EXAMPLE 1.4. CONTINUOUS FUNCTIONS. Let

$$N_c(x) := \int_{1-}^{x} (\delta_1 + du) = \begin{cases} x, & x \geq 1, \\ 0, & x < 1, \end{cases}$$

where δ_1 is Dirac point measure at 1 and du is Lebesgue measure on $[1, \infty)$. This function is continuous on $(1, \infty)$ and approximates $N(x) = \lfloor x \rfloor$, the counting function of the natural numbers. The "zeta function" associated with N_c is

$$\zeta_c(s) = \int_{1-}^{\infty} u^{-s}\,(\delta_1 + du) = 1 + \frac{1}{s-1} = \frac{s}{s-1}, \qquad \Re s > 1.$$

This function is close to the Riemann zeta function on the half line $\{s : s > 1\}$. (However, the two functions are not close on the complex half-plane $\{s : \Re s > 1\}$; e.g. the Riemann zeta function is unbounded as $s \to 1 + i\infty$ while $\zeta_c(s) \to 1$.)

To determine a continuous function that corresponds to the classical weighted prime counting function $\Pi(\cdot)$, we express $\log \zeta_c(s)$ as a Mellin transform:

THEOREM 1.5. *For* $\Re s > 1$,

$$\int_1^{\infty} u^{-s} \frac{1-u^{-1}}{\log u} du = \log \frac{s}{s-1}. \tag{1.8}$$

PROOF. Let $F(s)$ denote the integral on the left side of the last formula. Then

$$F'(s) = -\int_1^{\infty} u^{-s}\,(1-u^{-1})\,du = -\frac{1}{s-1} + \frac{1}{s} = \left\{\log \frac{s}{s-1}\right\}'.$$

Thus F agrees with $\log s/(s-1)$ to within a constant. The constant is 0 since

$$\lim_{s \to +\infty} F(s) = 0 = \lim_{s \to +\infty} \log \frac{s}{s-1}. \qquad \square$$

Recalling the relation

$$\int_1^{\infty} u^{-s}\, d\Pi(u) = \log \zeta(s)$$

connecting the weighted prime counting function with the logarithm of the Riemann zeta function, the choice

$$\Pi_c(x) := \int_1^{x} \frac{1-u^{-1}}{\log u} du \tag{1.9}$$

yields an analogous "continuous prime counting function" associated with N_c. The function $\Pi_c(x)$ is close to the logarithmic integral $\mathrm{li}(x)$, which in turn approximates the counting function of the rational primes. This example suggests the plausibility of the PNT, but of course it is not a proof of that result. Continuous analogues of prime number theory will occur repeatedly in this work, and they will provide an essential ingredient for the constructions of Chapter 17.

1.4. π, Π, and an extended notion of g-numbers

Example 1.4 is continuous, which perhaps seems unusual, but it has many of the distribution properties of discrete g-primes and g-integers. In the sequel, we shall make use of that example and some others that are not discrete. Using the notion of $\exp^*$, to be defined in Chapter 3, we expand our definition of g-numbers to encompass these cases:

A g-number system is a pair of right continuous increasing functions Π *and* N *on* $[1, \infty)$ *satisfying* $dN = \exp^* d\Pi$ *and* $\Pi(1) = 0$.

This usage differs from that of T. Hilberdink [**Hi12**], who calls such a system *outer*; he applies the name *g-prime system* to one for which there exists a function π that is connected to the weighted counting function Π by

$$\Pi(x) = \pi(x) + \frac{1}{2}\pi(x^{1/2}) + \frac{1}{3}\pi(x^{1/3}) + \dots \tag{1.10}$$

with the property that π is *increasing.* (Here, of course, Π also is increasing.)

Suppose we are given an increasing function Π. We can determine a function π that formally satisfies (1.10) by Moebius inversion:

$$\pi(x) := \sum_{k=1}^{\infty} \frac{\mu(k)}{k}\, \Pi(x^{1/k}). \tag{1.11}$$

For given x, this series is terminating (and hence convergent) if $\Pi(u) = 0$ for all $u < a$ for some $a > 1$, for in this case $x^{1/k} < a$ for all sufficiently large k. If, on the other hand, $\Pi(x) > 0$ for all $x > 1$, we can easily show that the preceding series still converges if Π satisfies a Lipschitz condition at $x = 1$.

LEMMA 1.6. *Suppose* $\Pi(x)$ *is increasing on* $[1, \infty)$ *and* $\Pi(x) \ll x-1$ *for* x *near* $1+$. *Then the series* (1.11) *is absolutely convergent and is uniformly convergent on any given interval* $(1, X)$. *Moreover, the function defined by* (1.11) *satisfies* (1.10).

PROOF. Suppose $x > 1$ is given. For $k > \log x$ we have

$$\Pi(x^{1/k}) \ll x^{1/k} - 1 = e^{(\log x)/k} - 1 \ll \frac{\log x}{k}\,.$$

Thus the series (1.11) is dominated by

$$\Pi(x) \sum_{k \le \log x} \frac{1}{k} + \sum_{k > \log x} \frac{\log x}{k^2}\,,$$

so the series converges absolutely. For $x \le X$, the series is dominated by the number $\Pi(X)\log X + O(1)$, so it converges uniformly there by the Weierstrass M-test.

If we insert $\pi(x)$ as defined by (1.11) into the right side of (1.10) we get an absolutely summable series: indeed

$$\sum_{n=1}^{\infty} \frac{1}{n} \sum_{k=1}^{\infty} \Bigl|\frac{\mu(k)}{k}\, \Pi(x^{1/nk})\Bigr| \le \sum_{\ell=1}^{\infty} \frac{1}{\ell}\Pi(x^{1/\ell})\, d(\ell),$$

where $d(\ell) = \sum_{kn=\ell} 1$, the "divisor function." If we repeat the argument of the first part and then apply summation by parts using the famous estimate $\sum_{\ell \le t} d(\ell) \sim t\log t$, we find that the double sum is dominated by

$$\Pi(x) \sum_{\ell \le \log x} \frac{d(\ell)}{\ell} + \sum_{\ell > \log x} \frac{d(\ell)\log x}{\ell^2} \ll \Pi(x)(\log\log x)^2.$$

It follows that Moebius inversion is valid, and the function π satisfies (1.10). □

For $\Pi(x)$ to be increasing, it is not necessary that $\pi(x)$ given by (1.11) also be increasing. Indeed, a simple example is provided by setting $\Pi(x) = 0$ for $x < 2$ and $\Pi(x) = 1$ for $x \geq 2$. For $x < 8$, our formula gives

$$\pi(x) = \Pi(x) - (1/2)\,\Pi(x^{1/2}) \mp \cdots = \begin{cases} 0, & x < 2, \\ 1, & 2 \leq x < 4, \\ \frac{1}{2}, & 4 \leq x < 8. \end{cases}$$

In general, we shall require only of our g-number systems that $\Pi(x)$ be monotone, but it is interesting to show that the important Example 1.4 satisfies the extended condition that $\pi(x)$ is monotone.

Proposition 1.7. *Let $\Pi = \Pi_c$ given by* (1.9) *be the (weighted) continuous "prime-counting" function of Example* 1.4 *and let $\pi = \pi_c$ be the associated function given by* (1.11)*. Then π is well-defined and is increasing on* $(1, \infty)$.

Proof. The integrand of Π, $(1 - t^{-1})/\log t$, tends to 1 as $t \to 1+$. Thus $\Pi(x) \ll x - 1$ for x near $1+$, and hence the series defining $\pi(x)$ converges by Lemma 1.6.

Formally differentiating the series for $\pi(x)$ yields

$$\pi'(x) = \sum_{k=1}^{\infty} \frac{\mu(k)}{k} \frac{x^{1/k} - 1}{x \log x}\,.$$

The differentiated series converges absolutely and also uniformly on any given interval $(1, X)$ by the argument used in proving the last lemma. Thus the termwise differentiation is valid ([**Ap74**], Theorem 9.14; [**BS00**], Theorem 9.4.4).

It remains to show that $\pi(x)$ is increasing. Writing $x^{1/k}$ as $\exp(\{\log x\}/k)$ and expanding exp as a Maclaurin series, we find

$$\pi'(x)\, x \log x = \sum_{k=1}^{\infty} \frac{\mu(k)}{k}\,(x^{1/k} - 1) = \sum_{k=1}^{\infty} \frac{\mu(k)}{k} \sum_{l=1}^{\infty} \frac{(\log x)^l}{l!\, k^l}\,.$$

We can exchange the order of the last two sums by the absolute convergence of the series. This yields

$$\begin{aligned} \pi'(x) \log x &= \frac{1}{x} \sum_{l=1}^{\infty} \frac{(\log x)^l}{l!} \sum_{k=1}^{\infty} \frac{\mu(k)}{k^{l+1}} \\ &= \frac{1}{x} \sum_{l=1}^{\infty} \frac{(\log x)^l}{l!} \frac{1}{\zeta(l+1)} > \frac{6}{\pi^2} \frac{x-1}{x} > 0 \end{aligned}$$

for all $x > 1$. Thus π is increasing. □

1.5. Notes

A survey of g-numbers by the first author and P. T. Bateman is given in [**BD69**]. Some of our arguments come from this article.

§1.4. In [**Hi12**], it is shown that $N(x) := 1 + c(x-1)$ determines a g-prime system in the sense of Hilberdink for $0 < c < 2$, while for some $c > 2$, the prime counting function $\pi(x)$ is not always monotonic increasing. Proposition 1.7 is the case $c = 1$ of Hilberdink's result.

CHAPTER 2

Analytic Machinery

Are we there yet?

SUMMARY. A medley of functions and associated measures. The Mellin transform. A family of norms. Convergence of measures. Convolutions. A derivation operator on measures.

2.1. A function class

We begin by setting out conditions on functions that justify Mellin transforms, convergence, and other analytic techniques. Happily, these conditions are not very restrictive, and nearly all the functions that occur in our study, e.g. $N(\cdot)$, $\Pi(\cdot)$, satisfy them. Readers eager to get to g-numbers might skim this chapter for its definitions and conventions and return for details as needed.

Here are the requirements for our functions, along with brief comments:

(1) have support contained in $[1, \infty)$, the reasonable "home" for g-integers.
(2) have total variation function of polynomial growth. We shall make essential use of Mellin transforms and integrations on $[1, \infty)$, for which we want limited growth at infinity. This condition also implies local bounded variation, so we can associate a Borel-Stieltjes measure dF with each function F as well as integrate by parts. (For "elementary" methods, transforms are not used, so growth is not an issue and local bounded variation alone would suffice.)
(3) be right continuous. This choice will be convenient in associating a Borel measure with each function.

Let $\mathcal{V}$ denote the class of right continuous real or complex valued functions F with support contained in $[1, \infty)$ whose associated total variation function on the interval $[1, x]$, called $F_v(x)$, is of polynomial growth at infinity. Here are two instances of functions in $\mathcal{V}$ arising in Example 1.4: one is N_c, the continuous approximation to the integer counting function; the other is

$$F(x) := \begin{cases} 0, & x < 1 \\ 1, & x \geq 1. \end{cases} \tag{2.1}$$

2.2. Measures

We may regard Borel sets on $\mathbb{R}$ as generated by the half open intervals of the form $(a, b]$ for $a < b$, and we introduce a signed Borel measure dF corresponding to each element $F \in \mathcal{V}$ by setting

$$dF((a, b]) := F(b) - F(a).$$

We shall often write "$dF \in d\mathcal{V}$" to say that dF is a measure associated with $F \in \mathcal{V}$. (An exception to our convention of connecting function and measure names is the function given in (2.1), for which it is customary to represent the associated measure as δ_1, Dirac measure at 1.)

With these conventions, we can express a function F in terms of a measure dF by

$$\int_{1-}^{x} dF = F(x), \quad 1 \le x < \infty.$$

We shall call F the cumulative distribution function (briefly: distribution function) associated with dF. For $b > a \ge 1$, we shall understand

$$\int_a^b dF := \int_{(a,\,b]} dF = F(b) - F(a).$$

(Note that $\int_a^b dF$ includes any point mass that dF may have at b, but not at a.)

We associate with the total variation function F_v the total variation measure $|dF| = d(F_v)$, briefly dF_v, and we have

$$F_v(x) = \int_{1-}^{x} |dF|, \quad 1 \le x < \infty.$$

2.3. Mellin transforms

Given $F \in \mathcal{V}$ with $F(x) = O(x^\tau)$ for some $\tau \ge 0$, integration by parts shows that the *Mellin integral*

$$\int_{1-}^{\infty} u^{-s} dF(u) = s \int_{1}^{\infty} u^{-s-1} F(u)\, du$$

exists for all $s \in \mathbb{C}$ with $\Re s > \tau$. (Note that the integrated terms vanish at $1-$ and at infinity.) The integral defines a function of s that we call the *Mellin (or Mellin-Stieltjes) transform* of dF, and we set

$$\widehat{F}(s) := \int_{1-}^{\infty} u^{-s} dF(u), \quad \Re s > \tau.$$

For example, the Riemann zeta function is expressible on $\{s : \Re s > 1\}$ as the Mellin transform of the counting measure $dN(x)$ of positive rational integers.

Since the g-zeta function is expressible by a generalized Dirichlet series, as in (1.3), the reader may ask, Why not restrict ourselves to series? The main reason for the generality is that we shall often approximate discrete counting functions by continuous ones, and Mellin transforms enable us to treat continuous as well as discrete measures.

We assume that the basic facts, such as the definition of abscissa of convergence (denoted σ_c), abscissa of absolute convergence (denoted σ_a), conditions justifying differentiation, etc., are known for Dirichlet series; similar relations hold for Mellin transforms. Here we note only that the region of convergence of a Mellin integral is a half plane and that the Mellin transform has the important property of carrying convolutions into pointwise products (see §2.6). We establish or give references for further information as we need it.

We note here an inequality that yields two useful estimates for Mellin transforms with real arguments.

LEMMA 2.1. *Suppose that $F, G \in \mathcal{V}$ with $F(u) \le G(u)$ for all $u \ge 1$ and suppose that H is a left continuous, positive, decreasing function on $[1, \infty)$. Then, for all $x > 1$,*

$$\int_{1-}^{x} H(u)\, dF(u) \le \int_{1-}^{x} H(u)\, dG(u).$$

PROOF. Integration by parts gives

$$\int_{1-}^{x} H(u)\,\{dG(u) - dF(u)\} = H(x)\{G(x) - F(x)\} + \int_{1}^{x} \{G(u) - F(u)\}\{-dH(u)\} \ge 0. \qquad \square$$

Specializing to $H(u) := u^{-\sigma}$ and letting $x \to \infty$, we obtain an inequality between Mellin transforms:

LEMMA 2.2. *Suppose that $F, G \in \mathcal{V}$ with $F(x) \le G(x)$ for all $x \ge 1$. Then for all $\sigma > \sigma_c$ we have $\widehat{F}(\sigma) \le \widehat{G}(\sigma)$.*

2.4. Norms

We introduce a family of norms for measures via Mellin transforms. This will be useful for the next section and in later analysis. If $F \in \mathcal{V}$, $F_v(x) \ll x^{\tau}$ for some $\tau \ge 0$, and $\sigma > \tau$, then, by integration by parts,

$$\widehat{F_v}(\sigma) = \int_{1-}^{\infty} u^{-\sigma}\, |dF(u)| = \sigma \int_{1}^{\infty} u^{-\sigma-1} F_v(u)\, du < \infty.$$

In this case, we say that the Mellin integral is *absolutely convergent* at σ. (We say that a Mellin integral is *absolutely convergent at a complex point s* if it is absolutely convergent at σ, where $\sigma = \Re s$.) We define $d\mathcal{V}_\sigma$ to be the set of measures with this property and note that $d\mathcal{V}_\sigma$ is a linear space, since it admits addition and scalar multiplication.

We define $\|dF\|_\sigma := \widehat{F_v}(\sigma)$ and call this the σ-norm of dF. Let $\mathcal{V}_\sigma$ denote the collections of functions $F \in \mathcal{V}$ whose associated measure has finite σ-norm. It is easy to see that $\|\cdot\|_\sigma$ satisfies the expected properties of a norm. Also, $\|dF\|_{\sigma'} \le \|dF\|_\sigma$ for any $\sigma' > \sigma$, and in particular $\|dF\|_{\sigma'}$ is finite whenever $\|dF\|_\sigma$ is.

For the opposite relation between the growth of F_v and the norm, note that if $\|dF\|_\sigma < \infty$, then, for any $x \ge 1$,

$$F_v(x) = \int_{1-}^{x} |dF(u)| \le x^{\sigma} \int_{1-}^{x} u^{-\sigma} |dF(u)| \le x^{\sigma} \|dF\|_\sigma \ll x^{\sigma}. \tag{2.2}$$

It is important to have the restriction $\sigma \ge 0$ for norms, since some of the preceding inequalities are false for negative values of σ.

2.5. Convergence

We are going to use limits of sequences of measures, and for this we set out a notion of convergence. For $dF_i \in d\mathcal{V}$, $i = 1, 2, \ldots$, we say that the sequence *converges to a limit* (in $d\mathcal{V}$) if there exists a measure $dF \in d\mathcal{V}$ and some $\sigma \ge 0$ for which

$$\|dF_i - dF\|_\sigma \to 0, \quad i \to \infty. \tag{2.3}$$

Of course, if (2.3) holds for some σ, then it holds for any larger value of the parameter. In a related vein, we have the following observation:

LEMMA 2.3. *Suppose that a sequence $\{dF_i\}$ converges and g is a continuous function of polynomial growth on $[1, \infty)$. Then $\{g\,dF_i\}$ converges.*

The proof consists of noting that, if $g(u) \ll u^c$, then

$$\|g\,dF_i\|_{\sigma+c} \ll \|dF_i\|_\sigma.$$

LEMMA 2.4. *If a sequence of measures in $d\mathcal{V}$ converges, then the associated sequence of distribution functions converges locally uniformly on $\mathbb{R}$.*

PROOF. We can assume that dF_i converges to 0, i.e. $\|dF_i\|_\sigma \to 0$ for some σ, and let $X > 1$ be given. Then, by (2.2), we have, uniformly for $x \in [1, X]$,

$$|F_i(x)| \le (F_i)_v(x) \le x^\sigma \|dF_i\|_\sigma \le X^\sigma \|dF_i\|_\sigma \to 0, \quad i \to \infty. \qquad \square$$

The space of measures $d\mathcal{V}_\sigma$ is complete (in the usual sense that a Cauchy sequence in $d\mathcal{V}_\sigma$ has a limit in this space): if $\sum_i \|dF_i\|_\sigma < \infty$ for some σ, then, by a simple argument, there exists a measure $dF \in d\mathcal{V}_\sigma$ to which $\sum_i dF_i$ converges.

2.6. Convolution of measures

We define multiplicative convolution, a multiplication on $d\mathcal{V}$, by setting

$$\int_E dF * dG := \iint_{uv \in E} dF(u)\, dG(v) \tag{2.4}$$

first for $E = (a, b]$, a half open interval, with the set $\{uv \in (a, b]\}$ understood as

$$\{(u, v) \in (1/2, \infty) \times (1/2, \infty) : a < uv \le b\}.$$

(The 1/2's serve only as numbers smaller than 1, in order to include possible contributions of dF, dG at 1.) The convolution measure is then extended to all Borel sets E in $\mathbb{R}$. This operation is commutative and associative. Rudin [**Ru87**] gives an elegant proof of associativity of convolutions by transferring the relation into the Mellin transform space, where we can appeal to the associativity of $\mathbb{C}$.

The measure δ_1, Dirac measure at 1, is the unity element of convolutions: for any $dF \in d\mathcal{V}$ we have $\delta_1 * dF = dF$. We shall occasionally use repeated convolutions – for $dF \in d\mathcal{V}$, set $dF^{*0} := \delta_1$ and $dF^{*n} := dF * dF^{*n-1}$ for $n \ge 1$.

The following integration formulas will be used often. Let dF, $dG \in d\mathcal{V}$ and f be a Borel function that is bounded on an interval $[1, x]$. Then

$$\int_{1-}^{x} dF * dG = \int_{1-}^{x} F(x/v)\, dG(v) = \int_{1-}^{x} G(x/u)\, dF(u) \tag{2.5}$$

and

$$\begin{aligned} \int_{1-}^{x} f\,(dF * dG) &= \iint_{uv \le x} f(uv)\, dF(u)\, dG(v) \\ &= \int_{1-}^{x} \left\{ \int_{1-}^{x/v} f(uv)\, dF(u) \right\} dG(v)\,. \end{aligned} \tag{2.6}$$

The first formula follows from the definition of convolution and Fubini's Theorem on iterated integration. Formula (2.6) holds for $f = \chi_S$, the indicator function of some bounded Borel measurable set S, by (2.4) with $S \cap [1, x]$ in place of E. By linearity the formula holds for f a simple function, and finally it holds for a general bounded measurable function by uniform approximation by simple functions.

LEMMA 2.5. *Suppose that F, $G \in \mathcal{V}_\sigma$, and define $H(x) = 0$ for $x < 1$ and*

$$H(x) := \int_{1-}^{x} dF * dG, \qquad x \geq 1.$$

Then $H \in \mathcal{V}_\sigma$.

PROOF. By construction, H has support in $[1, \infty)$.
Next, we show the growth condition for H_v. By the triangle inequality,

$$H_v(x) = \int_{1-}^{x} |dF * dG| \leq \int_{1-}^{x} |dF| * |dG|.$$

The last inequality and (2.5) together imply that

$$H_v(x) \leq \int_{1-}^{x} F_v(x/u)\,|dG(u)| \leq \int_{1-}^{x} O((x/u)^\sigma)\,|dG(u)| = O(x^\sigma),$$

since $\widehat{G_v}(\sigma) < \infty$.
It remains to show that H is right continuous. We have

$$\lim_{\epsilon \to 0+} H(x+\epsilon) - H(x) = \lim_{\epsilon \to 0+} \int_{1-}^{x+\epsilon} \left\{ F\Big(\frac{x+\epsilon}{u}\Big) - F\Big(\frac{x}{u}\Big) \right\} dG(u).$$

(This formula is valid, since $F(x/u) = 0$ for $x < u \leq x + \epsilon$.) The F difference in the last integral goes to zero pointwise as $\epsilon \to 0$ by right continuity of F, and hence the integral goes to zero by dominated convergence. (Here $2F_v(\{x+1\}/u)$ is a dominating function for the integrand.) □

The following formula shows how the Mellin transform converts convolutions into pointwise multiplications. This property is the reason the Mellin (rather than, say, Laplace) transform is useful in studying multiplicative properties.

LEMMA 2.6. *Let F, $G \in \mathcal{V}$ and suppose $s \in \mathbb{C}$ is a number for which the Mellin transform of dF converges and that of dG converges absolutely. Then*

$$\int_{1-}^{\infty} u^{-s}\,(dF * dG)(u) = \widehat{F}(s)\,\widehat{G}(s).$$

(Of course, the roles of F and G can be exchanged here.)

PROOF. Let Z be any number exceeding 1. By (2.6),

$$\begin{aligned}
\int_{1-}^{Z} u^{-s}(dF * dG)(u) &= \int_{y=1-}^{Z} \left\{ \int_{x=1-}^{Z/y} (xy)^{-s} dF(x) \right\} dG(y) \\
&= \left\{ \int_{1-}^{\sqrt{Z}} x^{-s} dF(x) \right\} \left\{ \int_{1-}^{\sqrt{Z}} y^{-s} dG(y) \right\} \\
&\quad + \int_{y=1-}^{\sqrt{Z}} y^{-s} \left\{ \int_{x=\sqrt{Z}}^{Z/y} x^{-s} dF(x) \right\} dG(y) \\
&\quad + \int_{y=\sqrt{Z}}^{Z} y^{-s} \left\{ \int_{x=1-}^{Z/y} x^{-s} dF(x) \right\} dG(y) \\
&=: I + II + III, \quad \text{say.}
\end{aligned}$$

As $Z \to \infty$ we have $I \to \widehat{F}(s)\,\widehat{G}(s)$;

$$|II| \le \int_{y=1-}^{\sqrt{Z}} y^{-\sigma} o(1)\, dG_v(y) = o(1),$$

since $\int_{\sqrt{Z}}^{Z/y} x^{-s} dF(x) \to 0$ uniformly for $1 \le y \le \sqrt{Z}$ as $Z \to \infty$; and

$$|III| \le \int_{y=\sqrt{Z}}^{Z} O(1) y^{-\sigma} dG_v(y) = o(1). \qquad \square$$

EXAMPLE 2.7. DIVISOR FUNCTION. The divisor function d of elementary number theory corresponds to the convolution of the counting function of the (rational) integers with itself in the sense that

$$\sum_{n\le x} d(n) = \sum_{mn\le x} 1 = \iint_{st\le x} d\lfloor s\rfloor\, d\lfloor t\rfloor = \int_{1-}^{x} d\lfloor\cdot\rfloor^{*2}.$$

EXAMPLE 2.8. CONVOLUTION POWERS OF LEBESGUE MEASURE. For dt Borel–Lebesgue measure on $[1, \infty)$, we have

$$dt^{*n} = \log^{n-1} t\; dt/(n-1)!\,, \quad n \ge 1.$$

(This relation is most easily verified via the cumulative distribution function or Mellin transforms.)

We conclude this section with a useful inequality.

LEMMA 2.9. *Suppose that F and G are nondecreasing functions in $\mathcal{V}$ with $F(u) \le G(u)$ for all $u \ge 1$. Then, for all positive integers n and real $x > 1$,*

$$\int_{1-}^{x} dF^{*n} \le \int_{1-}^{x} dG^{*n}.$$

PROOF. We start with the algebraic identity

$$dG^{*n} - dF^{*n} = (dG - dF) * (dG^{*n-1} + dG^{*n-2} * dF + \cdots + dF^{*n-1})$$

and integrate each side:

$$\int_{1-}^{x} dG^{*n} - dF^{*n}$$
$$= \int_{1-}^{x} \{G(x/u) - F(x/u)\}(dG^{*n-1} + dG^{*n-2} * dF + \cdots + dF^{*n-1})(u) \ge 0$$

since $G - F$ and the measures are each nonnegative. $\square$

We shall have occasion to convolve measures involving fractional powers of the argument, such as $d\pi(u^{1/2})$, which occurs in $d\Pi(u)$ (see (1.6) and (1.10)). For this we have the following formula.

LEMMA 2.10. *Let n, k be positive integers and*

$$f(u) = f_n(u) := \phi(u^{1/n}), \quad r(u) = r_n(u) := \rho(u^{1/n})$$

for functions ϕ, $\rho \in \mathcal{V}$. We have

$$\int_{1-}^{x} df * dr = \int_{1-}^{x^{1/n}} d\phi * d\rho, \tag{2.7}$$

$$\int_{1-}^{x} df^{*k} = \int_{1-}^{x^{1/n}} d\phi^{*k}. \tag{2.8}$$

With a small abuse of notation, we sometimes write the last formula as

$$\int_{1-}^{x} d\phi(u^{1/n})^{*k} = \int_{1-}^{x^{1/n}} d\phi^{*k}.$$

PROOF. We use iterated integration and a change of variable. Write

$$\begin{aligned}\int_{1-}^{x} df * dr &= \int_{1-}^{x} f(x/t)\, dr(t) = \int_{1-}^{x} \phi(x^{1/n}/t^{1/n})\, d\rho(t^{1/n}) \\ &= \int_{1-}^{x^{1/n}} \phi(x^{1/n}/u)\, d\rho(u) = \int_{1-}^{x^{1/n}} d\phi * d\rho.\end{aligned}$$

To show (2.8), we proceed inductively, taking $dr = df^{*(k-1)}$. □

2.7. Convolution of functions

The preceding notion of convolution has a useful specialization to functions. Let $\mathcal{R}_{\text{loc}}$ be the class of functions on $[1, \infty)$ that are locally Riemann integrable (rather than Lebesgue integrable, to avoid some unneeded complexity). We define $f * g$, the multiplicative convolution of functions f, $g \in \mathcal{R}_{\text{loc}}$, by

$$f * g\,(x) := \int_{1}^{x} f\left(\frac{x}{u}\right) g(u)\, \frac{du}{u}.$$

In the case of absolutely continuous measures (where we can write $dF(x) = f(x)\, dx$) this definition essentially coincides with the previous notion of convolution, as we show now.

LEMMA 2.11. *Suppose that f, $g \in \mathcal{R}_{\text{loc}}$ and F, G are their integrals, i.e.*

$$F(x) = \int_{1}^{x} f(u)\, du, \quad G(x) = \int_{1}^{x} g(u)\, du.$$

Then

$$f * g(x)\, dx = (dF * dG)\,(x), \quad x \geq 1.$$

PROOF. It suffices to show that

$$\int_{1}^{x} f * g(v)\, dv = \int_{1}^{x} (dF * dG)\,(v), \quad x \geq 1. \tag{2.9}$$

Each side of the last formula is a continuous function (in each case $\int_x^y \to 0$ as $y - x \to 0$), so the desired result follows from (2.9) upon differentiating.

We show (2.9) by first changing the order of iterated integration and then changing the integration variables:

$$\begin{aligned}&\int_{1}^{x} f * g\,(v)\, dv \\ &= \int_{1}^{x} \left(\int_{1}^{v} f\left(\frac{v}{u}\right) g(u)\, \frac{du}{u} \right) dv = \int_{1}^{x} \left(\int_{u}^{x} f\left(\frac{v}{u}\right) dv \right) g(u)\, \frac{du}{u} \\ &= \int_{1}^{x} \left(\int_{1}^{x/u} f(w)\, dw \right) g(u)\, du = \int_{1}^{x} F\left(\frac{x}{u}\right) dG(u) = \int_{1}^{x} (dF * dG)\,(v).\end{aligned}$$

□

Also, by making a change of variable in the defining relation, we see that $f * g = g * f$, i.e., convolution of functions is commutative.

Suppose that f and g are of polynomial growth. Then the Mellin transform of dF exists in some half plane, and it is expressed as

$$\widehat{F}(s) = \int_1^\infty x^{-s} dF(x) = \int_1^\infty x^{-s} f(x)\, dx.$$

The last integral is the classical definition of the Mellin transform of a function f, and it is usually denoted also by $\hat{f}(s)$ (sorry!).

By the last lemma and Lemma 2.6 the product formula for the Mellin transform of functions is

$$\left(\int_1^\infty x^{-s} f(x)\, dx\right)\left(\int_1^\infty x^{-s} g(x)\, dx\right) = \int_1^\infty x^{-s}\,(f * g)(x)\, dx, \tag{2.10}$$

or, $\hat{f}(s)\hat{g}(s) = (f * g)\hat{}(s)$. We shall also occasionally use repeated convolutions f^{*n}, $n = 2, 3, \ldots$, and for these

$$(\hat{f}(s))^n = (f^{*n})\hat{}(s).$$

EXAMPLE 2.12. EXAMPLE 2.8 REVISITED. Define a function

$$1(x) := \begin{cases} 1, & \text{if } x \geq 1, \\ 0, & \text{if } x < 1. \end{cases}$$

We have the convolution formula

$$1^{*2}(x) = 1 * 1(x) = \int_1^x u^{-1}\, du = \log x$$

and, by induction,

$$1^{*n}(x) = \int_1^x \frac{(\log x/u)^{n-2}}{(n-2)!} u^{-1}\, du = \frac{(\log x)^{n-1}}{(n-1)!}.$$

2.8. The L and T operators

In this section we introduce two mappings of the class of measures $d\mathcal{V}$ to itself. These will appear many times in the sequel, particularly in connection with Mellin transforms and their derivatives.

For k any nonnegative integer, define L^k on $d\mathcal{V}$ by

$$L^k : dF(t) \mapsto (\log t)^k\, dF(t);$$

for $k = 1$, we write L instead of L^1.

For $\alpha \in \mathbb{C}$, define T^α on $d\mathcal{V}$ by

$$T^\alpha : dF(u) \mapsto t^\alpha\, dF(t),$$

and for $\alpha = 1$, we write T instead of T^1.

Let us look at some properties of L. Differentiation of a Mellin transform yields the formula

$$\frac{d}{ds}\int_{1-}^\infty u^{-s}\, dF(u) = -\int_{1-}^\infty u^{-s} \log u\, dF(u) = -\int_{1-}^\infty u^{-s}(L\, dF)(u),$$

which suggests that L is connected with differentiation. We call a linear operator on a ring that satisfies the product rule of differentiation a *derivation*. We claim that L has this property. L is clearly linear, and we now show that it satisfies the product rule.

LEMMA 2.13. *For dF, $dG \in d\mathcal{V}$, we have*

$$L\,(dF * dG) = (L\,dF) * dG + dF * (L\,dG). \tag{2.11}$$

PROOF. It suffices to establish the relation for the associated distribution functions. By (2.6), we have

$$\begin{aligned}\int_1^x L\,(dF * dG) &= \iint_{uv \le x} \log(uv)\,dF(u)\,dG(v)\\ &= \iint_{uv \le x} \log u\,dF(u)\,dG(v) + dF(u)\log v\,dG(v)\\ &= \int_1^x (L\,dF) * dG + dF * (L\,dG).\end{aligned}$$

(Note that it is irrelevant whether these integrals start at $1-$ or at 1, since the logarithm vanishes at 1.) □

If we define $L^0\,dF := dF$ and $L^n\,dF := L(L^{n-1}\,dF)$ for $n \ge 1$, then the preceding formula implies that

$$L(dF^{*n}) = n\,dF^{*n-1} * L\,dF, \tag{2.12}$$

$$L^n(dF * dG) = \sum_{j=0}^{n} \binom{n}{j} L^j dF * L^{*n-j} dG. \tag{2.13}$$

We are going to apply L term-by-term to infinite series. This is justified by the following observation.

LEMMA 2.14. *L is a continuous operator on $d\mathcal{V}$.*

The proof consists in noting that $\{L\,dF_i\}$ converges whenever $\{dF_i\}$ does, by applying Lemma 2.3.

Now we look at T.

LEMMA 2.15. *For dF, $dG \in d\mathcal{V}$ and any $\alpha \in \mathbb{C}$, we have*

$$T^\alpha\,(dF * dG) = (T^\alpha\,dF) * (T^\alpha\,dG). \tag{2.14}$$

PROOF. Using the first part of (2.6), we get

$$\begin{aligned}\int_{1-}^x T^\alpha\,(dF * dG) &= \iint_{st \le x} (st)^\alpha\,dF(s)\,dG(t)\\ &= \iint_{st \le x} (s^\alpha\,dF(s))\,(t^\alpha\,dG(t)) = \int_{1-}^x (T^\alpha\,dF) * (T^\alpha\,dG).\end{aligned}$$

Since the integrated version of (2.14) holds, so does the measure form. □

Some simple properties of T are that it is one-to-one and onto; also, the composition $T^\alpha \circ T^{-\alpha} = T^0$, the identity mapping on $d\mathcal{V}$.

The T and L maps also have obvious interpretations as operators on $\mathcal{R}_{\text{loc}}$, the algebra of all locally Riemann integrable functions supported in $[1, \infty)$:

$$(T^\alpha f)(t) := t^\alpha f(t), \quad (L^k f)(t) := (\log t)^k f(t),$$

with the understanding that $(Lf)(t) = (\log t)f(t)$.

L is a derivation, i.e.

$$L(f * g) = (Lf) * g + f * Lg. \tag{2.15}$$

Indeed, if we use Lemma 2.11 to interpret $f * g(x)\,dx$ as $dF * dG$, then Lemma 2.13 shows the derivation property at once.

Each T operator is an isomorphism on the algebra, satisfying

$$T^\alpha(f * g) = T^\alpha f * T^\alpha g. \tag{2.16}$$

This is verified by noting that

$$x^\alpha \int_1^x f\left(\frac{x}{u}\right) g(u)\,\frac{du}{u} = \int_1^x \left\{f\left(\frac{x}{u}\right)\left(\frac{x}{u}\right)^\alpha\right\}\{g(u)u^\alpha\}\,\frac{du}{u}.$$

EXAMPLE 2.16. A T PRODUCT. The formula

$$\left(\int_1^\infty f(x)x^{-s-1}\,dx\right)^n = \int_1^\infty f^{*n}(x)x^{-s-1}\,dx$$

will be used later. This equality can be established in terms of the operator T^{-1} by writing

$$\begin{aligned}\left(\int_1^\infty x^{-s}(T^{-1}f)(x)\,dx\right)^n &= \int_1^\infty x^{-s}(T^{-1}f)^{*n}(x)\,dx\\ &= \int_1^\infty x^{-s}T^{-1}(f^{*n})(x)\,dx.\end{aligned}$$

2.9. Notes

References for this chapter include [**Ro88**], Ch. 6 and 7 of [**Ru87**], Ch. 9 of [**Ti39**], [**Di70a**], as well as the works cited in the Preface.

CHAPTER 3

dN as an Exponential and Chebyshev's Identity

Can exponentials compound our interest?

SUMMARY. The counting measure of integers as an exponential of the weighted prime and prime power counting measure. Connection with Chebyshev's elementary prime number identity.

3.1. Goals and plan

This chapter has two main goals. The first is to express dN, the integer counting measure, as the exponential of the measure $d\Pi$, specifically

$$dN = \delta_1 + d\Pi + \frac{1}{2!}d\Pi * d\Pi + \frac{1}{3!}d\Pi * d\Pi * d\Pi + \cdots =: \exp^* d\Pi, \tag{3.1}$$

with $*$ denoting multiplicative convolution. This relation provides the fundamental connection between the counting functions of g-primes and the associated g-integers, and it is the reason for introducing $\Pi(x)$, which counts primes and prime powers.

Chebyshev showed in the 1850s that the rational primes satisfy the inequalities

$$\pi(x) \le \Pi(x) \ll x/\log x. \tag{3.2}$$

By analogy with the classical case, we say that a g-number system satisfies a Chebyshev O-bound if (3.2) holds. We shall study these estimates in Chapters 9 and 11. The argument Chebyshev used was in essence based on the identity

$$L\,dN = dN * d\psi, \tag{3.3}$$

where L denotes the log operator, introduced in §2.8, and $d\psi = L\,d\Pi$ is the weighted prime and prime power counting measure.

Our second goal for this chapter is to show that (3.3) holds for generalized number systems, and moreover this relation is intimately connected with the exponential representation above. If we recall the derivation property of L, the connection between the formulas is suggested by the calculus relation for differentiable functions

$$y(x) = Ce^{f(x)} \iff y'(x) = y(x)\,f'(x).$$

We shall show that the power series (3.1) is convergent and that the analogues of familiar analytic operations are valid. Then we show that exp has the expected properties of an exponential, that (3.3) and (3.1) are equivalent, and that these relations hold for g-numbers.

At first glance the exponential representation (3.1) looks "far-out." However, the Mellin formulas for $\zeta(s)$ and $\log\zeta(s)$ and the Mellin transform's homomorphic property of taking convolutions to products suggest how (3.1) might make sense:

starting with (1.5), expanding the exponential as a series, and using Lemma 2.6 for each term, for s in some half plane, we find

$$\begin{aligned}\zeta(s) &= \exp\log\zeta(s) = \exp\int_1^\infty u^{-s}d\Pi(u)\\ &= 1 + \int u^{-s}\,d\Pi(u) + \frac{1}{2!}\Big\{\int u^{-s}d\Pi(u)\Big\}^2 + \frac{1}{3!}\Big\{\int u^{-s}d\Pi(u)\Big\}^3 + \dots\\ &= \int u^{-s}\,\{\delta_1 + d\Pi + \frac{1}{2!}d\Pi^{*2} + \frac{1}{3!}d\Pi^{*3} + \dots\}(u)\,.\end{aligned}$$

(For compactness in the preceding formula, we have suppressed the limits on the integrals and used $d\Pi^{*i}$ to denote an i-fold convolution of $d\Pi$ with itself.) That is, we have shown

$$\int_{1-}^\infty u^{-s}dN(u) = \int_1^\infty u^{-s}\exp^*\{d\Pi\}(u).$$

The uniqueness theorem for Mellin transforms (see [**BD04**] §6.6, or for the Dirichlet series version, [**Ch68**] Ch. X, Th. 7; [**MV07**] Th. 1.6) implies

THEOREM 3.1. *The formula* $dN = \exp^* d\Pi$ *holds for any g-number system of polynomial growth.*

REMARK 3.2. The polynomial growth condition arises from our analytic proof, which requires the zeta function to be convergent in some half plane. This is not a serious restriction, for we shall study only g-number systems that are reasonably similar to the natural numbers. An unconditional version of (3.1) holds as well (cf. [**Di70a**]) if we use a weaker notion of convergence than that given in §2.5.

The preceding formulas together yield the interesting Mellin transform identity for the zeta function

$$\exp\Big\{\int_1^\infty u^{-s}\,d\Pi(u)\Big\} = \int_{1-}^\infty u^{-s}\exp^*\{d\Pi\}(u). \tag{3.4}$$

The last identity leads to some useful observations about the zeta function.

LEMMA 3.3. *Suppose that the integer counting function N of a g-number system $\mathcal{N}$ satisfies the O-density condition*

$$\limsup_{x\to\infty} N(x)/x < \infty. \tag{1.2 bis}$$

Then $\zeta_{\mathcal{N}}(s) = \zeta(s)$ converges in the open half-plane $\{\Re s > 1\}$ and is nonzero there.

PROOF. First, for any $\sigma > 1$,

$$\zeta(\sigma) = \int_{1-}^\infty u^{-\sigma}dN(u) = \sigma\int_1^\infty u^{-\sigma-1}N(u)\,du \ll \sigma\int_1^\infty u^{-\sigma}\,du = \frac{\sigma}{\sigma-1} < \infty.$$

Thus the zeta function converges absolutely on $\{\Re s > 1\}$.

Next, we have for $\sigma > 1$,

$$|\log\zeta(\sigma+it)| = \Big|\int_1^\infty u^{-\sigma-it}d\Pi(u)\Big| \le \int_1^\infty u^{-\sigma}d\Pi(u) = \log\zeta(\sigma) < \zeta(\sigma) < \infty,$$

and since $\log\zeta(s)$ is bounded on this half plane, $\zeta(s) \neq 0$ here. □

Relations (3.1) and (3.3) are at the heart of our treatment of g-numbers. They apply to continuous "counting functions" as well as discrete ones. Thus it is useful to study the properties of convolution exponentials.

3.2. Power series in measures

Convolution exponentials are one case of power series in measures. We shall presently meet some other series as well, so we consider here the general notion of $g(dF)$ for measures dF and functions g that are holomorphic at the origin.

Let $\mathcal{V}(c)$ denote the collection of functions $F \in \mathcal{V}$ satisfying $F(1) = c$, and let $d\mathcal{V}(c)$ represent the associated measures. Since elements of $\mathcal{V}$ satisfy $F(x) = 0$ for all $x < 1$, we have $dF(\{1\}) = F(1)$, and elements of $d\mathcal{V}(c)$ have a point mass c at the point 1. Our main cases will be $c = 0$ and 1. The convergence of a series in a measure dF depends on the size of $|dF(\{1\})|$, as we now show.

PROPOSITION 3.4. *Let $g(z)$ be a holomorphic function at 0 whose Maclaurin series has radius of convergence ρ. Let $dF \in d\mathcal{V}(c)$ with $|c| < \rho$. Then the series*

$$g(dF) := \sum_{n=1}^{\infty} a_n \, dF^{*n}$$

converges to an element of $d\mathcal{V}(c)$. Further, g is continuous on $\cup_{|c|<\rho}\{d\mathcal{V}(c)\}$.

PROOF. Since $dF \in d\mathcal{V}$, there exists some positive number σ such that

$$\|dF\|_\sigma := \int_{1-}^{\infty} u^{-\sigma} |dF|(u) < \infty .$$

Further, since $|dF|(\{1\}) = |c|$ and $u^{-\sigma} \to 0$ as $\sigma \to \infty$ for all $u > 1$, by Lebesgue's convergence theorem we have $\|dF\|_\sigma \to |c|$ as $\sigma \to \infty$. Thus there exists a number σ_1 such that $\|dF\|_\sigma < \rho$ for $\sigma > \sigma_1$.

For such a value of σ, we show that

$$\Big\| g(dF) - \sum_{n=0}^{N} a_n \, dF^{*n} \Big\|_\sigma = \Big\| \sum_{n>N} a_n \, dF^{*n} \Big\|_\sigma \to 0 , \qquad N \to \infty .$$

By the triangle inequality, it suffices to show that, as $N \to \infty$,

$$\sum_{n>N} |a_n| \int_{1-}^{\infty} u^{-\sigma} |dF|^{*n}(u) \to 0.$$

By Lemma 2.6, the last sum is

$$\sum_{n>N} |a_n| \Big(\int_{1-}^{\infty} u^{-\sigma} |dF|(u) \Big)^n = \sum_{n>N} |a_n| (\|dF\|_\sigma)^n ,$$

and this converges to 0 as $N \to \infty$, since $\|dF\|_\sigma < \rho$ and the last expression is the tail of a convergent series.

To show continuity, suppose $\|dF\|_\sigma < \rho$ and $\|dF_\nu - dF\|_\sigma \to 0$ as $\nu \to \infty$. Let $M_\nu := \max(\|dF_\nu\|_\sigma, \|dF\|_\sigma)$, and say ν is large enough that $M_\nu \le \rho' < \rho$. Then

$$\begin{aligned}
\|g(dF_\nu) - g(dF)\|_\sigma &\le \sum_{n=1}^{\infty} |a_n|\, \|dF_\nu{}^n - dF^n\|_\sigma \\
&\le \sum_{n=1}^{\infty} |a_n| \left\| \sum_{j=0}^{n-1} dF_\nu{}^{n-1-j} * dF^j * (dF_\nu - dF) \right\|_\sigma \\
&\le \left(\sum_{n=1}^{\infty} |a_n|\, n M_\nu{}^{n-1} \right) \|dF_\nu - dF\|_\sigma \\
&\le \left(\sum_{n=1}^{\infty} |a_n|\, n \rho'^{\,n-1} \right) \|dF_\nu - dF\|_\sigma \to 0, \quad \nu \to \infty. \qquad \square
\end{aligned}$$

Most of our applications will be to measures $dF \in \mathcal{V}(0)$; in this case $g(dF)$ will exist for any function g that is holomorphic at 0.

3.3. Inverses

We say that $dF \in d\mathcal{V}$ is invertible if there exists $dG \in d\mathcal{V}$ such that $dF * dG = \delta_1$. Obviously dG is invertible as well. The usual argument from group theory shows that if an inverse exists, it is unique; we will generally refer to dG as *the* inverse of dF and denote it by dF^{*-1}. One example of an inverse relation, from classical number theory, is $dN * dM = \delta_1$, where M is the summatory function of the Moebius function. Another example is

$$(\delta_1 + dt) * (\delta_1 - dt/t) = \delta_1. \tag{3.5}$$

Lemma 2.6 implies that the Mellin transforms of a pair of invertible measures are reciprocals. One way to verify the last formula is to use Mellin transforms.

Evaluation at the set $\{1\}$ shows that $dF(\{1\}) \neq 0$ is a necessary condition for dF to be invertible. Now we show this condition is sufficient as well.

PROPOSITION 3.5. *If $dF \in d\mathcal{V}$ and $dF(\{1\}) \neq 0$, then dF is invertible. If $dF(\{1\}) = 1$, then*

$$dF^{*-1} = \sum_{n=0}^{\infty} (-1)^n (dF - \delta_1)^{*n}. \tag{3.6}$$

PROOF. First, suppose $dF(\{1\}) = 1$, and set $dF_1 := dF_1 - \delta_1$ with $dF_1 \in d\mathcal{V}(0)$. The function $g(z) := (1+z)^{-1}$ has the series

$$g(z) = \sum_{n=0}^{\infty} (-1)^n z^n$$

in the disc $\{z \in \mathbb{C} : |z| < 1\}$. We claim that $g(dF_1)$ is the inverse of dF.

First, note that $g(dF_1) \in d\mathcal{V}$. Indeed, for σ large enough, we have $\|dF_1\|_\sigma < 1$ (as shown in the proof of the last proposition), and so

$$\Big\| \sum_{n=0}^{\infty} (-1)^n (dF_1)^{*n} \Big\|_\sigma \le \sum_{n=0}^{\infty} \|dF_1\|_\sigma^n < \infty.$$

Next, we have

$$\begin{aligned}\big\|(\delta_1 + dF_1)*g(dF_1) - \delta_1\big\|_\sigma &\le \Big\|(\delta_1 + dF_1) * \sum_{n=0}^{N-1}(-1)^n\big(dF_1\big)^{*n} - \delta_1\Big\|_\sigma \\ &\quad + \Big\|(\delta_1 + dF_1) * \sum_{n=N}^{\infty}(-1)^n\big(dF_1\big)^{*n}\Big\|_\sigma \\ &\le \big\|dF_1\big\|_\sigma^N + \|dF\|_\sigma \cdot \sum_{n=N}^{\infty}\big\|dF_1\big\|_\sigma^n \to 0\end{aligned}$$

as $N \to \infty$. Thus $dF * g(dF_1) = \delta_1$, so $dF \in d\mathcal{V}(1)$ is invertible.

Finally, suppose more generally that $dF(\{1\}) = c \neq 0$. Let $dF_2 = c^{-1}dF$. Now $dF_2(\{1\}) = 1$, so there exists a measure dG_2 such that $dF_2 * dG_2 = \delta_1$. Now $dG = c^{-1}dG_2$ is the inverse of dF. □

It is immediate that the measures of $d\mathcal{V}(0)$ form a group under addition, with 0 the neutral element and $-dG$ the inverse of dG. We show now that $d\mathcal{V}(1)$ forms a group under convolution. It is clear that the convolution of two measures in $d\mathcal{V}(1)$ also lies there and that δ_1 is the neutral element. As we noted in §2.6, convolution is associative. The preceding proposition implies that elements of $d\mathcal{V}(1)$ have an inverse in $d\mathcal{V}$. Finally, since

$$dF(\{1\})\, dF^{-1}(\{1\}) = (dF * dF^{-1})(\{1\}) = \delta_1(\{1\}) = 1,$$

the inverse of each element of $d\mathcal{V}(1)$ also lies in this set.

We show next that there is a homomorphic mapping from the group $(d\mathcal{V}(0), +)$ to $(d\mathcal{V}(1), *)$. As in the case of numbers, the exponential provides this map.

3.4. The exponential on $\mathcal{V}$

Given $F \in \mathcal{V}$, define

$$\exp^* dF := \delta_1 + dF + \frac{1}{2!}dF * dF + \cdots = \sum_{i=0}^{\infty} \frac{1}{i!}\,(dF)^{*i}.$$

As usual, $dF^{*0} := \delta_1$, the Dirac measure at 1, and for $i \ge 1$, $dF^{*i} := dF * dF^{*i-1}$, the i-fold convolution of dF. By Proposition 3.4, this series is convergent regardless of the value of $dF(1)$, since the exponential is an entire function.

EXAMPLES 3.6. TWO EXPONENTIAL RELATIONS. In §3.1 we showed that $dN = \exp^* d\Pi$ holds for any g-number system. Another exponential relation is provided by the continuous analogue of dN, discussed in §1.4:

$$\delta_1 + dt = \exp^* d\Pi_c \ \text{ with } \ \Pi_c(x) := \int_1^x \frac{1 - u^{-1}}{\log u}\,du. \tag{3.7}$$

We shall establish this formula by a generalization of (3.4).

PROPOSITION 3.7. *Suppose $F \in \mathcal{V}$ and $\widehat{F}$ is absolutely convergent at σ_a. Then, for all s with $\Re s \ge \sigma_a$,*

$$\exp \int_1^\infty u^{-s} dF(u) = \int_{1-}^\infty u^{-s} \exp^* dF(u).$$

PROOF. Expanding the exponential as a series and applying Lemma 2.6, we find

$$\begin{aligned}\exp \widehat{F}(s) &= \sum_{n=0}^{\infty} \frac{1}{n!}\Big\{ \int_1^{\infty} u^{-s} dF(u)\Big\}^n = \sum_{n=0}^{\infty} \frac{1}{n!} \int_{1-}^{\infty} u^{-s} dF^{*n}(u) \\ &= \int_{1-}^{\infty} u^{-s} \sum_{n=0}^{\infty} \frac{1}{n!} dF^{*n}(u) = \int_{1-}^{\infty} u^{-s} \exp^* dF(u).\end{aligned}$$

The exchange of the sum and integral is justified by absolute convergence. □

PROOF OF (3.7). By Theorem 1.5 and the last proposition we have

$$\int_{1-}^{\infty} u^{-s}(\delta_1 + du) = \frac{s}{s-1} = \exp \int_1^{\infty} u^{-s} d\Pi_c(u) = \int_{1-}^{\infty} u^{-s} \exp^* d\Pi_c(u).$$

Now $\delta_1 + du = \exp^* d\Pi_c$ by the identity theorem for Mellin transforms. □

We shall revisit (3.7) in the next section.

As another application of the proposition, we show

LEMMA 3.8. *Let $F \in \mathcal{V}$ and $\alpha \in \mathbb{R}$. Then*

$$T^{\alpha} \exp^* dF = \exp^*(T^{\alpha} dF).$$

PROOF. One can apply Lemma 2.15 termwise followed by a convergence argument. As an alternative, we use Proposition 3.7 twice to obtain

$$\int_{1-}^{\infty} u^{-s} u^{\alpha} (\exp^* dF)(u) = \int_{1-}^{\infty} u^{-(s-\alpha)} (\exp^* dF)(u) = \exp \int_1^{\infty} u^{-(s-\alpha)} dF(u),$$

$$\int_{1-}^{\infty} u^{-s} (\exp^* u^{\alpha} dF)(u) = \exp \int_1^{\infty} u^{-s} u^{\alpha} dF(u) = \exp \int_1^{\infty} u^{-(s-\alpha)} dF(u).$$

Since the Mellin transforms are the same, so are the measures. □

We showed in Proposition 3.7 how the exponential "commutes" with Mellin transforms. The following inequality involves analogous exponentials and integrals, but with a positive measure and taken over a bounded interval.

LEMMA 3.9. *Let $F \in \mathcal{V}$ be nondecreasing. For $x \geq 1$ we have*

$$\exp\Big\{ \int_{1-}^{x} dF(u) \Big\} \geq \int_{1-}^{x} \exp^*\{dF\}(u).$$

PROOF. Remember that dF is zero to the left of 1. Now the left side of the last equation equals

$$1 + \int_{1-}^{x} dF(u) + \frac{1}{2!} \iint_{u \leq x,\, v \leq x} dF(u)\, dF(v) + \dots,$$

while the right side equals

$$1 + \int_{1-}^{x} dF(u) + \frac{1}{2!} \iint_{uv \leq x} dF(u)\, dF(v) + \dots .$$

The multiple integrals are taken over subregions in the second expression, so the claimed inequality holds. □

In our use of the exponential, we shall generally assume that $dF \in d\mathcal{V}(0)$, in which case $\exp^* dF \in d\mathcal{V}(1)$. If $F(1) = c \neq 0$, we can express $dF = c\delta_1 + dF_1$ where $F_1(1) = 0$ and apply the following proposition to obtain

$$\exp^* dF = \exp^*\{c\delta_1\} * \exp^* dF_1 = e^c \exp^* dF_1.$$

The exp operator on $d\mathcal{V}$ has the algebraic properties of an exponential; in particular it takes sums into (convolution) products.

PROPOSITION 3.10. *For $dF, dG \in d\mathcal{V}$, we have*

$$\exp^*(dF + dG) = \exp^* dF * \exp^* dG.$$

PROOF. For K a positive integer, we add the elements of the square matrix

$$\left\{\frac{1}{i!} dF^{*i} * \frac{1}{j!} dG^{*j}\right\}_{0 \le i,j \le K}$$

as a double sum in two ways:

First, sum by rows and then by columns, yielding

$$\sum_{0 \le i \le K} \frac{1}{i!} dF^{*i} * \sum_{0 \le j \le K} \frac{1}{j!} dG^{*j}, \tag{3.8}$$

an approximation to $\exp^* dF * \exp^* dG$,

Next, sum over points (i, j) on fixed diagonals of the form $i + j =$ constant and lying within the square $[0, K] \times [0, K]$, yielding

$$\sum_{0 \le k \le K} \frac{1}{k!} \sum_{i+j=k} \frac{k!}{i!j!} dF^{*i} * dG^{*j} + dT,$$

where dT represents the sum of the elements of the $(K + 1) \times (K + 1)$ matrix lying outside the triangular region $\{i + j \le K\}$. The contribution from the first region is

$$\sum_{k \le K} \frac{1}{k!} (dF + dG)^{*k}, \tag{3.9}$$

an approximation to $\exp^*(dF + dG)$.

Let σ be sufficiently large that $\|dF\|_\sigma$ and $\|dG\|_\sigma$ are each finite. We next show that dT, the difference of the expressions in (3.8) and (3.9), has arbitrarily small σ-norm for large K. Replace dF and dG by their total variation measures and extend the dT sum beyond the $(K + 1) \times (K + 1)$ square to get

$$\|dT\|_\sigma \le \sum_{k>K} \frac{1}{k!} \int_{1-}^{\infty} u^{-\sigma} (|dF| + |dG|)^{*k}(u) = \sum_{k>K} \frac{1}{k!} \left(\|dF\|_\sigma + \|dG\|_\sigma\right)^k.$$

The last series goes to 0 as $K \to \infty$. The preceding argument implies also that (3.9) converges to $\exp^*(dF + dG)$.

It remains to show that (3.8) converges to $\exp^* dF * \exp^* dG$. We have

$$\begin{aligned} dH_K &:= \exp^* dF * \exp^* dG - \sum_{0 \le i \le K} \frac{1}{i!} dF^{*i} * \sum_{0 \le j \le K} \frac{1}{j!} dG^{*j} \\ &= \exp^* dF * \sum_{j>K} \frac{1}{j!} dG^{*j} + \sum_{0 \le j \le K} \frac{1}{j!} dG^{*j} \sum_{i>K} \frac{1}{i!} dF^{*i}. \end{aligned}$$

Using the bound

$$\Big| \sum_{0\le j\le K} \frac{1}{j!} dG^{*j} \Big| \le \exp^*|dG|,$$

and Proposition 3.7 (connecting exponentials and Mellin transforms) we get

$$\|dH_K\|_\sigma \le \exp^*(\|dF\|_\sigma) \sum_{j>K} \frac{1}{j!}(\|dG\|_\sigma)^j + \exp^*(\|dG\|_\sigma) \sum_{i>K} \frac{1}{i!}(\|dF\|_\sigma)^i$$

which converges to 0 as $K \to \infty$. □

COROLLARY 3.11. *If* $dF \in d\mathcal{V}$ *and* $\exp^* dF = dG$, *then* $\exp^*\{-dF\} = dG^{*-1}$

The proof consists of noting that $\exp^* dF * \exp^*(-dF) = \exp^* 0 = \delta_1$.

EXAMPLES 3.12. dN^{*2} AND dN^{*-1} AS EXPONENTIALS. In classical number theory, $dN * dN$ corresponds to the divisor function and dN^{*-1} to the Moebius function. We have

$$dN * dN = \exp^*(2\, d\Pi), \qquad dN^{*-1} = \exp^*(-d\Pi).$$

EXAMPLE 3.13. SQUARE ROOTS. Recall from (3.1) and (1.9) that

$$dN = \exp^* d\Pi, \qquad \delta_1 + dt = \exp^*\Big(\frac{1-t^{-1}}{\log t} dt\Big) =: \exp^* d\Pi_c.$$

Set

$$dN^{*1/2} := \exp^*(\, d\Pi/2), \quad \big(\delta_1 + dt\big)^{*1/2} := \exp^*(d\Pi_c/2).$$

We have $dN^{*1/2} * dN^{*1/2} = dN$ and $\big(\delta_1 + dt\big)^{*1/2} * \big(\delta_1 + dt\big)^{*1/2} = \delta_1 + dt$.

The PNT (for both rational and g-numbers) asserts that $\Pi(x) - \Pi_c(x)$ is "small," where $\Pi_c(x) := \int_1^x (1-t^{-1})\, dt/(\log t)$. As an application of several of our relations, here are two parallel formulas involving $\Pi(x) - \Pi_c(x)$ that will be useful in estimating the integer counting function $N(x)$.

LEMMA 3.14. *For* $x \ge 1$ *we have*

$$N(x) - \int_1^x N(u)u^{-1}\, du = \int_{1-}^x \exp^*(d\Pi - d\Pi_c), \tag{3.10}$$

$$N(x)/x = \int_{1-}^x T^{-1}\{dN * (\delta_1 - dt/t)\} = \int_{1-}^x T^{-1} \exp^*(d\Pi - d\Pi_c), \tag{3.11}$$

where T *is defined in* §2.8.

PROOF. For the first relation, write

$$dN * (\delta_1 - dt/t) = dN * (\delta_1 + dt)^{-1} = \exp^*(d\Pi - d\Pi_c).$$

Thus

$$\int_{1-}^x \exp^*(d\Pi - d\Pi_c) = N(x) - \int_1^x N\Big(\frac{x}{t}\Big)\frac{dt}{t} = N(x) - \int_1^x N(u)\,\frac{du}{u}.$$

For the second relation, use (3.5) to write

$$N(x) = \int_{1-}^x dN * (\delta_1 - dt/t) * (\delta_1 + dt) = \int_{1-}^x \frac{x}{t}\,\{dN * (\delta_1 - dt/t)\}.$$

Thus $N(x)/x$ has the first claimed form. By writing $dN = \exp^* d\Pi$ and $\delta_1 + dt = \exp^* d\Pi_c$, we obtain the second claimed formula for $N(x)/x$. □

3.5. Three equivalent formulas

The exponential relation is equivalent to two other formulas. By this equivalence, Chebyshev's identity (3.3) follows at once from the representation (3.1).

PROPOSITION 3.15. *Let $F, G \in \mathcal{V}$ and suppose $F(1) = 1$ and $G(1) = 0$. Then following three expressions are equivalent:*

$$dF = \exp^* dG \tag{3.12}$$

$$L\,dF = dF * L\,dG \tag{3.13}$$

$$dG = \sum_{i=1}^{\infty} (-1)^{i-1}(dF - \delta_1)^{*i}/i. \tag{3.14}$$

Here L is the log operator and δ_1 Dirac measure at 1.

(The last expression is formally similar to Taylor's series for the logarithm.)

PROOF. (3.12) $\Rightarrow$ (3.13): Apply L termwise to the exponential series. This is valid by Lemma 2.14. We then use (2.12) to evaluate $L\,dG^{*n}$ to get

$$\begin{aligned} L\,dF &= L\{\delta_1 + dG + \frac{1}{2!}dG^{*2} + \frac{1}{3!}dG^{*3} + \dots\} \\ &= L\,dG * \big\{0 + \delta_1 + dG + \frac{1}{2!}dG^{*2} + \dots\big\} = L\,dG * dF. \end{aligned}$$

(3.13) $\Rightarrow$ (3.12): Consider $dH := dF^{*-1} * \exp^* dG$. We show $dH = \delta_1$. First, $dH(\{1\}) = dF^{*-1}(\{1\})\,(\exp^* dG)(\{1\}) = e^{dG(\{1\})}/dF(\{1\}) = e^0/1 = 1 = \delta_1(\{1\})$. Now it suffices to show that $LdH = 0$.

We have

$$LdH = (LdF^{*-1}) * \exp^* dG + dF^{*-1} * L\exp^* dG.$$

We showed just above that $L\exp^* dG = (\exp^* dG) * LdG$. Next,

$$0 = L\delta_1 = L(dF * dF^{*-1}) = (LdF) * dF^{*-1} + dF * (LdF^{*-1}).$$

Thus

$$LdF^{*-1} = -(LdF) * dF^{*-2},$$

and we have

$$\begin{aligned} LdH &= (\exp^* dG) * \big\{ -(LdF) * dF^{*-2} + dF^{*-1} * LdG\big\} \\ &= (\exp^* dG) * dF^{*-2} * \big\{dF * LdG - LdF\big\} = 0. \end{aligned}$$

Thus $dF^{*-1} * \exp^* dG = \delta_1$, i.e. (3.12) holds.

(3.13) $\Leftrightarrow$ (3.14): Formula (3.14) is equivalent to

$$LdG = L\sum_{i=1}^{\infty}(-1)^{i-1}(dF - \delta_1)^{*i}/i\,, \tag{3.15}$$

since both sides of (3.14) are 0 on the set $\{1\}$, the only place where L is 0. Now L is a continuous operator, so we can apply it termwise to the last sum. Using the series representation (3.6) for dF^{*-1}, we obtain

$$LdG = LdF * \sum_{i=1}^{\infty}(-1)^{i-1}(dF - \delta_1)^{*(i-1)} = LdF * dF^{*-1}.$$

The last formula is equivalent to (3.13). □

THE EXPONENTIAL REPRESENTATION (3.7) (AGAIN). This time, we prove the formula using (3.13), specifically, we show that

$$L(\delta_1 + dt) = (\delta_1 + dt) * (1 - t^{-1})\, dt. \tag{3.16}$$

It suffices to verify that the integrals of the two sides are equal. On the one hand,

$$\int_{1-}^{x} L(\delta_1 + dt) = \int_{1-}^{x} \log t\, dt = x \log x - x + 1,$$

and on the other, by iterated integration,

$$\int_{1-}^{x} (\delta_1 + dt) * (1 - t^{-1})\, dt = \int_{1}^{x} \frac{x}{u} (1 - u^{-1})\, du = x \log x - x + 1.$$

Thus (3.16) holds, and so $\delta_1 + dt = \exp^* d\Pi_c$ by the preceding proposition.

3.6. Notes

Exponentials of arithmetic functions on $\mathbb{N}$ are treated in [**Di70a**] and [**BD04**], §2.4.

§3.1 The interested reader might use (3.1) or (3.3) to show that $\int_1^x L dN \geq \psi(x)$.

§3.4. Exponentials are presented via Dirichlet series in [**Ap76**], §11.9.

§3.5. By Proposition 3.15 the exponential mapping from $\mathcal{V}(0)$ to $\mathcal{V}(1)$ is one-to-one and onto; this insures that g-prime and g-integer systems determine one other uniquely.

CHAPTER 4

Upper and Lower Estimates of $N(x)$

Size matters

SUMMARY. Renormalizing g-integer counting functions having polynomial growth to make linear growth. Growth rate of type $x \log^\alpha x$. Upper and lower logarithmic densities. An example of g-primes and associated g-integers having *the same* growth rate. Regular growth.

4.1. Normalization and restriction

In §1.1 it was convenient to assume that a g-number system $\mathcal{N}$ satisfies the O-density condition (1.2), i.e. that $N(x)/x$ is bounded. Suppose instead that our g-number system satisfies the more general polynomial growth condition

$$\limsup_{x\to\infty} N(x)/x^\alpha < \infty \tag{4.1}$$

for some positive $\alpha \neq 1$. While systems of such growth are less usual, they lie in the function classes described in §2.1 and allow analysis by Mellin transform techniques.

However, as we show now, a simple mapping takes such a g-number system $\mathcal{N}$ into another one, $\mathcal{N}'$, for which (4.1) holds with $\alpha = 1$. Let $\mathcal{P}'$ denote a new set of g-primes created from $\mathcal{P}$ by the mapping $p_i \mapsto p_i^\alpha =: p_i'$. The g-integers n' of the associated g-number system $\mathcal{N}'$ are related to those of $\mathcal{N}$ by $n_i' = n_i^\alpha$. The counting function of $\mathcal{N}'$ satisfies

$$N'(x) = \#\{n_i' \leq x\} = \#\{n_i^\alpha \leq x\} = \#\{n_i \leq x^{1/\alpha}\} = N(x^{1/\alpha}) \ll x,$$

i.e. the O-density condition (1.2) holds for $\mathcal{N}'$. Thus, to treat a system whose counting function is of polynomial growth, we may as well assume from the outset that $N(x)/x$ is bounded.

We now give another example of a g-number system $\mathcal{N}$ whose integer counting function has "excessive" size. Recall from classical number theory the divisor function $\tau(n)$ and its counting function $D(x)$. The latter satisfies

$$dD = dN * dN = \exp^*(2\, d\Pi).$$

Here N is the counting function of the rational integers and Π that of the weighted primes and prime powers. The g-primes of $\mathcal{N}$ are the rational primes, each with multiplicity 2, and the g-integers are the rational integers, with n having multiplicity $\tau(n)$. By Dirichlet's well-known approximation, the counting function of $\mathcal{N}$ has size $D(x) \sim x \log x$. *In this book, we generally treat cases in which the g-prime counting function is near $x/\log x$ and the counting function of the g-integers is $O(x)$.*

In this chapter, we shall consider upper and lower logarithmic densities, which are cruder measures of the growth of $N(\cdot)$ than density. (In Chapters 6 and 7 we shall take up g-number systems satisfying various notions of density.) At the end of the chapter we examine two anomalous cases.

4.2. O-log density

We define $\overline{ld}\,(\mathcal{N})$, the upper logarithmic density of $\mathcal{N}$, to be

$$\limsup_{x\to\infty}\left\{\frac{1}{\log x}\int_{1-}^{x} u^{-1}\,dN(u)\right\}.$$

If this quantity is finite, we say that $\mathcal{N}$ has *O-log density.* We define $\underline{ld}\,(\mathcal{N})$, lower logarithmic density, to be the limit inferior of the corresponding expression. We say that $\mathcal{N}$ has *logarithmic density* in case the two quantities are finite and equal. Lower log density is discussed in the following section and log density in §5.2.

When a number system satisfies an O-density condition (cf. §§1.2, 6.2), it satisfies an O-log density condition as well, as shown by summation or integration by parts. The converse implication is false, as we illustrate in Example 6.1 below; thus we say that O-log density is a "weaker" condition than O-density. In the rest of this section we give simple criteria for generalized integers to have O-log density. These involve the behavior of either the associated zeta function at $s = 1+$ or (a weighted version of) the prime counting function. Generally similar relations hold for lower densities.

LEMMA 4.1. *Suppose that* $F \in \mathcal{V}$ *is a nondecreasing function with a Mellin transform*

$$\widehat{F}(s) := \int_{1-}^{\infty} u^{-s} dF(u).$$

Then $\widehat{F}(s) = O(1/s)$ *as* $s \to 0+$ *if and only if* $F(x) = O(\log x)$ *as* $x \to \infty$.

PROOF. Suppose first that there is a number B such that $\widehat{F}(s) \le B/s$ for $0 < s < 1$. Given $x > e$, take $s = 1/\log x \in (0,\,1)$. We find

$$(1/e)F(x) \le \int_{1-}^{x} u^{-s} dF(u) \le \int_{1-}^{\infty} u^{-s} dF(u) \le B\log x.$$

In the other direction, suppose that

$$F(x) \le C + C\log x, \quad x \ge 1.$$

For $s > 0$, we have by Lemma 2.2 (comparison of integrators),

$$\widehat{F}(s) = \int_{1-}^{\infty} u^{-s} dF(u) \le C\int_{1-}^{\infty} u^{-s}(\delta_1 + du/u) = C + C/s,$$

i.e. $\widehat{F}(s) \ll 1/s, \quad 0 < s < 1.$ □

Taking $\widehat{F}(s) = \zeta_{\mathcal{N}}(s+1)$ in the preceding lemma, we obtain

PROPOSITION 4.2. *A g-number system* $\mathcal{N}$ *has O-log density if and only if*

$$\limsup_{s\to 1+}\,(s-1)\,\zeta_{\mathcal{N}}(s) < \infty. \tag{4.2}$$

We call the value of the left side of (4.2) the upper right hand residue of $\zeta_{\mathcal{N}}(s)$ at $s = 1$, or more briefly, the *upper residue of* $\zeta_{\mathcal{N}}(s)$ there. Adapting the proof of Lemma 4.1, we obtain

COROLLARY 4.3. *A g-number system $\mathcal{N}$ has log density 0 if and only if $\zeta_{\mathcal{N}}(s)$ has upper residue 0 at $s = 1$.*

What role do small primes play in determining whether a g-number system has O-log density? Not surprisingly, the answer is None, as we show next. Thus we can alter the initial part of a g-prime distribution in checking whether the associated g-integers have O-log density.

We treat differently a g-number system generated by discrete primes from one arising from a general increasing function $\Pi(x)$. In the first case, say $\mathcal{P}$ is a discrete g-prime system with counting function $\pi(x)$ and a number $B > 1$ is given. Partition $\mathcal{P}$ into $\mathcal{P}_1$, the g-primes not exceeding B, and $\mathcal{P}_2$, the remaining g-primes. For $i = 1, 2$, let π_i denote the counting function of $\mathcal{P}_i$, and Π_i the weighted g-prime and g-prime power counting function of $\mathcal{P}_i$. Of course, powers of a prime $p \in \mathcal{P}_1$ can exceed B and will be counted in Π_1.

In the second case, write $d\Pi = d\Pi_1 + d\Pi_2$ with

$$d\Pi_1(u) = \chi_{(1,\,B]}(u)\,d\Pi(u) \quad\text{and}\quad d\Pi_2(u) = \chi_{(B,\,\infty)}(u)\,d\Pi(u),$$

with χ_E the indicator function of a set E.

In both cases let N_i denote the counting function of $\mathcal{N}_i$. We have

$$dN = \exp^*\{d\Pi_1 + d\Pi_2\} = dN_1 * dN_2 \quad\text{and}\quad \zeta(s) = \zeta_1(s)\,\zeta_2(s),$$

where ζ_i is the zeta function associated with $\mathcal{N}_i$.

PROPOSITION 4.4. *Let $\mathcal{N}$ be a g-number system generated by $d\Pi = d\Pi_1 + d\Pi_2$ as above. Then $\mathcal{N}$ has O-log density if and only if $\mathcal{N}_2$ has the same.*

PROOF. It is obvious that if $\mathcal{N}$ has O-log density, so does $\mathcal{N}_2$. Now suppose that $\mathcal{N}_2$ has O-log density. By Proposition 4.2, $(s-1)\zeta_{\mathcal{N}_2}(s)$ is bounded as $s \to 1+$.

We now show that $\zeta_1(1)$ is finite. For $\mathcal{P}$ discrete with smallest prime $q > 1$,

$$\begin{aligned}\log\zeta_1(1) &= \int_1^\infty u^{-1}d\Pi_1(u) = \int_1^B \Big\{u^{-1} + \frac{1}{2}u^{-2} + \frac{1}{3}u^{-3} + \dots\Big\}\,d\pi(u)\\ &\le \int_1^B u^{-1}\Big\{1 + \frac{1}{q} + \frac{1}{q^2} + \dots\Big\}\,d\pi(u) \le \frac{q}{q-1}\,\pi(B) < \infty,\end{aligned}$$

and for a general increasing function Π,

$$\log\zeta_1(1) = \int_1^\infty u^{-1}d\Pi_1(u) = \int_1^B u^{-1}d\Pi(u) \le \Pi(B) < \infty.$$

For $s \in (1, 2)$, we have $\zeta(s) \le \zeta_1(1)\,\zeta_2(s) \ll 1/(s-1)$. Thus $\mathcal{N}$ has O-log density, by another application of Proposition 4.2. □

The reader is encouraged to give an elementary proof of this proposition.

PROPOSITION 4.5. *Let $\mathcal{N}$ be a g-number system for which*

$$\int_1^x \frac{d\Pi(u)}{u} = \log\log ex + E(x), \quad x \ge 1,$$

holds for some function $E(x)$. If $E(x)$ is bounded above, then $\mathcal{N}$ has O-log density. On the other hand, if $E(x)$ is unbounded above but bounded from below, then $\mathcal{N}$ does not have O-log density.

PROOF. Since

$$\int_1^x \frac{1-u^{-1}}{u\log u}\,du - \log\log ex$$

is bounded, the formula of the proposition can be changed to

$$\int_1^x \frac{d\Pi(u)}{u} = \int_1^x \frac{1-u^{-1}}{u\log u}\,du + E_1(x), \quad x \geq 1, \tag{4.3}$$

where E and E_1 differ by at most a bounded quantity.

Suppose $E_1(x)$ is bounded above by some positive number K. Then (4.3) and Lemma 2.2 give the Mellin inequality

$$\int_1^\infty u^{-s} d\Pi(u) < \int_1^\infty u^{-(s-1)} \frac{1-u^{-1}}{u\log u} du + K.$$

Recalling formulas (1.5) and (1.8), we find

$$\log\zeta(s) < \log\frac{s}{s-1} + K, \quad s > 1.$$

Thus $\limsup_{s\to 1+}(s-1)\zeta(s)$ is finite, and so $\mathcal{N}$ has O-log density by Proposition 4.2.

Next, suppose (4.3) holds with $E_1(x) \geq -M$ for all $x \geq 1$ and $E_1(x)$ is unbounded above. If the zeta function of $\mathcal{N}$ has abscissa of convergence $\sigma_c > 1$, then $\mathcal{N}$ does not have O-log density and we are done; thus it suffices to assume that $\sigma_c = 1$. Since $d\Pi \geq 0$, we have $\log\zeta(s) < \zeta(s)$ for $s > 1$, and hence the Mellin integral for $\log\zeta(s)$ also is convergent there. Thus,

$$\int_1^\infty u^{-(s-1)}\, dE_1(u) = \int_1^\infty u^{-s}\, d\Pi(u) - \int_1^\infty u^{-s} \frac{1-u^{-1}}{\log\, u}\, du$$

holds for $s > 1$. Integrating by parts and writing $s = 1 + \epsilon$, we then obtain

$$I(\epsilon) := \epsilon \int_1^\infty u^{-1-\epsilon} E_1(u)\, du = \log\{\epsilon\,\zeta(1+\epsilon)/(1+\epsilon)\}, \quad \epsilon > 0. \tag{4.4}$$

Given a large positive number L, we have $E_1(u_0) \geq L$ for some (large!) number u_0. Then $E_1(u)$ remains large for a rather long interval beyond u_0: we have

$$\begin{aligned} E_1(u) &= E_1(u_0) + \int_{u_0}^u \left\{\frac{d\Pi(t)}{t} - \frac{1-t^{-1}}{t\log t}\,dt\right\} \\ &> L - \int_{u_0}^u \frac{dt}{t\log t} = L - \log\frac{\log u}{\log u_0} \geq \frac{L}{2} \end{aligned}$$

for

$$\log u_0 \leq \log u \leq (\log u_0)\exp(L/2) := \log u_1.$$

It follows that, for a suitable small ϵ, $I(\epsilon)$ is large: indeed,

$$I(\epsilon) > -M\epsilon \int_1^\infty u^{-1-\epsilon}\, du + \frac{L\epsilon}{2}\int_{u_0}^{u_1} u^{-1-\epsilon}\, du = -M + \frac{L}{2}({u_0}^{-\epsilon} - {u_1}^{-\epsilon}).$$

Now choose $\epsilon = (\log 2)/(\log u_0)$, so that ${u_0}^{-\epsilon} = 1/2$. A small calculation shows that ${u_1}^{-\epsilon} < 1/10$ for any $L \geq 3$, say. Thus

$$I(\epsilon) > -M + L/5.$$

Since M is fixed and L can be arbitrarily large (and hence ϵ small), we have

$$\limsup_{\epsilon\to 0+} \log\{\epsilon\zeta(1+\epsilon)\} = \limsup_{\epsilon\to 0+} I(\epsilon) = \infty.$$

By another application of Proposition 4.2, $\mathcal{N}$ does not have O-log density. □

EXAMPLE 4.6. PNT $\not\Rightarrow$ O-LOG DENSITY. Suppose that the prime counting function of a g-number system $\mathcal{N}$ satisfies

$$\Pi(x) - \int_1^x \frac{1-v^{-1}}{\log v}\,dv = \frac{x+o(x)}{\log ex\,\log\log 10x}, \quad x \geq 1. \tag{4.5}$$

The PNT clearly holds here. However, with the notation of the last proposition,

$$E_1(x) \sim \int_1^x \frac{dt}{t\log et\,\log\log 10t},$$

which diverges to infinity as $x \to \infty$. By the last proposition, $\mathcal{N}$ does not have O-log density.

4.3. Lower log density

Recall that we defined the lower logarithmic density of $\mathcal{N}$ by

$$\underline{ld}(\mathcal{N}) = \liminf_{x\to\infty} \frac{1}{\log x}\int_{1-}^x u^{-1}\,dN(u).$$

Here we establish some conditions for $\underline{ld}(\mathcal{N})$ to be positive.

THEOREM 4.7. *If a g-number system $\mathcal{N}$ has positive lower logarithmic density, then*

$$\int_1^x u^{-1}d\Pi(u) \geq \log\log ex - M \tag{4.6}$$

holds for some positive number M and all $x \geq 1$. Conversely, if (4.6) holds, then $\mathcal{N}$ has either positive or infinite lower logarithmic density.

PROOF. Suppose first that $\underline{ld}(\mathcal{N}) > 0$. By Lemmas 3.9 and 2.15, we have

$$\exp\left\{\int_{1-}^x T^{-1}d\Pi\right\} \geq \int_{1-}^x \exp^*\{T^{-1}d\Pi\}(u) = \int_{1-}^x T^{-1}\exp^*\{d\Pi\}(u).$$

By (3.1) and the hypothesis, the last integral is

$$\int_{1-}^x T^{-1}dN = \int_{1-}^x u^{-1}dN(u) \geq c\log x$$

for all $x \geq 1$ and some positive number c. This establishes (4.6).

Next, assume that (4.6) holds. We have

$$\int_1^x u^{-1}d\Pi(u) + K \geq \int_1^x \frac{1-u^{-1}}{u\log u}du$$

for some positive K, by the boundedness of

$$\int_1^x \frac{1-u^{-1}}{u\log u}du - \log\log x.$$

Defining $d\Pi_+ := d\Pi + K\delta_1$, where δ_1 is Dirac measure at 1, we obtain the equivalent inequality

$$\int_{1-}^x T^{-1}d\Pi_+ = \int_{1-}^x u^{-1}d\Pi_+(u) \geq \int_{1-}^x u^{-1}\frac{1-u^{-1}}{\log u}du = \int_{1-}^x T^{-1}d\Pi_c\,.$$

Using Lemmas 2.9 and 2.15, we find for each nonnegative integer n

$$\int_{1-}^{x} T^{-1} d\Pi_+^{*n} \geq \int_{1-}^{x} T^{-1} d\Pi_c^{*n},$$

and summing terms of the exponential series, we obtain

$$\int_{1-}^{x} T^{-1} \exp^* d\Pi_+ \geq \int_{1-}^{x} T^{-1} \exp^* d\Pi_c. \tag{4.7}$$

Now

$$\exp^* d\Pi_+ = \exp^*\{d\Pi + K\delta_1\} = (\exp^* d\Pi) * (\exp^*\{K\delta_1\}) = e^K dN$$

and $\exp^* d\Pi_c = \delta_1 + dt$, by the homomorphic property of exp and the identifications of $\exp^* d\Pi$ and $\exp^* d\Pi_c$ provided by (3.1) and (3.7) respectively.

Thus (4.7) translates to

$$e^K \int_{1-}^{x} T^{-1} dN \geq \int_{1-}^{x} T^{-1}(\delta_1 + dt) = 1 + \log x,$$

so the lower logarithmic density of $\mathcal{N}$ is either positive or infinite. □

PROPOSITION 4.8. *Suppose that a g-number system $\mathcal{N}$ has positive lower logarithmic density. Then*

$$\liminf_{s \to 1+} (s-1)\zeta_{\mathcal{N}}(s) > 0. \tag{4.8}$$

We call the value of the left side of (4.8) the *lower residue of $\zeta_{\mathcal{N}}(s)$ at $s = 1$.* The proposition is proved by integration by parts, as was done for the corresponding part of Proposition 4.2.

Does the converse of Proposition 4.8 hold? The situation here is different from Proposition 4.2: without further conditions, the answer is No. In the next section we shall give an example in which $\underline{ld}(\mathcal{N}) = 0$ but the zeta function has an infinite residue at $s = 1+$. However, if the zeta function also has a finite *upper* residue at 1, then, as we show in Corollary 4.10, $\underline{ld}(\mathcal{N}) > 0$. In the remainder of this section, we give some sufficient conditions for positive lower logarithmic density.

PROPOSITION 4.9. *Suppose that $\mathcal{N}$ satisfies* (4.8) *and the bound*

$$\zeta_{\mathcal{N}}(1+\epsilon)/\zeta_{\mathcal{N}}(1+2\epsilon) \ll 1 \tag{4.9}$$

for $0 < \epsilon < 1$. Then $\mathcal{N}$ has positive lower logarithmic density.

PROOF. By hypothesis, there exist numbers m, B, and ϵ_0 such that

$$\epsilon\,\zeta(1+\epsilon) \geq m > 0 \quad \text{and} \quad \frac{2\zeta(1+\epsilon)}{\zeta(1+2\epsilon)} \leq B$$

for $0 < \epsilon < \epsilon_0$. Note that $B \geq 2$, since $\zeta(s)$ decreases on the half-line $\{s : s > 1\}$. Let $\epsilon \in (0, \epsilon_0]$, to be specified. We have

$$B\zeta(1+2\epsilon) - \zeta(1+\epsilon) = \int_{1-}^{\infty} u^{-1-\epsilon}(Bu^{-\epsilon} - 1)\, dN(u).$$

The last integrand is positive for $1 < u < B^{1/\epsilon} =: U$, and it is negative for all $u > U$. It follows that

$$(2-1)\zeta(1+\epsilon) < \int_{1-}^{U} (Bu^{-\epsilon} - 1)u^{-1-\epsilon} dN(u) < \int_{1-}^{U} Bu^{-1} dN(u)$$

(the last inequality is immediate) or

$$\frac{m}{\epsilon} \leq \zeta(1+\epsilon) < B \int_{1-}^{U} u^{-1} dN(u).$$

For x sufficiently large, take $\epsilon = \log B / \log x$. Then $U = x$ and we obtain finally

$$\int_{1-}^{x} u^{-1} dN(u) > \frac{m \log x}{B \log B} \gg \log x. \qquad \square$$

An immediate consequence of this proposition is

COROLLARY 4.10. *If* $\epsilon\zeta_{\mathcal{N}}(1+\epsilon)$ *has positive finite bounds from above and below as* $\epsilon \to 0+$, *then* $\mathcal{N}$ *has positive lower logarithmic density.*

The interested reader is encouraged to give a direct proof of the corollary.

Could we have proved the result of Proposition 4.9 without assuming condition (4.8)? By Proposition 4.8, this condition is necessary in order for $\mathcal{N}$ to have positive lower logarithmic density. Thus, our question is, Does (4.9) imply (4.8)? The answer is No, as we now show.

EXAMPLE 4.11. CONDITION (4.9) $\not\Rightarrow$ CONDITION (4.8). Define the g-integer counting function $F(x)$ of a (continuous) system by taking dF to be the convolution square root of $\delta_1 + dt$ (see Example 3.13). Note that $dF \geq 0$, since it is the exponential of a nonnegative measure. By Lemma 2.6, we have

$$\widehat{F}(s) := \int_{1-}^{\infty} x^{-s} dF(x) = \Big(\frac{s}{s-1}\Big)^{1/2}, \quad s > 1,$$

and hence $\lim_{s\to 1+}(s-1)\widehat{F}(s) = 0$; i.e. (4.8) is not satisfied. On the other hand, (4.9) holds, as

$$\frac{\widehat{F}(1+\epsilon)}{\widehat{F}(1+2\epsilon)} \sim \Big\{\frac{2\epsilon}{\epsilon}\Big\}^{1/2} = \sqrt{2} \ \text{ as } \epsilon \to 0+\,.$$

Another version of Proposition 4.9 can be deduced from the following

PROPOSITION 4.12. *Formula* (4.9) *holds for* $\mathcal{N}$ *if and only if*

$$\frac{-\zeta_{\mathcal{N}}'}{\zeta_{\mathcal{N}}}(1+\epsilon) \ll \frac{1}{\epsilon}, \quad 0 < \epsilon < 1. \tag{4.10}$$

PROOF. By the mean value theorem,

$$\log\zeta(1+\epsilon) - \log\zeta(1+2\epsilon) = \frac{-\zeta'}{\zeta}(1+c\,\epsilon)\,\epsilon$$

for some $c \in (1,\, 2)$. Also, by (1.6),

$$\frac{-\zeta'}{\zeta}(s) = \int_{1}^{\infty} u^{-s} d\psi(u),$$

where $\psi\uparrow$, so $(-\zeta'/\zeta)(s)$ is a decreasing function of s. These relations yield

$$\frac{-\zeta'}{\zeta}(1+2\epsilon)\,\epsilon \leq \log\{\zeta(1+\epsilon)/\zeta(1+2\epsilon)\} \leq \frac{-\zeta'}{\zeta}(1+\epsilon)\,\epsilon.$$

The right hand inequality and (4.10) imply that $\zeta(1+\epsilon)/\zeta(1+2\epsilon) \ll 1$. Conversely, the left hand inequality and (4.9) give an upper bound for $2\epsilon\,(-\zeta'/\zeta)(1+2\epsilon)$. $\square$

COROLLARY 4.13. *Suppose that* (4.8) *and* (4.10) *both hold. Then* $\mathcal{N}$ *has positive lower logarithmic density.*

4.4. An example with infinite residue but 0 lower log density

EXAMPLE 4.14. Define a very lacunary g-prime system $\mathcal{P}$ by setting $p_1 = e$ and for $m \geq 2$, setting $p_m = \exp\{m \exp p_{m-1}\}$ with multiplicity $b_m = \lfloor {p_m}^{1+1/m} \rfloor$. Let $\mathcal{N} = \{n_i\}$ denote the associated g-integers with $n_0 = 1$ and, as usual, $\zeta(s)$ the g-zeta function. We show first that

$$\epsilon\zeta(1+\epsilon) \to \infty, \quad \epsilon \to 0+, \tag{4.11}$$

and then that

$$\liminf_{x\to\infty} \frac{1}{\log x} \sum_{n_i \leq x} \frac{1}{n_i} = 0. \tag{4.12}$$

To begin, we show that $\zeta(s)$ converges for $s > 1$. Without loss of generality, we can assume that $s \in (1, 2)$. Partition $\mathcal{P}$ into

$$\mathcal{P}_1 := \{p_m \text{ with multiplicity } b_m,\ m < 2/(s-1)\},$$
$$\mathcal{P}_2 := \{p_m \text{ with multiplicity } b_m,\ m \geq 2/(s-1)\}.$$

Let $\pi_1(x)$ and $\pi_2(x)$ denote the g-prime counting functions and $\Pi_1(x)$ and $\Pi_2(x)$ the weighted g-prime and g-prime power counting functions of $\mathcal{P}_1$ and $\mathcal{P}_2$ respectively. Then we have

$$\exp\{d\Pi\} = \exp\{d\Pi_1 + d\Pi_2\} = \exp\{d\Pi_1\} * \exp\{d\Pi_2\}.$$

It follows that

$$\zeta(s) = \int_{1-}^{\infty} x^{-s}(\exp\{d\Pi_1\} * \exp\{d\Pi_2\})(x) = \zeta_1(s) \cdot \zeta_2(s),$$

where

$$\zeta_\nu(s) := \int_{1-}^{\infty} x^{-s} \exp\{d\Pi_\nu\}(x), \quad \nu = 1, 2.$$

Now, we have

$$\zeta_1(s) = \prod_{m<2/(s-1)} (1 - {p_m}^{-s})^{-b_m},$$

which is a finite product and hence convergent. Next,

$$\log\zeta_2(s) = \int_1^\infty x^{-s} d\Pi_2(x) = \sum_1^\infty \frac{1}{k} \int_1^\infty x^{-s} d\pi_2(x^{1/k}) = \sum_1^\infty \frac{1}{k} \int_1^\infty x^{-ks} d\pi_2(x).$$

Let $I = I(k, s)$ denote the last integral. We have

$$\begin{aligned} I = &\sum_{m\geq 2/(s-1)} {p_m}^{-ks} \lfloor {p_m}^{1+1/m} \rfloor \leq \sum_{m\geq 2/(s-1)} {p_m}^{-k-(s-1)} {p_m}^{1+1/m} \\ \leq &\sum_{m\geq 2/(s-1)} {p_m}^{-(k-1)-1/m} = \sum_{m\geq 2/(s-1)} {p_m}^{-(k-1)} \exp\{-\exp p_{m-1}\} \\ \leq &\, {p_M}^{-(k-1)} \sum_{m\geq 2/(s-1)} \exp\{-\exp p_{m-1}\}, \end{aligned}$$

where $M = \lceil 2/(s-1) \rceil$. We have

$$\log\zeta_2(s) \leq \sum_{k\geq 1} {p_M}^{-(k-1)} \sum_{m\geq 2/(s-1)} \exp\{-\exp p_{m-1}\};$$

thus the integral for $\zeta_2(s)$ converges for $s > 1$, and hence so does that of $\zeta(s)$.

Next, we use the high multiplicity of integers to show that the right hand residue of $\zeta(s)$ at $s = 1$ is infinite. For $s > 1$, choose m such that $1/(2m+2) < s - 1 \le 1/(2m)$. Using a single term of the series for $\log \zeta_2(s)$, we have

$$\begin{aligned}
(s-1)\zeta(s) &> (s-1)\log\zeta_2(s) \ge (s-1)\, p_m{}^{-s} \lfloor p_m{}^{1+1/m} \rfloor \\
&\ge (s-1)\, p_m{}^{-s+1+1/m} (1 - p_m{}^{-1-1/m}) \\
&> \frac{1}{2m+2}\, p_m{}^{1/(2m)} \frac{1}{2} = \frac{1}{4m+4} \exp\{(1/2)\exp p_{m-1}\},
\end{aligned}$$

which goes to ∞ as $s \to 1+$, proving (4.11).

Finally, we show $\underline{ld}(\mathcal{N}) = 0$. For μ a positive integer, take $x = p_{\mu+1} - 1$. If a g-integer n_i satisfies $n_i \le x$, then n_i has only g-prime divisors p_m with $m \le \mu$. Hence we have

$$\int_{1-}^{x} t^{-1} dN(t) \le \prod_{m \le \mu} \left(1 - p_m{}^{-1}\right)^{-b_m},$$

and so

$$\begin{aligned}
\log \int_{1-}^{x} t^{-1} dN(t) &\le \sum_{m \le \mu} \lfloor p_m{}^{1+1/m} \rfloor \sum_{k=1}^{\infty} \frac{1}{k} p_m{}^{-k} \\
&\le 2 \sum_{m \le \mu} p_m{}^{1+1/m} p_m{}^{-1} \le 2\mu\, p_\mu{}^{1/\mu}.
\end{aligned}$$

Now

$$\log x \ge \frac{1}{2} \log p_{\mu+1} = \frac{\mu+1}{2} \exp p_\mu,$$

and hence

$$\int_{1-}^{x} t^{-1} dN(t) \le \frac{2}{\mu+1} \exp\{-p_\mu + 2\mu\, p_\mu{}^{1/\mu}\} \log x = o(\log x).$$

for $x = p_{\mu+1} - 1 \to \infty$. This establishes (4.12).

4.5. Extreme thinness is inherited

The rational primes are distributed "thinly" in the rational integers in the sense that $\pi_{\mathbb{N}}(x) = o(N_{\mathbb{N}}(x))$. In this section we exhibit a condition under which the counting function of a system of g-integers actually has the *same* order of growth as that of the associated g-primes.

PROPOSITION 4.15. *Suppose that the counting function of a g-prime system $\mathcal{P}$ satisfies*

$$\pi(x) = \pi_{\mathcal{P}}(x) \ll x/\log^\lambda x,$$

for a constant $\lambda > 1$. Then the g-integer counting function of $\mathcal{N}_{\mathcal{P}}$ also satisfies

$$N(x) \ll x/\log^\lambda x.$$

PROOF. Deleting the first few g-primes from $\mathcal{P}$ changes $N(x)$ asymptotically only by a fixed factor. Take $\alpha = \alpha(\lambda)$ so large that

$$\alpha > e^\lambda \quad \text{and} \quad (\log \alpha)^{\lambda-1} > 2/(\lambda - 1).$$

The function $x \mapsto x/\log^\lambda x$ is increasing for $x > e^\lambda$, so it is convenient to assume, without loss of generality, that $\pi(x) = 0$ for $x \le \alpha$.

By Lemma 1.1, the weighted g-prime and g-prime power counting function $\Pi(x)$ satisfies the same O-bound as does $\pi(x)$. Let $B > 0$ be a constant such that

$$\Pi(x) \leq Bx/\log^\lambda x, \quad x > 1. \tag{4.13}$$

Of course, $\Pi(x) = 0$ for $x \leq \alpha$.

We show by induction that, for all $k \in \mathbb{N}$ and $x > \alpha$,

$$\Pi_k(x) := \int_1^x d\Pi^{*k} \leq 2^{\lambda(k-1)} B^k x/\log^\lambda x\,. \tag{4.14}$$

Indeed, (4.14) holds for $k = 1$ by (4.13). Next, by Dirichlet's hyperbola method,

$$\Pi_{k+1}(x) \leq \int_1^{\sqrt{x}} \Pi_k(x/t)\, d\Pi(t) + \int_1^{\sqrt{x}} \Pi(x/t)\, d\Pi_k(t) =: I_1 + I_2$$

say, since the contribution of the overlapped square $[1, \sqrt{x}] \times [1, \sqrt{x}]$ is nonnegative. We have

$$I_2 \leq B \int_1^{\sqrt{x}} \frac{x/t}{\log^\lambda x/t}\, d\Pi_k(t) \leq \frac{2^\lambda B\, x}{\log^\lambda x} \int_1^{\sqrt{x}} t^{-1} d\Pi_k(t).$$

Let c_k denote the coefficient on the right side of (4.14). Extend the range of the last integral to infinity and apply integration by parts. We find

$$\int_1^\infty t^{-1} d\Pi_k(t) = \int_1^\infty t^{-2} \Pi_k(t)\, dt \leq c_k \int_\alpha^\infty \frac{dt}{t \log^\lambda t} = \frac{c_k/(\lambda - 1)}{(\log \alpha)^{\lambda - 1}} < \frac{c_k}{2},$$

by the definition of α.

Similarly,

$$I_1 \leq c_k \int_1^{\sqrt{x}} \frac{x/t}{\log^\lambda x/t}\, d\Pi(t) \leq \frac{2^\lambda c_k\, x}{\log^\lambda x} \int_1^{\sqrt{x}} t^{-1} d\Pi(t),$$

and using the bound for $\int_1^\infty t^{-1} d\Pi_k(t)$ with $k = 1$, we find that the last integral is at most $B/2$. Combining the bounds for I_1 and I_2, we find

$$\Pi_{k+1}(x) \leq \frac{2^\lambda B\, c_k\, x}{\log^\lambda x} = \frac{c_{k+1}\, x}{\log^\lambda x},$$

proving (4.14).

Now we insert (4.14) into the exponential formula for dN. We get

$$\begin{aligned} N(x) &= \int_{1-}^x \left\{\delta_1 + \Pi + \frac{1}{2!} d\Pi * d\Pi + \dots\right\} \\ &< 1 + 2^{-\lambda} \sum_{k=1}^\infty \frac{1}{k!} (2^\lambda B)^k \frac{x}{\log^\lambda x} \ll \frac{x}{\log^\lambda x}, \end{aligned}$$

since the infinite series is convergent. Thus the claimed O-bound for $N(x)$ holds. □

EXAMPLE 4.16. A THIN SET OF NUMBERS. Let $\mathcal{P}$ be a g-prime system with

$$p_n := n(\log(2 + n))^2, \quad n = 1, 2, \dots.$$

A small calculation shows that

$$\pi_{\mathcal{P}}(x) := \#\{n : p_n \leq x\} \sim x/\log^2 x.$$

By the last proposition, we have also $N_{\mathcal{P}}(x) \ll x/\log^2 x$.

4.6. Regular growth

We say that a positive nondecreasing function f on $[1, \infty)$ has *regular growth* if there exists a nonnegative exponent α such that, for all $c > 0$, $f(cx)/f(x) \sim c^\alpha$ as $x \to \infty$. One can show ([**BD69**], §4) that if f has this property, then for any $\epsilon > 0$,

$$c_\epsilon x^{\alpha-\epsilon} \leq f(x) \leq C_\epsilon x^{\alpha+\epsilon}$$

holds with suitable constants c_ϵ and C_ϵ for all $x \geq 1$. In particular, if the g-integer counting function of a system $\mathcal{N}$ is of regular growth with exponent $\alpha \neq 1$, then the abscissa of convergence of the corresponding zeta function is not 1. As noted in § 4.1, systems we consider are normalized to make $\alpha = 1$.

It is immediate from the definitions of regular growth and density that

$$\text{positive density} \Rightarrow \text{regular growth}.$$

However, regular growth does not have connections with other of the properties we treat, so it is not going to play an important role in this work.

Example 4.17. Regular growth $\not\Rightarrow$ logarithmic density (and *a fortiori* regular growth does not imply density.) Let $N(x) = \lfloor x \rfloor$ and take $dN_2 := dN * dN$. By the Dirichlet divisor estimate, $N_2(x) \sim x \log x$. Easy calculations show that $N_2(cx)/N_2(x) \sim c$, while

$$\int_{1-}^{x} \frac{dN_2(u)}{u} = \frac{N_2(x)}{x} + \int_1^x \frac{N_2(u)}{u^2}\, du \sim \frac{1}{2} \log^2 x.$$

Thus, N_2 has regular growth but not logarithmic density.

Example 4.18. Logarithmic density $\not\Rightarrow$ regular growth. Set

$$\Pi(x) := \sum_{k \leq \log x/(\log 2)} 2^k/k\,, \quad x \geq 1.$$

We have

$$\log \zeta(s) = \sum_{k=1}^{\infty} 2^{k(1-s)}/k = \log \frac{1}{1 - 2^{1-s}}\,, \quad \Re s > 1,$$

so

$$\zeta(s) = \left(1 - 2^{1-s}\right)^{-1} = 1 + 2 \cdot 2^{-s} + 4 \cdot 4^{-s} + \dots,$$

and hence (by the identity theorem for Dirichlet series)

$$N(x) = \sum_{0 \leq \nu \leq \log x/(\log 2)} 2^\nu.$$

This g-number system has logarithmic density, for

$$\int_{1-}^{x} u^{-1} dN(u) = 1 + \sum_{2^n \leq x} 1 = 1 + \left\lfloor \frac{\log x}{\log 2} \right\rfloor \sim \frac{1}{\log 2} \log x.$$

On the other hand, this system is not of regular growth, for as $k \to \infty$,

$$\frac{N(\frac{2}{3}\, 2^k)}{N(2^k)} = \frac{2^k - 1}{2^{k+1} - 1} \sim \frac{1}{2} \neq \frac{2}{3}\,.$$

4.7. Notes

§4.1. Beurling [**Be37**] considered also number systems having an integer counting function satisfying $N(x) \sim cx \log^k x$. We do not treat this extension here.

§4.2. Here and below we discuss upper, lower, and right hand residues of a generating function $\widehat{F}(s)$ at a point, but the term *residue* is reserved for an analytic function that is meromorphically continuable at the point.

§4.5. It is likely that the integer counting function of Example 4.16 has regular growth, but we have not shown this.

§4.6. The notion of regular growth is closely connected with Karamata's theory of regular variation. These topics are discussed in detail in the encyclopedic works of [**BGT87**] and [**Ko04**].

CHAPTER 5

Mertens' Formulas and Logarithmic Density

Euler's number is really *constant*

SUMMARY. Connection of Mertens' sum and product formulas with logarithmic density and the right hand residue of the zeta function at 1. Theorems of A. Tauber and of Hardy-Littlewood-Karamata.

5.1. Introduction

In this chapter we examine for g-numbers

- O- and asymptotic relations for the Mertens product

$$\prod_{p\leq x}(1-1/p)^{-1}$$

- O- and asymptotic relations for the Mertens sum

$$\psi_1(x) := \int_1^x \frac{d\psi(u)}{u}$$

- O-log density and logarithmic density of the associated g-integers.

We shall show the intimate connections of these notions.

We shall assume throughout this chapter that the g-primes are discrete, so that the Mertens product has a clear interpretation. Also, we assume that the counting function of g-integers satisfies $N(x) \ll x$, so $\zeta(s)$ has abscissa of convergence 1.

5.2. Logarithmic density

Recall that we say $\mathcal{N}$ has *logarithmic density* if

$$\lim_{x\to\infty}\left\{\frac{1}{\log x}\int_{1-}^x \frac{1}{u}\,dN(u)\right\}$$

exists. Let $\zeta(s)$ be the zeta function of a g-number system $\mathcal{N}$, and suppose that $\lim_{s\to 1+}(s-1)\zeta(s)$ exists and equals α. We say in this case that zeta has a *right hand residue* α at $s=1$. We begin by showing the equivalence of these properties.

PROPOSITION 5.1. *$\mathcal{N}$ has logarithmic density if and only if $\zeta(s)$ has a right hand residue at $s=1$. In case these quantities exist, they are equal.*

PROOF. Assume $\mathcal{N}$ has logarithmic density α. Then, for large x,

$$\ell(x) := \int_{1-}^x u^{-1}\,dN(u) \sim \alpha\log x.$$

For $\Re s > 1$, integration by parts yields

$$\begin{aligned}\zeta(s) &= \int_{1-}^{\infty} u^{-(s-1)}\, d\ell(u) = (s-1)\int_1^\infty u^{-s}\ell(u)\, du \\ &= (s-1)\int_1^\infty u^{-s}(\alpha + o(1))\log u\, du = \frac{\alpha + o(1)}{s-1}\end{aligned}$$

as $s \to 1+$, i.e. $(s-1)\zeta(s)$ has limit α.

In the other direction, the Hardy-Littlewood-Karamata (HLK) Tauberian Theorem, proved in the next section, applies directly: if $\zeta = \zeta_{\mathcal{N}}$ satisfies the asymptotic relation

$$\zeta(s) \sim \alpha/(s-1), \quad s \to 1+,$$

then $\mathcal{N}$ has logarithmic density α. □

We noted in §4.2 that validity of the PNT does not guarantee that a g-number system satisfies even an O-logarithmic density condition. However, a slightly different hypothesis yields logarithmic density.

COROLLARY 5.2. *$\mathcal{N}$ has a positive logarithmic density if and only if*

$$\lim_{s\to 1+}\int_1^\infty u^{-s-1}\Big\{\Pi(u) - \int_1^u \frac{1-v^{-1}}{\log v}dv\Big\}\, du$$

exists.

PROOF. The displayed expression equals

$$\lim_{s\to 1+} \log\{\zeta(s)(s-1)/s\},$$

the logarithm of the right hand residue of zeta, as we see by integrating by parts (1.5) and (1.8). By the last proposition, the latter expression is a real number if and only if $\mathcal{N}_{\mathcal{P}}$ has positive logarithmic density. □

Another criterion is given in terms of $\int d\psi(u)/u$.

THEOREM 5.3. *Suppose that the Chebyshev function $\psi(x)$ of a g-number system $\mathcal{N}$ satisfies*

$$\psi_1(x) := \int_1^x \frac{d\psi(u)}{u} = \log x + O\Big(\int_1^x \frac{du}{u\, f(u)}\Big), \quad x > 1, \tag{5.1}$$

where $f > 0$ and $\int_2^\infty du/\{uf(u)\log u\} < \infty$. Then $\mathcal{N}$ has logarithmic density.

On the other hand, if

$$\psi_1(x) \ge \log x + \int_1^x \frac{du}{u\, f(u)}, \quad x > 1, \tag{5.2}$$

holds with f instead satisfying $\int_2^\infty du/\{uf(u)\log u\} = \infty$, then $\mathcal{N}$ does not have logarithmic density.

PROOF. Let

$$F(s) := (s-1)\int_1^\infty u^{-s}(\psi_1(u) - \log\, u)\, du.$$

The integral is convergent for $s > 1$ under our standing assumption that $N(x) \ll x$, because of the (weak!) estimate

$$\psi_1(x) = \int_1^x T^{-1}d\psi \le \int_1^x T^{-1}dN * T^{-1}d\psi = \int_1^x T^{-1}LdN \ll \log^2 x.$$

Now we have

$$\begin{aligned}
-\frac{\zeta'}{\zeta}(s) &= \int_1^\infty u^{-(s-1)} d\psi_1(u) \\
&= (s-1)\left(\int_1^\infty u^{-s}\log u\, du + \int_1^\infty u^{-s}(\psi_1(u) - \log u)\, du\right) \\
&= \frac{1}{s-1} + F(s).
\end{aligned}$$

Hence, with an integration, for $1 < s < 2$,

$$\begin{aligned}
\log \zeta(s) - \log \zeta(2) &= \int_s^2 -\frac{\zeta'}{\zeta}(w)\, dw = \int_s^2 \Big(\frac{1}{w-1} + F(w)\Big)\, dw \\
&= \log\frac{1}{s-1} + \int_s^2 F(w)\, dw.
\end{aligned}$$

By Proposition 5.1, it suffices to show that the last term on the right side has a finite limit as $s \to 1+$. Indeed, for $1 < s \le 2$, we have

$$\begin{aligned}
|F(s)| &\le C(s-1)\int_1^\infty u^{-s}\left(\int_1^u \frac{dv}{vf(v)}\right) du \\
&= C(s-1)\int_1^\infty \frac{1}{vf(v)}\left(\int_v^\infty u^{-s}\, du\right) dv = C\int_1^\infty \frac{v^{-s}}{f(v)}\, dv.
\end{aligned}$$

Hence, for $1 < s_1 < s_2$,

$$\begin{aligned}
\left|\int_{s_1}^{s_2} F(w)\, dw\right| &\le \int_{s_1}^{s_2} |F(w)|\, dw \le C\int_{s_1}^{s_2}\left(\int_1^\infty \frac{v^{-w}}{f(v)}\, dv\right) dw \\
&= C\int_1^\infty \frac{v^{-(s_1-1)} - v^{-(s_2-1)}}{vf(v)\log v}\, dv \to 0
\end{aligned}$$

as $s_1, s_2 \to 1+$, by the dominated convergence theorem, since $v^{-(s_1-1)} - v^{-(s_2-1)}$ goes to 0 pointwise. By Cauchy's criterion, $\int_s^2 F(w)\, dw$ has a finite limit as $s \to 1+$.

On the other hand, if (5.2) holds with $\int_2^\infty du/\{uf(u)\log u\} = \infty$, then the preceding calculations show

$$\int_s^2 F(w)\, dw = \int_s^2\int_1^\infty \frac{v^{-w}}{f(v)}\, dv\, dw = \int_1^\infty \frac{v^{-(s-1)} - v^{-1}}{vf(v)\log v}\, dv,$$

and the last integral goes to infinity as $s \to 1+$. Thus $(s-1)\zeta(s) \to \infty$, and $\mathcal{N}$ does not have logarithmic density. □

5.3. The Hardy-Littlewood-Karamata Theorem

If a real-valued function $F \in \mathcal{V}$ satisfies $F(x) = \{c + o(1)\}\log x$ as $x \to \infty$, then a small calculation shows that its Mellin transform satisfies

$$\widehat{F}(s) := \int_{1-}^\infty u^{-s} dF(u) = \frac{c + o(1)}{s} \qquad s \to 0+. \tag{5.3}$$

We call a result of this type *abelian*, because the key step in its proof is typically an averaging argument based on Abel's method of summation (or integration) by parts. The HLK Theorem asserts that the converse holds for each nondecreasing function $F(x)$. Note that Lemma 4.1 is an "O-" version of the theorem.

THEOREM 5.4 (HLK). *Suppose $F \in \mathcal{V}$ is real and nondecreasing and its Mellin transform satisfies* (5.3). *Then $F(x) = \{c + o(1)\} \log x$ as $x \to \infty$.*

PROOF. We may assume that $c > 0$; otherwise, replace $F(x)$ by $F(x) + c \log x$ with some $c > 0$. Also, we may assume that $F(x) = 0$ near 1, which does not affect (5.3) or the result of the theorem. For any polynomial

$$P(z) = \sum_{n=0}^{N} a_n \, z^n$$

with $\int_0^1 P(u)\, du > 0$ and number $s > 0$, we have

$$(5.4)\quad \int_{1-}^{\infty} u^{-s} P(u^{-s})\, dF(u) = \sum_{n=0}^{N} a_n \int_{1-}^{\infty} u^{-s(n+1)}\, dF(u) = \sum_{n=0}^{N} \frac{a_n}{n+1} \frac{c + o_n(1)}{s}$$
$$= \frac{c + o_P(1)}{s} \sum_{n=0}^{N} \frac{a_n}{n+1}, \qquad s \to 0+.$$

It is OK to take the error term $o_P(1)$ outside the sum since

$$\sum_{n=0}^{N} \frac{a_n}{n+1} = \int_0^1 P(u) > 0.$$

If we change the integration variable on the left side of (5.4), we find

$$-\int_0^1 tP(t)\, dF(t^{-1/s}) = \frac{c + o_P(1)}{s} \int_0^1 P(u)\, du\,.$$

Note that $c + o_P(1) > 0$ for sufficiently small positive s.

For any given Riemann integrable function G on $[0,\, 1]$ with $\int G > 0$ and any positive number ϵ, we can choose polynomials P_1 and P_2, also with positive integrals, such that $P_1(x) \le G(x) \le P_2(x)$ for $0 \le x \le 1$ with

$$\int_0^1 \{P_2(t) - G(t)\}\, dt < \epsilon, \qquad \int_0^1 \{G(t) - P_1(t)\}\, dt < \epsilon.$$

The function $F(t^{-1/s})$ is non-increasing in t, hence

$$-\int_0^1 tP_1(t)\, dF(t^{-1/s}) \le -\int_0^1 tG(t)\, dF(t^{-1/s}) \le -\int_0^1 tP_2(t)\, dF(t^{-1/s}),$$

or

$$\frac{c + o_{P_1}(1)}{s} \int_0^1 P_1(u)\, du \le -\int_0^1 tG(t)\, dF(t^{-1/s}) \le \frac{c + o_{P_2}(1)}{s} \int_0^1 P_2(u)\, du.$$

Since $\int_0^1 \{P_2(t) - P_1(t)\}\, dt$ can be taken arbitrarily small and G lies between P_1 and P_2, we have

$$-\int_0^1 tG(t)\, dF(t^{-1/s}) = \frac{c + o(1)}{s} \int_0^1 G(u)\, du, \qquad s \to 0+.$$

Now take

$$G(t) := \begin{cases} 1/t, & 1/e \le t \le 1 \\ 0, & 0 \le t < 1/e. \end{cases}$$

With this choice, we find

$$F(e^{1/s}) - 0 = -\int_{1/e}^{1} dF(t^{-1/s}) = \frac{c + o(1)}{s} \int_{1/e}^{1} u^{-1}\, du = \frac{c + o(1)}{s},$$

and, with one last change of variable, $F(x) = \{c + o(1)\} \log x$ as $x \to \infty$. □

5.4. Mertens' sum formula

A weak form of Mertens' classical sum formula asserts that

$$\int_1^x \frac{d\psi(t)}{t} \sim \log x. \tag{5.5}$$

Here we show the connection of this relation with logarithmic density for a g-number system.

As a warm-up, we give an O- estimate of $\psi_1(x)$ that is sharper than the auxiliary estimate in the proof of Theorem 5.3.

PROPOSITION 5.5. *Suppose a g-number system has finite upper logarithmic density and positive lower logarithmic density. Then*

$$\psi_1(x) \ll \log x.$$

PROOF. The Chebyshev identity (3.1) and Lemma 2.15 with $\alpha = -1$ together yield

$$\sum_{n \le y} \frac{\Lambda(n)}{n} * \frac{1}{n} = \sum_{n \le y} \frac{\log n}{n}.$$

The right side of the last equation is at most

$$\log y \sum_{n \le y} \frac{1}{n} \ll \log^2 y,$$

while the left side equals

$$\sum_{mn \le y} \frac{\Lambda(m)}{m} \frac{1}{n} \ge \sum_{m \le \sqrt{y}} \frac{\Lambda(m)}{m} \sum_{n \le \sqrt{y}} \frac{1}{n} \gg \psi_1(\sqrt{y}) \log y.$$

Taking $y = x^2$, we get the desired estimate. □

THEOREM 5.6. *Suppose that a g-number system has positive logarithmic density. Then* (5.5) *holds.*

PROOF. By integration by parts,

$$\zeta(s) = \int_{1-}^{\infty} u^{-(s-1)} dN(u)/u = (s-1) \int_1^{\infty} u^{-s} \Big\{ \int_{1-}^{u} dN(x)/x \Big\}\, du$$
$$= (s-1) \int_1^{\infty} u^{-s} \{A + o(1)\} \log u\, du = \frac{A}{s-1} + o\Big(\frac{1}{s-1}\Big), \quad s \to 1+,$$

for some positive constant A.

Similarly, again for $s \to 1+$,

$$-\zeta'(s) = \int_1^{\infty} u^{-(s-1)} \log u\, dN(u)/u = A(s-1)^{-2} + o\big((s-1)^{-2}\big).$$

Divide the two expressions and let $s \to 1+$ to obtain

$$\int_1^{\infty} u^{-s} d\psi(u) = -\zeta'(s)/\zeta(s) \sim \frac{1}{s-1}.$$

Since $\psi \uparrow$, the result follows from the Hardy-Littlewood-Karamata Tauberian Theorem of the last section. □

The converse assertion, "(5.5) $\Longrightarrow$ positive logarithmic density" is not true. This is easily seen by reconsidering Example 4.6: here

$$\int_1^x \frac{d\psi(t)}{t} = \log x + \frac{\{1+o(1)\}\log x}{\log\log 10x} \sim \log x\,, \tag{5.6}$$

and in this case $\mathcal{N}$ does not have finite upper logarithmic density. However, as we showed in Theorem 5.3, a hypothesis slightly stronger than (5.6) insures that $\mathcal{N}$ does have positive logarithmic density.

We shall show in Theorem 8.5 that $\pi(x) = o(N(x))$ under the assumption of a regular growth condition. Here we prove another o-estimate using logarithmic density. (Recall from §4.6 that log density is independent of regular growth.) Also, we give a lower bound for $\pi(x)$.

COROLLARY 5.7. *Suppose that $\mathcal{N}$ is a g-number system with positive logarithmic density. Then $\pi(x) = o(x)$ and, for any $\epsilon > 0$, $\pi(x) \gg x^{1-\epsilon}$.*

PROOF. By the last theorem

$$\psi_1(x) := \int_1^x \frac{d\psi(u)}{u} = \{1+o(1)\}\log x.$$

Then, by integration by parts,

$$\begin{aligned}\pi(x) \le \Pi(x) &= \int_{p_1-}^x \frac{u\,d\psi_1(u)}{\log u}\\ &= \frac{\psi_1(x)x}{\log x} - \int_{p_1}^x \psi_1(u)\Big(\frac{1}{\log u} - \frac{1}{\log^2 u}\Big)\,du\\ &= \{1+o(1)\}x - \int_{p_1}^x \{1+o(1)\}\Big(1 - \frac{1}{\log u}\Big)\,du = o(x).\end{aligned}$$

For the lower bound estimate, we difference the result of (5.5) and write

$$\epsilon\log x \sim \int_{x^{1-\epsilon}}^x \frac{d\psi(t)}{t} < x^{\epsilon-1}\{\psi(x) - \psi(x^{1-\epsilon})\} < x^{\epsilon-1}\psi(x).$$

Thus $\psi(x) \gg x^{1-\epsilon}\log x$ and so $\pi(x) \gg x^{1-\epsilon}$. □

REMARK 5.8. Theorem 5.6 is proved under the assumption that $\mathcal{N}$ has positive logarithmic density. Can the condition be weakened to allow logarithmic density 0? The answer is No: in Example 6.1 (below), $\pi(a_k)/a_k > \log k \to \infty$ as $k \to \infty$, and on the other hand the g-number system constructed has logarithmic density 0.

5.5. Mertens' product formula

Mertens' classical product formula asserts that

$$\prod_{p\le x}\Big(1 - \frac{1}{p}\Big)^{-1} \Big/ \log x \to e^{\gamma}, \quad x \to \infty. \tag{5.7}$$

Here p ranges over the natural primes and γ denotes Euler's number. We shall establish an analogous formula for g-primes under the hypothesis that the g-integers have a logarithmic density.

EXAMPLE 5.9. AN ODD MERTENS PRODUCT. As a hint of what to expect, consider the Mertens product for the g-prime system $\mathcal{P}_1$ which consists of the natural primes from 3 onwards. $\mathcal{N}_1$ is the collection of odd numbers and has density 1/2. Formula (5.7) gives

$$\prod_{3\le p\le x}\Big(1-\frac{1}{p}\Big)^{-1}\Big/\log x \to e^{\gamma}/2, \qquad x\to\infty.$$

We showed in Proposition 5.1 that a g-number system has a logarithmic density precisely when it has a right hand residue at $s = 1$, and in this case, the two quantities are the same. Now we show further that a g-number system $\mathcal{N}$ has these properties if and only if Mertens' product formula holds in $\mathcal{N}$.

THEOREM 5.10. *Suppose that $\mathcal{N}$ is a g-number system generated by g-primes $\mathcal{P} = \{p_i\}$. We have*

$$\prod_{p_i\le x}\Big(1-\frac{1}{p_i}\Big)^{-1}\Big/\log x \to A\,e^{\gamma}, \quad x\to\infty, \tag{5.8}$$

if and only if $\mathcal{N}$ has logarithmic density A.

5.6. A remark on γ

It seems natural enough that A should appear in the last formula, but it is perhaps surprising also to have Euler's constant γ, particularly since this number does not seem connected with Beurling numbers: if we use the definition

$$\gamma := \lim_{x\to\infty}\Big\{\sum_{n\le x}\frac{1}{n}-\log x\Big\} = \int_{1-}^{\infty} u^{-1}(d\lfloor u\rfloor - du),$$

then the most obvious definition of the corresponding quantity for a Beurling system $\mathcal{N} = \{n_i\}$ with density (or logarithmic density) A would seem to be

$$\gamma_{\mathcal{N}} = \lim_{x\to\infty}\Big\{\sum_{n_i\le x}\frac{1}{n_i}-A\log x\Big\} = \int_{1-}^{\infty}\frac{1}{u}\{dN_{\mathcal{N}}(u)-A\,du\}.$$

This number, however, is generally different from Euler's constant. For example, for $\mathcal{N}$ the odd numbers, the last formula gives

$$\begin{aligned}\gamma_{\text{odds}} &= \lim_{x\to\infty}\Big\{\sum_{\substack{n\le x\\ n\text{ odd}}}\frac{1}{n}-\frac{1}{2}\log x\Big\}\\ &= \lim_{x\to\infty}\Big\{\sum_{n\le x}\frac{1}{n}-\log x-\Big(\sum_{n\le x/2}\frac{1}{2n}-\frac{1}{2}\log x\Big)\Big\}\\ &= \gamma-\frac{\log(1/2)+\gamma}{2} = \frac{\gamma+\log 2}{2} > \gamma.\end{aligned}$$

PROOF OF THE THEOREM. The argument proceeds via the following steps:

- an equivalent form of the theorem
- a generating function
- a tauberian theorem for Mellin transforms.

5.7. An equivalent form and proof of "only if"

Taking the log of each side of formula (5.8) and expanding $\log(1-1/p)$ as a series gives

$$\sum_{p\le x}\sum_{\alpha\ge 1}\frac{1}{\alpha p^{\alpha}}-\log\log x\to\log A+\gamma \tag{5.9}$$

as an expression equivalent to (5.8). In the following lemma, we replace the double sum and the iterated log in (5.9) by functions that are better suited for analysis.

LEMMA 5.11. *The formula*

$$\int_1^x\frac{1}{u}\Big\{d\Pi(u)-\frac{1-u^{-1}}{\log u}du\Big\}\to\log A,\quad x\to\infty, \tag{5.10}$$

is equivalent to (5.9).

PROOF. As a preliminary, note that each of (5.9) and (5.10) implies the crude bound

$$\sum_{p\le x}\frac{1}{p}\ll\log\log x. \tag{5.11}$$

Now, we change the double sum in (5.9) into $\int d\Pi(u)/u$; the two expressions differ only in some higher prime powers, and we show the difference goes to 0 as $x\to\infty$. We have

$$0\le\sum_{p\le x}\sum_{\alpha\ge 1}\frac{1}{\alpha p^{\alpha}}-\int_1^x\frac{d\Pi(u)}{u}=\sum_{p\le x}\sum_{\alpha\colon p^{\alpha}>x}\frac{1}{\alpha p^{\alpha}}.$$

Note that $\alpha\ge\max\{2,\log x/(\log p)\}$. Thus the last sum is at most

$$\sum_{p\le x}\frac{\log p}{\log x}\sum_{\alpha\ge 2}p^{-\alpha}\ll\frac{1}{\log x}\sum_{p\le x}\frac{\log p}{p}\frac{1}{p}\ll\frac{\log\log x}{\log x}=o(1),$$

by the boundedness of $\log p/p$ and (5.11).

Next, we replace $\log\log x$ by a function having a better Mellin transform. Integration by parts and a change of variable give

$$\lim_{x\to\infty}\Big\{\int_1^x\frac{1-u^{-1}}{u\log u}du-\log\log x\Big\}=-\int_1^{\infty}u^{-2}\log\log u\,du=-\int_0^{\infty}e^{-v}\log v\,dv.$$

We interpret the last integral using two formulas for the derivative of Euler's gamma function (cf. [**MV07**], Appendix C). On the one hand,

$$\Gamma'(1)=\frac{d}{ds}\int_0^{\infty}e^{-v}v^{s-1}\,dv\Big|_{s=1}=\int_0^{\infty}e^{-v}\log v\,dv,$$

and on the other, the logarithmic derivative of the product form of Γ gives

$$\Gamma'(1)=\Gamma'(1)/\Gamma(1)=-\gamma.$$

Thus, as $x\to\infty$,

$$\int_1^x\frac{1-u^{-1}}{u\log u}du-\log\log x\to\gamma.$$

Together, the two expressions show the equivalence of (5.10) and (5.9). □

We now prove the "only if" assertion of the theorem by showing that the relation

$$(s-1)\zeta(s) \to A, \quad s \to 1+ \tag{5.12}$$

follows from (5.10), and appealing to Proposition 5.1. We start by noting that (1.5) and (1.8) together yield

$$\log \zeta(s) - \log \frac{s}{s-1} = \int_1^\infty u^{-s} \left\{ d\Pi(u) - \frac{1-u^{-1}}{\log u}\, du \right\}. \tag{5.13}$$

Next, let $\ell(x)$ denote the left side of (5.10). If we express the last formula in terms of ℓ, integrate by parts, and apply Lemma 5.11, we find

$$\begin{aligned}\log\{(s-1)\zeta(s)/s\} &= \int_1^\infty u^{-(s-1)}\, d\ell(u) = (s-1)\int_1^\infty u^{-s}\, \ell(u)\, du \\ &= (s-1)\int_1^\infty u^{-s}\, \{\log A + o(1)\}\, du = \log A + o(1)\end{aligned}$$

as $s \to 1+$. Also, $\log s \to 0$ as $s \to 1$ in the last formula, so upon exponentiating, (5.12) holds, which proves one implication of the theorem.

5.8. Tauber's Theorem and conclusion of the argument

Applying Proposition 5.1 again, we complete the proof of Theorem 5.10 by deducing (5.10) from (5.12). We use the Mellin version of the following tauberian theorem.

THEOREM 5.12 (Tauber). *Let f be a right continuous function on $[1, \infty)$ that is locally of bounded variation. Further, suppose that*

$$\widehat{f}(s) := \int_{1-}^\infty u^{-s} df(u)$$

exists for all $s > 1$, that $\widehat{f}(s) \to \ell$ as $s \to 1+$, *and that*

$$\int_1^x \frac{\log u}{u}\, df(u) = o(\log x), \quad x \to \infty. \tag{5.14}$$

Then

$$\int_{1-}^x \frac{1}{u}\, df(u) \to \ell \qquad \text{as } x \to \infty.$$

(We remark that (5.14) is also a necessary condition for this theorem.)

In Tauber's theorem take $\ell = \log A$ and

$$df(u) = d\Pi(u) - \frac{1-u^{-1}}{\log u}\, du.$$

By (5.12), we have $\zeta(s) \sim A/(s-1)$ as $s \to 1+$. Thus

$$\int_1^\infty u^{-s}\, df(u) = \log\{\zeta(s)(s-1)/s\} \to \log A = \ell.$$

Recalling that $\log u\, d\Pi(u) = d\psi(u)$, we have from Theorem 5.6,

$$\int_1^x \frac{\log u}{u}\, df(u) = \int_1^x \frac{1}{u}\{d\psi(u) - (1-u^{-1})\, du\} = o(\log x).$$

Thus, the condition of the theorem is satisfied and so, as $x \to \infty$,

$$\int_{1-}^{x} \frac{1}{u}\, df(u) = \int_{1}^{x} \frac{1}{u}\, d\Pi(u) - \int_{1}^{x} \frac{1-u^{-1}}{u \log u}\, du \to \log A\,. \qquad \square$$

PROOF OF TAUBER'S THEOREM. We suppose without loss of generality that $f(u) = 0$ for all $u \le e$; this avoids having factors such as $1/\log u$ near $u = 1$. Set

$$G(x) := \int_{e}^{x} \frac{\log u}{u}\, df(u).$$

By the hypothesis on $\widehat{f}$, we have

$$I(\epsilon) := \int_{e}^{\infty} u^{-\epsilon} \frac{dG(u)}{\log u} = \int_{e}^{\infty} u^{-1-\epsilon}\, df(u) \to \ell, \quad \epsilon \to 0+\,.$$

Integrating by parts and recalling that $G(x) = o(\log x)$ by (5.14), we find

$$I(\epsilon) = I_1(\epsilon) + I_2(\epsilon) := \epsilon \int_{e}^{\infty} \frac{G(u)\, du}{u^{1+\epsilon} \log u} + \int_{e}^{\infty} \frac{G(u)\, du}{u^{1+\epsilon} \log^2 u} \to \ell, \quad \epsilon \to 0+\,.$$

To proceed, we show first that $I_1(\epsilon) \to 0$ as $\epsilon \to 0+$. For given $\delta > 0$, we have $|G(u)/\log u| < \delta$ for all u exceeding some $U = U(\delta)$ and $|G(u)/\log u| < M$ for some M and $e \le u \le U$. Thus

$$|I_1(\epsilon)| \le \epsilon \int_{e}^{U} \frac{M\, du}{u^{1+\epsilon}} + \epsilon \int_{U}^{\infty} \frac{\delta\, du}{u^{1+\epsilon}} \le M\{e^{-\epsilon} - U^{-\epsilon}\} + \delta,$$

whence $\limsup_{\epsilon \to 0+} |I_1(\epsilon)| \le \delta$. Since δ is arbitrarily small, $I_1(\epsilon) \to 0$ as claimed. It follows that $I_2(\epsilon) \to \ell$ as $\epsilon \to 0+$.

Our next goal is to show that

$$\int_{e}^{x} \frac{G(u)\, du}{u \log^2 u} \to \ell, \quad x \to \infty. \tag{5.15}$$

Given a (large) positive x, set $\epsilon := 1/\log x$ and

$$\begin{aligned} I_3(\epsilon) &:= \int_{e}^{x} \frac{G(u)\, du}{u \log^2 u} - \int_{e}^{\infty} \frac{G(u)\, du}{u^{1+\epsilon} \log^2 u} \\ &= \int_{e}^{x} (1 - u^{-\epsilon}) \frac{G(u)\, du}{u \log^2 u} - \int_{x}^{\infty} \frac{G(u)\, du}{u^{1+\epsilon} \log^2 u}\,; \end{aligned}$$

we show $I_3(\epsilon) \to 0$ as $\epsilon \to 0+$.

The last integral has absolute value at most

$$\int_{x}^{\infty} \frac{o(1)\, du}{u^{1+\epsilon} \log u} \le \frac{o(1)}{\log x} \int_{x}^{\infty} u^{-1-\epsilon} du = \frac{o(1) x^{-\epsilon}}{\epsilon \log x} = o(1).$$

Now, $0 < 1 - u^{-\epsilon} \le \epsilon \log u$ for $\epsilon > 0$ and $u > 1$; thus

$$\left| \int_{e}^{x} (1 - u^{-\epsilon}) \frac{G(u)\, du}{u \log^2 u} \right| \le \int_{e}^{x} \epsilon \log u \frac{|G(u)|\, du}{u \log^2 u} = \epsilon \int_{e}^{x} \frac{o(1)\, du}{u}\,.$$

By reasoning used above, for given $\delta > 0$, this is at most

$$\epsilon M \int_{e}^{U} \frac{du}{u} + \epsilon \delta \int_{U}^{x} \frac{du}{u} \le \epsilon M \log U + \delta \epsilon \log x,$$

and so $I_3(\epsilon) \to 0$. It follows that (5.15) holds.

Finally, by integration by parts,

$$\int_e^x \frac{1}{u}\, df(u) = \int_e^x \frac{dG(u)}{\log u} = \frac{G(x)}{\log x} + \int_e^x \frac{G(u)\, du}{u \log^2 u} \to \ell, \quad x \to \infty. \qquad \square$$

5.9. Notes

§5.5. The product result, Theorem 5.10, is based on an article [**Ol11**] by R. Olofsson. This theme is treated also by P. Pollack [**Pl13**].

§5.8. Tauber's original paper from 1897 treated power series with radius of convergence 1. Our Mellin transform version follows that for Laplace transforms by J. Korevaar [**Ko04**] §I-14.

CHAPTER 6

O-Density of g-integers

How weighty are the matters contained in this space

SUMMARY. O-density and criteria. Independence of O-density and log density. Irrelevance of small primes. More on the residue of $\zeta(s)$ at $s = 1$.

6.1. Non-relation of log-density and O-density

Recall that we say that a g-number system $\mathcal{N}$ has O-density (finite upper density) if its integer counting function satisfies $N(x) \ll x$. Also, we say that $\mathcal{N}$ has positive lower density if $N(x) \gg x$. Here we give some conditions for these densities that are related to estimates of classical prime number theory.

Integration by parts shows that if a g-number system has O-density, then it also has O-log density. Similarly, positive lower density implies positive lower log density. Example 4.18 shows that density is a stronger condition than log density. In this section we give two examples showing that if a g-number system has log density or has O-density, we cannot conclude the existence of the other.

EXAMPLE 6.1. LOG DENSITY DOES NOT IMPLY O-DENSITY. Suppose that a g-number system has log density. We show that this does not insure that the system has O-density or that its primes satisfy a Chebyshev O-bound $\pi(x) \ll x/\log x$.

For $k = 2, 3, \ldots$, let $a_k := e^{k^k}$ be a g-prime of multiplicity $\lfloor a_k \log k \rfloor$. This lacunary collection of primes, call it $\mathcal{P}_\ell$, violates the Chebyshev condition, for the number of primes at *the single point* $x = a_k$ is about $x \log\log\log x$, and *a fortiori*, $\mathcal{N}_\ell$, the g-integers generated by $\mathcal{P}_\ell$, do not satisfy an O-density condition.

On the other hand, $\mathcal{N}_\ell$ does have a log density, and it is 0. By Corollary 4.3, it suffices to show that, for some function $B(\epsilon) \to \infty$ as $\epsilon \to 0+$,

$$\log \zeta(1+\epsilon) \le \log(1/\epsilon) - B(\epsilon). \tag{6.1}$$

By (1.5), we have

$$\log \zeta(s) = \int_1^\infty u^{-s} d\Pi(u) = S_1 + \frac{1}{2} S_2 + \frac{1}{3} S_3 + \ldots,$$

where

$$S_\nu = S_\nu(s) := \sum_{k=2}^\infty \lfloor a_k \log k \rfloor {a_k}^{-\nu s}, \quad \nu = 1, 2, \ldots.$$

The series of terms of $\log \zeta(s) - S_1$ is uniformly bounded for $s > 1$:

$$\sum_{k=2}^\infty \frac{1}{\nu} S_\nu \le \sum_{k=2}^\infty \log k \sum_{\nu=2}^\infty {a_k}^{-(\nu-1)} \ll \sum_{k=2}^\infty (\log k)/a_k \ll 1. \tag{6.2}$$

Also, with $s = 1 + \epsilon$, we have

$$S_1 \le \sum_{k=2}^{\infty} (\log k)\, a_k{}^{-\epsilon}. \tag{6.3}$$

We show that there is a reasonably small integer $K = K(\epsilon)$ beyond which the numbers $a_k{}^{-\epsilon}$ go to zero so rapidly that the tail of the last series is dominated by its $K + 1$st term. For the initial range of the series it suffices to use the trivial estimate $a_k{}^{-\epsilon} < 1$ to get (by a weak form of Stirling's formula) the bound

$$\sum_{k=2}^{K} (\log k)\, a_k{}^{-\epsilon} < \sum_{k=2}^{K} \log k < (K + 1/2)\log K - K + 1. \tag{6.4}$$

We choose K to be able to apply the ratio test to the later terms of (6.3): for given $\epsilon > 0$, take K such that

$$K^K \le 1/\epsilon < (K+1)^{K+1}. \tag{6.5}$$

Let us suppose that $\epsilon \le 1/30$. The right inequality implies that $K \ge 3$. Next, a small calculation shows that

$$(a_{k+1})^{-\epsilon}/(a_k)^{-\epsilon} < \exp\{-2\epsilon k^{k+1}\}, \quad k \ge 3.$$

Now suppose that $k \ge K + 1 \ge 4$, so that $\epsilon k^k > 1$. We have

$$\frac{(a_{k+1})^{-\epsilon}\log(k+1)}{(a_k)^{-\epsilon}\log k} < \frac{\log(k+1)}{\log k}\exp\{-2k\} \le \frac{\log 5}{\log 4}\exp\{-8\} < \frac{1}{1000},$$

and a geometric series estimate gives

$$\sum_{k=K+1}^{\infty} (\log k)\, a_k{}^{-\epsilon} < \frac{1000}{999}\log(K+1)\, a_{K+1}{}^{-\epsilon} < \frac{3}{2}\log K.$$

This bound and (6.4) together give

$$S_1 < K\log K + 1 + 2\log K - K.$$

Recalling (6.2) and (6.5), there is an absolute constant c such that, for all $\epsilon < 1/30$,

$$\log\zeta(1+\epsilon) < K\log K + 2\log K - K + c \le \log 1/\epsilon - B(\epsilon),$$

where $B(\epsilon) = K - 2\log K - c \to \infty$ as $\epsilon \to 0+$. By Corollary 4.3, the log density of $\mathcal{N}_\ell$ is 0.

EXAMPLE 6.2. O-DENSITY DOES NOT IMPLY LOG DENSITY. We give a continuous example of a g-number system having O-density but not log-density. Define a sequence of real numbers $\{Y_\nu\}$ by taking $Y_\nu := \exp(100^\nu)$ for $\nu = 1, 2, \ldots$, and define a g-integer counting function by

$$N(x) := \int_{1-}^{x} \exp^*\Big\{\Big(1 - \sum_{\nu=1}^{\infty}\{\chi_{(Y_\nu, Y_\nu^2)}(u) - \chi_{(Y_\nu^2, Y_\nu^4)}(u)\}\Big)\frac{1-u^{-1}}{\log u}\,du\Big\}. \tag{6.6}$$

We first show that $N(x) \ll x$ by applying Theorem 6.5 (see below). We have

$$\begin{aligned}\pi(x) \le \Pi(x) &= \int_1^x \Big(1 - \sum_{\nu=1}^{\infty}\{\chi_{(Y_\nu, Y_\nu^2)}(u) - \chi_{(Y_\nu^2, Y_\nu^4)}(u)\}\Big)\frac{1-u^{-1}}{\log u}\,du \\ &\le \int_1^x 2\,\frac{1-u^{-1}}{\log u}\,du \ll \frac{x}{\log x}.\end{aligned}$$

(The sequence $\{Y_\nu\}$ increases sufficiently rapidly that the intervals do not overlap.)

Also, for $s > 1$, we have

$$(6.7)\qquad \begin{aligned}\log\zeta(s) &= \int_1^\infty u^{-s}\Big(1 - \sum_{\nu=1}^\infty \{\chi_{(Y_\nu,Y_\nu^2)}(u) - \chi_{(Y_\nu^2,Y_\nu^4)}(u)\}\Big)\frac{1-u^{-1}}{\log u}\,du \\ &= \log\frac{s}{s-1} - \sum_{\nu=1}^\infty J_\nu(s) + \sum_{\nu=1}^\infty K_\nu(s),\end{aligned}$$

where

$$J_\nu(s) := \int_{Y_\nu}^{Y_\nu^2} u^{-s}\frac{du}{\log u} - \int_{Y_\nu^2}^{Y_\nu^4} u^{-s}\frac{du}{\log u}$$

and

$$K_\nu(s) := \int_{Y_\nu}^{Y_\nu^2} u^{-s}\frac{du}{u\log u} - \int_{Y_\nu^2}^{Y_\nu^4} u^{-s}\frac{du}{u\log u}.$$

We have at once for $s > 1$,

$$0 < \sum_{\nu=1}^\infty K_\nu(s) < \int_{Y_1}^\infty \frac{du}{u^2\log u} < \frac{1}{Y_1} < 10^{-43},$$

and each individual J term satisfies

$$J(s) = \int_Y^{Y^2} u^{-s}\frac{du}{\log u} - \int_{Y^2}^{Y^4} u^{-s}\frac{du}{\log u} = \int_Y^{Y^2} u^{-s}\left(1-u^{1-s}\right)\frac{du}{\log u} > 0.$$

Thus

$$(s-1)\zeta(s)/s < \exp Y_1^{-1} \ll 1, \quad s > 1,$$

i.e. $\zeta(s)$ has a finite upper residue at $s = 1$. Both the conditions of Theorem 6.5 hold, and so $N(x) \ll x$.

It remains to prove that this system does not have log-density. For this, we apply Proposition 5.1 and show that the zeta function does not have a right hand residue at $s = 1$. For each positive integer M, define $\epsilon_i = \epsilon_i(M)$ by

$$\epsilon_1 := 1/(2\log Y_M) = 0.5\cdot 100^{-M} \quad\text{and}\quad \epsilon_2 = 7\epsilon_1.$$

We show that the ratio $\epsilon_2\zeta(1+\epsilon_2)/\epsilon_1\zeta(1+\epsilon_1)$ is bounded away from 1.

By (6.7), we have the key relation

$$-\{\log \epsilon\zeta(1+\epsilon)/(1+\epsilon)\} = \sum_{\nu=1}^\infty J_\nu(1+\epsilon) + 10^{-43}\theta,$$

with $|\theta| < 1$. We show

- the contribution of $J_M(1+\epsilon_1)$ is reasonably large
- the contributions of *all* terms $J_\nu(1+\epsilon_2)$ are small,

which will yield the desired inequalities.

We need one numerical evaluation. Using either computer algebra or Table 5.1 of [**AS64**], we find

$$\begin{aligned}J_M(1+\epsilon_1) &= \int_{\epsilon_1\log Y_M}^{2\epsilon_1\log Y_M} e^{-v}\frac{dv}{v} - \int_{2\epsilon_1\log Y_M}^{4\epsilon_1\log Y_M} e^{-v}\frac{dv}{v} \\ &= \int_{1/2}^1 e^{-v}\frac{dv}{v} - \int_1^2 e^{-v}\frac{dv}{v} = 0.169906\ldots.\end{aligned}$$

Since all $J_\nu(1+\epsilon_1)$ are positive, we have

$$\sum_{\nu=1}^{\infty} J_\nu(1+\epsilon_1) > J_M(1+\epsilon_1) > 0.1699. \tag{6.8}$$

On the other hand, for $J_\nu(1+\epsilon_2)$, we give two simple upper estimates, depending upon the size of $\epsilon_2 \log Y_\nu$. For $\nu < M$, where $\epsilon_2 \log Y_\nu$ is small, we have

$$\begin{aligned} 0 < J_\nu(1+\epsilon_2) &= \int_{\epsilon_2 \log Y_\nu}^{2\epsilon_2 \log Y_\nu} \{e^{-v} - e^{-2v}\} \frac{dv}{v} \le \int_{\epsilon_2 \log Y_\nu}^{2\epsilon_2 \log Y_\nu} v \frac{dv}{v} \\ &= \epsilon_2 \log Y_\nu = \frac{7}{2} \frac{\log Y_\nu}{\log Y_M} = 3.5 \cdot 100^{\nu - M}; \end{aligned}$$

for $\nu \ge M$, $\epsilon_2 \log Y_\nu$ is large, and we have

$$\begin{aligned} 0 < J(1+\epsilon_2) &\le \int_{\epsilon_2 \log Y_\nu}^{2\epsilon_2 \log Y_\nu} e^{-v} \frac{dv}{v} \le \int_{\epsilon_2 \log Y_\nu}^{\infty} e^{-v}\, dv \\ &= \exp\{-\epsilon_2 \log Y_\nu\} = \exp\{-3.5 \cdot 100^{\nu - M}\}. \end{aligned}$$

Now

$$\begin{aligned} &\sum_{\nu<M} J_\nu(1+\epsilon_2) + \sum_{\nu \ge M} J_\nu(1+\epsilon_2) \\ &\qquad < 3.5\{10^{-2} + 10^{-4} + \ldots\} + \{e^{-3.5} + e^{-350} + \ldots\} < 0.0660. \end{aligned} \tag{6.9}$$

Comparing (6.8) and (6.9) and absorbing the other tiny errors from (6.7) into the round-offs, we find that

$$\frac{\epsilon_2 \zeta(1+\epsilon_2)}{\epsilon_1 \zeta(1+\epsilon_1)} > \exp\{0.10\} > 1.1$$

for a sequence of pairs of numbers $\{\epsilon_1(M), \epsilon_2(M)\}$ tending to $0+$ as $M \to \infty$. Thus $\zeta(s)$ does not have a right hand residue at $s = 1$, and by Proposition 5.1, the g-number system does not have log-density.

6.2. O-Criteria for O-density

In the remainder of this chapter, we give several conditions, some necessary, others sufficient, for O-density. The conditions are typically O-bounds on the prime counting function and/or the growth of the zeta function near its pole at $s = 1$.

We begin by showing, in analogy with Proposition 4.4, that the contribution of small primes plays no role in determining whether a g-number system has O-density or density. In the sequel, we shall freely remove initial primes when it is convenient.

PROPOSITION 6.3. *Let $B > 1$ and $d\Pi$ be the weighted prime and prime power counting measure of a g-number system $\mathcal{N}$, decomposed into $d\Pi_1 + d\Pi_2$ with Π_1 counting primes $p \le B$ and the powers of these primes, and Π_2 counting the remaining part of Π (as in Proposition 4.4). Also set $dN_i := \exp^* d\Pi_i$, $i = 1, 2$, and let $\mathcal{N}_2$ denote the g-number system associated with $d\Pi_2$. Then $\mathcal{N}$ has O-density (resp. density) if and only if $\mathcal{N}_2$ has the same.*

PROOF. We prove the proposition for density; for O-density, replace asymptotic estimates by O-bounds.

Suppose first that $\mathcal{N}$ has density A. We have

$$N_2(x) = \int_{1-}^{x} dN * \{\exp^*(-d\Pi_1)\} = \int_{1-}^{x} N(x/t)\exp^*(-d\Pi_1)(t)$$
$$= \int_{1-}^{x} \left\{\frac{Ax}{t} + o\Big(\frac{x}{t}\Big)\right\}\exp^*(-d\Pi_1)(t).$$

By the proof of Proposition 4.4, $\int_1^\infty T^{-1}d\Pi_1 < \infty$; thus $\int_{1-}^\infty T^{-1}\exp^*(-d\Pi_1)$ is (absolutely) convergent. This implies, first, that $\int_{1-}^x o(x/t)\exp^*(-d\Pi_1)(t) = o(x)$. Also, as $x \to \infty$,

$$\int_{1-}^{x} T^{-1}\exp^*(-d\Pi_1) \sim \int_{1-}^{\infty} T^{-1}\exp^*(-d\Pi_1) =: I.$$

Then, by Lemma 3.8 and Proposition 3.7,

$$I = \int_{1-}^{\infty} \exp^*(-T^{-1}d\Pi_1) = \exp\Big(-\int_{1-}^{\infty} T^{-1}d\Pi_1\Big) > 0.$$

Finally, as $x \to \infty$,

$$N_2(x) \sim Ax\int_{1-}^{x} T^{-1}\exp^*(-d\Pi_1) \sim Ax\exp\Big(-\int_{1}^{\infty} T^{-1}d\Pi_1\Big).$$

Next, suppose that $\mathcal{N}_2$ has density c_2. Write

$$N(x) = \int_{1-}^{x} dN_2 * \exp^*(d\Pi_1) = \int_{1-}^{x} N_2(x/t)\exp^*(d\Pi_1)(t),$$

and apply the preceding reasoning to conclude that

$$N(x) \sim c_2 x\exp\Big\{\int_{1}^{\infty} t^{-1}d\Pi_1(t)\Big\}.$$

Thus $\mathcal{N}$ has density as well. □

The following result follows immediately from the proof of Lemma 3.3 or from Proposition 4.2, after noting that O-density implies O-log density.

PROPOSITION 6.4. *If $N(x) \ll x$, then the associated zeta function has a finite upper residue at $s = 1$.*

Does Proposition 6.4 have a direct converse? The answer is No, for such an implication and Proposition 4.2 would together yield the relations

"O-log density $\iff$ finite upper residue $\iff$ O-density."

However, Example 6.1 implies that O-log density $\not\Rightarrow$ O-density. Thus Proposition 6.4 does not have an unconditional converse.

THEOREM 6.5. *If the zeta function of a g-number system has a finite upper residue at $s = 1$ and the counting function of g-primes satisfies the upper Chebyshev bound $\pi(x) \ll x/\log x$, then $N(x) \ll x$.*

PROOF. We use the Chebyshev upper bound in the equivalent form $\psi(x) \ll x$. By Chebyshev's identity (3.3) and Proposition 4.2,

$$T(x) := \int_{1-}^{x} \log u\, dN(u) = \int_{1-}^{x} \psi(x/u)\, dN(u) \ll \int_{1-}^{x} \frac{x}{u}\, dN(u) \ll x\log x.$$

Integration by parts gives

$$N(x) = 1 + \int_1^x (\log u)^{-1} dT(u) \ll x. \qquad \square$$

Could we have established the last theorem assuming only the existence of a Chebyshev upper bound? The answer is No; the g-integer system with counting function $D(x)$ of §4.1 satisfies $D(x) \sim x \log x \neq O(x)$ but $\pi(x) \sim 2x/\log x$. This example shows also that the condition $\pi(x) \ll x/\log x$ by itself does not imply that the associated g-integer system has a finite upper residue at $s = 1$.

Next, suppose that the counting function a g-number system can be resolved into a main part satisfying the conditions of the last theorem and a small remaining part. In this case, we show that the g-integers have O-density. It is convenient to handle separately the following convergence result.

LEMMA 6.6. *Suppose $\Pi_2(x) \in \mathcal{V}$ and there exists a nondecreasing function $A(x)$ on $[1, \infty)$ such that*

$$\left| \int_{1-}^x \exp^* d\Pi_2 \right| \leq A(x) \text{ and } \int_1^\infty A(t)\, t^{-2} dt < \infty. \tag{6.10}$$

Then

$$\int_{1-}^\infty T^{-1} \exp^* d\Pi_2 \tag{6.11}$$

converges.

PROOF. The convergence of $\int A(t)\, t^{-2} dt$ and the inequality

$$\frac{A(x)}{2x} = A(x) \int_x^{2x} \frac{dt}{t^2} \leq \int_x^{2x} A(t) \frac{dt}{t^2} = o(1)$$

imply that $A(x) = o(x)$. Set

$$E(u, x) := \int_u^x \exp^* d\Pi_2 , \quad u < x.$$

We have $|E(u, x)| \leq 2A(x)$. Then, by integration by parts,

$$\int_u^v t^{-1} (\exp^* d\Pi_2)(t) = \frac{E(u, v)}{v} + \int_u^v E(u, t) t^{-2} dt.$$

It follows that

$$\left| \int_u^v t^{-1} (\exp^* d\Pi_2)(t) \right| \leq \frac{2A(v)}{v} + 2 \int_u^v A(t) t^{-2} dt \to 0$$

as u, $v \to \infty$, and thus (6.11) holds. $\square$

COROLLARY 6.7. *Suppose that $\Pi(x)$ of a g-number system $\mathcal{N}$ can be written as a sum $\Pi_1(x) + \Pi_2(x)$, where $\Pi_1(x) \uparrow$, $\Pi_1(x) \ll x/\log x$, and*

$$\int_1^\infty x^{-s} d\Pi_1(x) - \log \frac{1}{s-1} \ll 1, \quad 1 < s < 2.$$

Further, suppose that the conditions of the last lemma hold for $\Pi_2(x)$. Then the integer counting function of $\mathcal{N}$ satisfies $N(x) \ll x$.

PROOF. Recall from (3.1) and Proposition 3.10 that

$$dN = \exp^* d\Pi = dN_1 * \exp^* d\Pi_2$$

with $dN_1 := \exp^* d\Pi_1$. The conditions for $\Pi_1(x)$ are those of the last theorem. Thus, $N_1(x) \ll x$. Combining this bound with the last lemma, we obtain

$$N(x) \leq \int_{1-}^{x} A\Big(\frac{x}{t}\Big)\, dN_1(t) = \int_{1-}^{x} N_1\Big(\frac{x}{t}\Big)\, dA(t) \ll x \int_{1-}^{x} \frac{dA(t)}{t}.$$

Now integrate by parts. We find

$$N(x) \ll A(x) + x \int_{1}^{\infty} A(t)\, t^{-2} dt \ll x. \qquad \square$$

COROLLARY 6.8. *Suppose that $\Pi(x) = \Pi_1(x) + \Pi_2(x)$, with $\Pi_1(x)$ as in the last theorem. In place of (6.10), assume that*

$$\int_{1-}^{\infty} T^{-1} \big|\exp^* d\Pi_2\big| < \infty$$

or that

$$\int_{1-}^{\infty} T^{-1} \big|d\Pi_2\big| < \infty.$$

In either case, $N(x) \ll x$ holds.

PROOF. Suppose the first integral condition holds. If we take

$$A(x) := \int_{1-}^{x} \big|\exp^* d\Pi_2\big|$$

in Lemma 6.6, we see by integration by parts that

$$\int_{1}^{x} A(t)\, t^{-2} dt = -\frac{A(x)}{x} + \int_{1-}^{x} t^{-1} \big|\exp^* d\Pi_2\big|(t) \leq \int_{1-}^{\infty} t^{-1} \big|\exp^* d\Pi_2\big|(t) < \infty.$$

Then, by the lemma, $N(x) \ll x$.

Now assume the second integral condition. By the triangle inequality and Lemma 2.15

$$\begin{aligned}\int_{1-}^{\infty} T^{-1} \big|\exp^* d\Pi_2\big| &\leq \int_{1-}^{\infty} T^{-1} \exp^* \big|d\Pi_2\big| \\ &= \int_{1-}^{\infty} \exp^* \big|T^{-1} d\Pi_2\big| = \exp \int_{1-}^{\infty} T^{-1} \big|d\Pi_2\big|,\end{aligned}$$

so we again have $N(x) \ll x$. $\square$

For a lower bound we have

THEOREM 6.9. *Suppose that a g-integer system $\mathcal{N}$ satisfies $N(x) \gg x$. Then* $\liminf_{\epsilon \to 0+} \epsilon\zeta(1+\epsilon) > 0$. *In the other direction, suppose that the lower Chebyshev bound $\psi(x) \gg x$ holds and also that $\epsilon\zeta(1+\epsilon)$ has positive finite bounds from above and below as $\epsilon \to 0+$. Then $N(x) \gg x$.*

PROOF. The first assertion is proved using integration by parts, as was done in showing the first part of Lemma 3.3. The argument for the converse statement is essentially that of Theorem 6.5, except that here Corollary 4.10 is used in place of Proposition 4.2 for the logarithmic density bound. $\square$

REMARKS 6.10. Note that both upper and lower bounds for $\epsilon\zeta(1+\epsilon)$ are conditions here and in the corollary. There is a further asymmetry between this theorem and Theorem 6.5: in the earlier one, we assumed the upper estimate $\pi(x) \ll x/\log x$ and used integration by parts to show $\psi(x) \ll x$; if we had assumed $\pi(x) \gg x/\log x$ here instead of $\psi(x) \gg x$, then we would not have been able to obtain the $\psi(x)$ bound without some further condition, such as $\int_1^x \Pi(t)t^{-1}dt = o(x)$.

6.3. Sharper criteria for O-density

Here, we show that a reasonably sharp Chebyshev upper bound yields an upper residue for $\zeta(s)$ which, in turn, leads to another O-density result (cf. Example 4.6 and Theorem 5.3).

PROPOSITION 6.11. *Suppose that the Chebyshev function $\psi(x)$ of a g-number system $\mathcal{N}$ satisfies the one-sided inequality*

$$\psi(x) \le x + x/f(x), \quad x \ge 1, \tag{6.12}$$

where $f(x) \ge 1$ and $\int_2^\infty du/\{uf(u)\log u\} < \infty$. Then $\mathcal{N}$ has O-density.

On the other hand, suppose $\psi(x) \ge x + x/f(x)$ holds with f instead satisfying $\int_2^\infty du/\{uf(u)\log u\} = \infty$. Then $\mathcal{N}$ does not have O-density.

PROOF. We have for $1 < s \le 2$,

$$\frac{-\zeta'}{\zeta}(s) = s\int_1^\infty u^{-s-1}\psi(u)\,du \le s\int_1^\infty u^{-s}\{1 + 1/f(u)\}\,du.$$

Thus

$$\frac{-\zeta'}{\zeta}(s) \le \frac{1}{s-1} + O(1) + 2\int_2^\infty u^{-s}\frac{du}{f(u)}.$$

Then, upon integrating, we have uniformly for $1 < s < 2$,

$$\begin{aligned}\log\zeta(s) - \log\zeta(2) &= \int_s^2 \frac{-\zeta'}{\zeta}(u)\,du \\ &\le \log\frac{1}{s-1} + O(1) + 2\int_s^2\Big\{\int_2^\infty u^{-v}\frac{du}{f(u)}\Big\}\,dv.\end{aligned}$$

The double integral is less than

$$\int_2^\infty\Big\{\int_s^\infty u^{-v}\,dv\Big\}\frac{du}{f(u)} = \int_2^\infty u^{-s}\frac{du}{f(u)\log u} < \int_2^\infty \frac{du}{uf(u)\log u} \ll 1,$$

whence $\zeta(s) \le C/(s-1)$ for some constant $C > 0$. Thus zeta has a finite upper residue at $s = 1$. Also, we have a Chebyshev bound for the g-number system by hypothesis. It follows that $N(x) \ll x$ by Theorem 6.5.

For the opposite estimate, we argue as follows. The contribution of small primes is irrelevant for O-density (Proposition 6.3), so we can assume that

$$\psi(x) = 0, \quad 1 \le x < 2, \qquad \psi(x) \ge x + cx/f(x), \quad x \ge 2,$$

for some $c > 0$. Supposing $\int_2^\infty du/\{uf(u)\log u\}$ is divergent, our path is first to show that the zeta function does not have a finite upper residue at $s = 1$, and then to apply Proposition 6.4.

We have for $s > 1$,

$$\frac{-\zeta'}{\zeta}(s) = s\int_2^\infty u^{-s-1}\psi(u)\,du \ge \int_2^\infty u^{-s}\{1 + c/f(u)\}\,du.$$

Once more integrating, we have for $1 < s < 2$,

$$\begin{aligned}\log \zeta(s) - \log \zeta(2) &= \int_s^2 \frac{-\zeta'}{\zeta}(v)\, dv \\ &\geq \int_s^2 \Big(\int_2^\infty u^{-v}\{1 + c/f(u)\}\, du \Big)\, dv.\end{aligned}$$

If we exchange the order of the last integrals and insert a negative term, we find

$$\begin{aligned}\log \zeta(s) - \log \zeta(2) &\geq \int_2^\infty \Big\{ \int_s^2 u^{-v} dv \Big\} \Big(1 - \frac{1}{u} + \frac{c}{f(u)} \Big)\, du \\ &= \int_2^\infty (u^{-s} - u^{-2}) \Big(\frac{1 - u^{-1}}{\log u} + \frac{c}{f(u) \log u} \Big)\, du \\ &\geq \int_1^\infty u^{-s} \frac{1 - u^{-1}}{\log u} du + \int_2^\infty u^{-s} \frac{c\, du}{f(u) \log u} - B\end{aligned}$$

for some number B, uniformly for $1 < s < 2$.

The last integral diverges to infinity as $s \to 1+$ by the monotone convergence theorem and the hypothesis, and the next to last integral equals $\log\{s/(s-1)\}$ by Theorem 1.5. It follows that

$$(s-1)\zeta(s) \gg \exp \int_2^\infty u^{-s} \frac{c\, du}{f(u) \log u} \to \infty$$

as $s \to 1+$. By the contrapositive form of Proposition 6.4, the g-number system does not have O-density. □

EXAMPLES 6.12. CRITICAL EXPONENT FOR O-DENSITY. If a g-number system satisfies $\psi(x) - x \leq x(\log\log x)^{-1.001}$, then it has O-density. On the other hand, if $\psi(x) - x \geq x(\log\log x)^{-1}$, then there is no O-density.

We next deduce an O-density result from a one-sided Mertens-type hypothesis (see §5.1) by an elementary argument.

THEOREM 6.13. *Suppose that the Chebyshev function $\psi(x)$ of a g-number system satisfies the one-sided Mertens-type inequality*

$$\psi_1(x) := \int_1^x u^{-1} d\psi(u) \leq \log x + O(1). \tag{6.13}$$

Then the associated g-integers satisfy $N(x) \ll x$.

PROOF. Suppose first that $\psi(x) = 0$ for $x \leq A$, where $A \geq 3$ and that

$$\int_A^x u^{-1} d\psi(u) \leq \log x - 1, \quad x > A. \tag{6.14}$$

We shall show in this case that the explicit inequality $N(x) \leq x$ holds for all $x \geq 1$.

Our argument proceeds by induction on intervals $[A^n, A^{n+1})$. On the initial interval, Chebyshev's identity gives

$$(L\, dN)(t) = \log t\, dN(t) = \{d\psi * dN\}(t) = 0, \quad 1 \leq t < A. \tag{6.15}$$

Thus $dN(t) = 0$ for $1 < t < A$, and since $dN\{1\} = 1$, we have

$$N(x) = 1 \leq x, \quad 1 \leq x < A.$$

Suppose that $N(u) \le u$ is known to hold for $1 \le u < A^n$, and let $A^n \le x < A^{n+1}$. For $t \ge A$ we have $x/t < A^n$, so $N(x/t) \le x/t$ holds by our assumption. Another application of Chebyshev's identity yields

$$T(x) := \int_1^x L\,dN = \int_1^x N(x/t)\,d\psi(t) \le \int_A^x \frac{x}{t}\,d\psi(t) \le x\,(\log x - 1) \tag{6.16}$$

for $A^n \le x < A^{n+1}$. Relation (6.15) together with the last inequality on each interval $[A^i,\, A^{i+1})$ for $1 \le i \le n$ imply that (6.16) holds on $A \le x < A^{n+1}$.

On this interval we have

$$\begin{aligned} N(x) &= 1 + \int_A^x \frac{dT(u)}{\log u} = 1 + \frac{T(x)}{\log x} + \int_A^x \frac{T(u)}{u\,\log^2 u}\,du \\ &\le 1 + \frac{x\,(\log x - 1)}{\log x} + \int_A^x \frac{\log u - 1}{\log^2 u}\,du = 1 + x - \frac{A}{\log A} < x, \end{aligned}$$

which completes the induction.

To handle the general case, suppose in (6.13) that

$$\int_1^x u^{-1} d\psi(u) \le \log x + B$$

for some constant B and all $x \ge 1$. Choose $A \ge 3$ such that

$$\int_1^A u^{-1} d\psi(u) \ge B + 1.$$

(If such a choice cannot be made, then $\int_1^\infty u^{-1} d\psi(u)$ is finite and (6.14) holds already for suitable A.) Define $\mathcal{P}_0 := \mathcal{P} \cap (A, \infty)$ and let ψ_0 and N_0 denote the Chebyshev and g-integer counting functions associated with $\mathcal{P}_0$. Now ψ_0 satisfies condition (6.14), so $N_0(x) \ll x$ for all $x \ge 1$. By Proposition 6.3, $N(x) \ll x$. $\square$

6.4. Notes

§4.6. A small calculation shows that the counting function of Example 4.18 has O-density. Thus we see also that O-density $\not\Rightarrow$ regular growth.

§§6.2, 4.3. In the proof of Theorem 6.9 and also that of Corollary 4.10, both lower bound estimates, we have used an upper bound hypothesis on $\epsilon\zeta(1+\epsilon)$. While *some* condition is needed, we wonder whether a weaker one would suffice.

CHAPTER 7

Density of g-integers

Will a number system sink if its density is greater than 1?

SUMMARY. Density and criteria. More on the residue of $\zeta(s)$ at $s = 1$. Axer's theorem on convolutions.

7.1. Densities and right hand residues

Recall that a g-number system $\mathcal{N}$ is said to have *density* δ if the counting function of its g-integers satisfies $\lim_{x\to\infty} N(x)/x = \delta$, and to have a *right hand residue* r if $\lim_{s\to 1+}\{(s-1)\,\zeta_{\mathcal{N}}(s)\}$ exists and equals r. We begin by giving the simple proof of the remark made at the end of Example 1.2 about evaluating a density that is known to exist

LEMMA 7.1. *Let $F \in \mathcal{V}$ and suppose $F(x)/x \to \delta$ as $x \to \infty$. Then the Mellin transform $\widehat{F}(s)$ converges for $\Re s > 1$ and*

$$\lim_{s\to 1+} (s-1)\,\widehat{F}(s) = \delta.$$

PROOF. Using integration by parts,

$$\widehat{F}(s) := \int_{1-}^{\infty} u^{-s} dF(u) = s\int_{1}^{\infty} u^{-s-1} F(u)\,du \sim \int_{1}^{\infty} u^{-s-1}\delta u\,du = \frac{\delta}{s-1},$$

as $s \to 1+$. Thus $(s-1)\,\widehat{F}(s) \to \delta$. □

The converse of the last lemma is not true: there exist Mellin transforms of nonnegative measures having a right hand residue for which the cumulative function does not have a density. One case is Example 4.18. Another is Example 6.1, where the right hand residue is 0 but there is no density.

7.2. Axer's Theorem

In the next section and in a later chapter we shall need versions of the following convolution estimate.

THEOREM 7.2 (Axer). *Let $A(x)$ and $B(x)$ be right continuous functions on $[1, \infty)$, each locally of bounded variation. Assume that $B(x) = o(x)$ and its total variation function satisfies $B_v(x) = O(x)$. Also, assume that $|A(x)| \le A_1(x)$, another right continuous function, with $\int_1^{\infty} A_1(u)u^{-2}du < \infty$. Finally, assume that either*

$$(A)\ A_1(x)\uparrow \quad or \quad (B)\ A_1(x)/x\downarrow.$$

Then $\int_{1-}^{x} dA * dB = o(x)$.

PROOF. We apply the Dirichlet hyperbola method. Let K be a large positive number and write

$$\begin{aligned}\int_{1-}^{x} dA * dB &= \int_{1-}^{x/K} A\Big(\frac{x}{t}\Big)\,dB(t) + \int_{1-}^{K} B\Big(\frac{x}{t}\Big)\,dA(t) - A(K)B\Big(\frac{x}{K}\Big).\\ &=: I + II - III, \quad \text{say.}\end{aligned}$$

In Case (A), we note first that $A_1(x) = o(x)$, by the first step in the proof of Lemma 6.6. Next, suppose that $B_v(t) \le Mt$ for all $t \ge 1$. We have, by Lemma 2.1,

$$|I| \le \int_{1-}^{x/K} A_1\Big(\frac{x}{t}\Big)\,dB_v(t) \le M\int_{1-}^{x/K} A_1\Big(\frac{x}{t}\Big)\,(\delta_1 + dt)$$

since $t \mapsto A_1(x/t)$ is positive, left continuous, and decreasing. Thus

$$|I| \le MA_1(x) + M\int_{1-}^{x/K} A_1\Big(\frac{x}{t}\Big)\,dt,$$

and if we change the variable in the last integral and extend the range of integration, we obtain

$$|I| \le o(x) + Mx\int_{K}^{\infty} A_1(u)u^{-2}du.$$

Choosing K large makes the last integral small; given $\epsilon > 0$, we have $|I| < \epsilon x$ for a large K. With this choice of K, we also have

$$|II| \le \int_{1-}^{K} o(x/t)\,dA_v(t) = o(x), \quad x \to \infty.$$

(Here $A_v(x)$ is the total variation function of $A(x)$; it is not connected with $A_1(x)$.) Also $III = O(1)\,o(x/K) = o(x)$, and thus $\int_{1-}^{x} dA * dB = o(x)$ holds in Case (A).

In Case (B), we have $II + III = o(x)$ as before, for any fixed positive K. It remains to show $I = o(x)$. The assumption here is that $A_2(x) := A_1(x)/x \downarrow$, and the integral hypothesis can be restated as $\int_1^\infty A_2(u)u^{-1}du < \infty$. It follows that

$$\frac{1}{2}A_2(x)\log x \le \int_{\sqrt{x}}^{x} t^{-1}A_2(t)\,dt = o(1), \quad x \to \infty,$$

which implies that $A_2(x) = o(\log^{-1} x)$.

Now, integrating by parts we find

$$\begin{aligned}|I| &\le \int_{1-}^{x/K} \frac{x}{t}A_2\Big(\frac{x}{t}\Big)\,dB_v(t)\\ &= B_v\Big(\frac{x}{K}\Big)\,KA_2(K) - x\int_{1}^{x/K} B_v(t)\Big\{-t^{-2}A_2\Big(\frac{x}{t}\Big)\,dt + t^{-1}\,d_tA_2\Big(\frac{x}{t}\Big)\Big\}.\end{aligned}$$

We can drop the last term, since $t \mapsto A_2(x/t)$ is increasing. Using the upper estimate for B_v, we find

$$\begin{aligned}|I| &\le MxA_2(K) + x\int_{1}^{x/K} Mt\,t^{-2}A_2\Big(\frac{x}{t}\Big)\,dt\\ &= O\Big(\frac{x}{\log K}\Big) + Mx\int_{1}^{x/K} A_2\Big(\frac{x}{t}\Big)\,t^{-1}dt.\end{aligned}$$

We change the integration variable in the last integral, obtaining

$$\int_K^x A_2(u)u^{-1}\,du \le \int_K^\infty A_2(u)u^{-1}\,du.$$

Thus I/x is arbitrarily small for sufficiently large K. With such a choice of K, each of I, II, $III = o(x)$, and so $\int_{1-}^x dA * dB = o(x)$ in this case as well. □

7.3. Criteria for density

Our first result is a direct refinement of Theorem 6.5.

THEOREM 7.3. *If $N(x) \sim Ax$, then the associated zeta function has a right hand residue c at $s = 1$. Conversely, if zeta has a right hand residue A at $s = 1$ and the counting function of g-primes satisfies the PNT, then $N(x) \sim Ax$.*

The first assertion can be deduced from Proposition 5.1 or directly by integration by parts of the Mellin integral for $\zeta(s)$. For the second assertion, replace the O- bounds of Theorem 6.5 with asymptotic estimates. (In place of Proposition 4.2, use Proposition 5.1.)

EXAMPLE 7.4. A SET OF *EVEN* PRIMES. Let $\mathcal{P} := \{p+1\}$ with $\{p\}$ the rational primes, i.e. $\mathcal{P} = \{3, 4, 6, 8, 12\ldots\}$. We show that the associated set of g-integers $\mathcal{N}$ has a density and that it equals $6/\pi^2$.

Clearly the PNT holds for $\mathcal{P}$. To apply the preceding theorem, it suffices to show that $\zeta_{\mathcal{P}}(s)$, the zeta function of $\mathcal{P}$, has a right hand limit of $6/\pi^2$ at $s = 1$. Recall that $(s-1)\zeta_0(s) \sim 1$ as $s \to 1+$ for $\zeta_0(s)$ the Riemann zeta function. Now

$$\frac{\zeta_{\mathcal{P}}(s)}{\zeta_0(s)} = \prod_p \frac{1-p^{-s}}{1-(p+1)^{-s}} = \prod_p \left\{1 - \frac{p^{-s}-(p+1)^{-s}}{1-(p+1)^{-s}}\right\},$$

and for $1 < s < 2$,

$$p^{-s} - (p+1)^{-s} \ll p^{-s-1} \ll p^{-2}.$$

Thus the last product is absolutely and uniformly convergent for $1 < s < 2$. It follows that we can take the limit pointwise in the product, and the density exists.

We find its value to be

$$\lim_{s\to 1+}(s-1)\zeta_{\mathcal{P}}(s) = \prod_p \left\{1 - \frac{p^{-1}-(p+1)^{-1}}{1-(p+1)^{-1}}\right\} = \prod_p (1-p^{-2}) = \frac{1}{\zeta_0(2)} = \frac{6}{\pi^2}.$$

An immediate consequence of the second assertion of the last theorem is

COROLLARY 7.5. *Suppose $\mathcal{N}$ is a g-number system having logarithmic density and for which the PNT holds. Then $\mathcal{N}$ has density.*

The last theorem is not of the form "if and only if," because we assumed the PNT for one implication. Example 4.18 shows that the theorem is false without this condition. Recall that we had

$$\Pi(x) := \sum_{1\le k\le \log x/(\log 2)} 2^k/k\,.$$

The PNT does not hold here, since

$$\Pi(2^n) - \Pi(2^n - 1) = 2^n/n \gg \Pi(2^n).$$

Also, since

$$(s-1)\zeta(s) = \frac{s-1}{1-2^{1-s}} \to \frac{1}{\log 2}, \quad s \to 1+,$$

$\zeta(s)$ has a right hand residue at $s = 1+$. On the other hand, we showed that $N(x)/x$ has no limit as $x \to \infty$.

Is there a variant of Theorem 7.3 that is valid without assuming the PNT? The answer is Yes, if we assume a little more than just the existence of a right hand residue of $\zeta(s)$. This residue can be represented as the limit as $s \to 1+$ of $\zeta(s)(s-1)/s$, and the last function is representable as the Mellin transform

$$\widehat{F}(s) := \int_{1-}^{\infty} u^{-s}\{dN * (\delta_1 - du/u)\}.$$

This observation and (3.11) immediately give the desired criterion for density:

PROPOSITION 7.6. *A g-number system $\mathcal{N}$ has a density if and only if $\widehat{F}(s)$, given above, converges at $s = 1$, i.e.*

$$\lim_{x\to\infty} \int_{1-}^{x} u^{-1}\{dN * (\delta_1 - du/u)\}$$

exists. If it exists, the density equals $\widehat{F}(1)$.

A direct proof consists in noting that the last integral equals

$$\int_{1-}^{x} \frac{dN(u)}{u} - \int_{1}^{x} \frac{N(u)\,du}{u^2} = \frac{N(x)}{x}$$

by integration by parts.

Another density criterion involving an integral of $N(x)$ is the following.

PROPOSITION 7.7. *Suppose $N(x)$ is the counting function of a g-number system $\mathcal{N}$ and A is a positive number. Then*

$$I_1 := \int_{1-}^{\infty} \frac{dN(t) - A\,dt}{t} \text{ exists} \iff I_2 := \int_{1}^{\infty} \frac{N(t) - At}{t^2}\,dt \text{ exists.}$$

If either integral exists, $\mathcal{N}$ has density A.

PROOF. Suppose first that I_1 exists, i.e.

$$F(x) := \int_{1-}^{x} \frac{dN(u) - A\,du}{u} = L + o(1), \quad x \to \infty.$$

Then

$$\begin{aligned} N(x) - Ax + A &= \int_{1-}^{x} u\,dF(u) = xF(x) - \int_{1}^{x} F(u)\,du \\ &= x\{L + o(1)\} - \int_{1}^{x} \{L + o(1)\}\,du = o(x). \end{aligned}$$

Thus $N(x) \sim Ax$ and $\mathcal{N}$ has density.

Integrating by parts, we have

$$\int_{1-}^{x} \frac{N(u) - Au}{u^2}\,du = -A - \frac{N(x) - Ax}{x} + \int_{1-}^{x} \frac{dN(u) - A\,du}{u}. \tag{7.1}$$

As $x \to \infty$, the last integral converges by hypothesis and $\{N(x) - Ax\}/x \to 0$ since $N(x) \sim Ax$. Thus I_2 exists as well.

Next, suppose that I_2 exists. Let $0 < \epsilon < 1$. We have

$$\frac{N(x)}{x} \le \frac{1+\epsilon}{\epsilon}\Big(\int_x^{x+\epsilon x} \frac{N(t)-At}{t^2}\,dt + A\log(1+\epsilon)\Big).$$

The integral goes to 0 as $x \to \infty$, and so

$$\limsup_{x\to\infty} \frac{N(x)}{x} \le \frac{1+\epsilon}{\epsilon} A\log(1+\epsilon).$$

Letting $\epsilon \to 0$, we find $\limsup_{x\to\infty} N(x)/x \le A$. Similarly, we deduce from

$$\frac{N(x)}{x} \ge \frac{1-\epsilon}{\epsilon}\Big(\int_{x-\epsilon x}^{x} \frac{N(t)-At}{t^2}\,dt + A\log\frac{1}{1-\epsilon}\Big)$$

that $\liminf_{x\to\infty} N(x)/x \ge A$. Thus $N(x)/x \to A$ as $x \to \infty$.

Appealing again to (7.1), we see that I_1 exists as well. □

Does Proposition 7.7 have a converse: if $\mathcal{N}$ is a g-number system having density A, does it follow that I_1 and I_2 exist? The answer is No. We shall show in Proposition 14.6 that the continuous g-number system with prime measure

$$d\Pi(u) := \frac{1-u^{-1}}{\log u}du + \Big(\frac{1-u^{-1}}{\log u}\Big)^2 du$$

has g-integer counting function

$$N(x) = 4x - \frac{4x}{\log x} + O\Big(\frac{x\log\log x}{\log^2 x}\Big).$$

It is clear that this system has density $A = 4$. On the other hand, the variant of I_2 having lower integration limit 10 satisfies

$$\int_{10}^{\infty} \frac{N(t)-At}{t^2}\,dt = \int_{10}^{\infty} \Big\{\frac{-4}{t\log t} + O\Big(\frac{\log\log t}{t\log^2 t}\Big)\Big\}\,dt = -\infty.$$

THEOREM 7.8. *Say that the function $\Pi(x)$ of a g-number system $\mathcal{N}$ can be written as a sum $\Pi_1(x) + \Pi_2(x)$, where $\Pi_1(x)\uparrow$, $\Pi_1(x) \sim x/\log x$, and*

$$\int_1^{\infty} x^{-s}d\Pi_1(x) - \log\frac{1}{s-1} \to \log A$$

as $s \to 1+$. Further, suppose that $\Pi_2(x)$ satisfies the conditions of Lemma 6.6. Then $\mathcal{N}$ has density $A\int_{1-}^{\infty} T^{-1}\exp^* d\Pi_2$.

PROOF. By (3.1) and Proposition 3.10, we have

$$dN = \exp^* d\Pi = dN_1 * \exp^* d\Pi_2$$

with $dN_1 := \exp^* d\Pi_1$. By Theorem 7.3,

$$N_1(x) := \int_{1-}^{x} \exp^* d\Pi_1 \sim Ax.$$

Subtracting $A(\delta_1 + dt)$ from dN_1 and adding the same, we find

$$N(x) = \int_{1-}^{x} \frac{Ax}{t}(\exp^* d\Pi_2)(t) + \int_{1-}^{x} \exp^* d\Pi_2 * (dN_1 - A\delta_1 - Adt). \tag{7.2}$$

We apply Axer's Theorem to the second integral in this formula with

$$dA := \exp^* d\Pi_2, \quad dB := dN_1 - A\delta_1 - Adt.$$

We have $|A(x)| \le A_1(x)$, where $A_1(x)$ satisfies the conditions (6.10), and by Theorem 7.3,

$$B(x) = N_1(x) - Ax = o(x), \quad B_v(x) \le N_1(x) + Ax \ll x.$$

Thus $\int_{1-}^{x} dA * dB = o(x)$,

Now the first term in (7.2) is asymptotic to

$$Ax \int_{1-}^{\infty} T^{-1} \exp^* d\Pi_2 =: A'x,$$

by Lemma 6.6, and hence $\mathcal{N}$ has density A'. □

We have, as analogues of Corollary 6.8,

COROLLARY 7.9. *Suppose that* $\Pi(x) = \Pi_1(x) + \Pi_2(x)$, *with* $\Pi_1(x)$ *as in the last theorem. In place of* (6.10) *assume that*

$$\int_{1-}^{\infty} T^{-1} \big| \exp^* d\Pi_2 \big| < \infty$$

or that

$$\int_{1-}^{\infty} T^{-1} \big| d\Pi_2 \big| < \infty.$$

In either case, $\mathcal{N}$ *has a density as before.*

A result that is related to but simpler than Theorem 7.8 is

PROPOSITION 7.10. *Suppose that the integer counting function* $N(x)$ *of a g-number system* $\mathcal{N}$ *satisfies*

$$\int_1^{\infty} \sup_{y \le x} \left| N(y) - \int_1^y N(u) u^{-1} du \right| \frac{dx}{x^2} < \infty .$$

Then $\mathcal{N}$ *has a density.*

PROOF. Write

$$dN = \exp^* d\Pi = \exp^* d\Pi_c * \exp^* d\Pi_2 ,$$

where, as usual, $d\Pi_c = (1 - t^{-1})\, dt/ \log t$.

We have, from (3.10),

$$N(x) - \int_1^x N(u) u^{-1}\, du = \int_{1-}^x \exp^*(d\Pi - d\Pi_c) = \int_{1-}^x \exp^* d\Pi_2 ,$$

If we take

$$A(x) := \sup_{y \le x} \left| N(y) - \int_1^y \frac{N(u)}{u}\, du \right| ,$$

then $A(x)$ is nondecreasing and satisfies (6.10).

Now,

$$N(x) = \int_{1-}^x \Big\{ \int_{1-}^{x/t} \exp^* d\Pi_c \Big\} (\exp^* d\Pi_2)(t) = x \int_{1-}^x T^{-1} \exp^* d\Pi_2 ,$$

and the last integral converges as $x \to \infty$ by Lemma 6.6. Thus $\mathcal{N}$ has a density, and it equals $\int_{1-}^{\infty} T^{-1} \exp^* d\Pi_2$. □

7.4. An L^1 criterion for density

In this section we show that a g-number system satisfying an L^1 condition on $\Pi(t) - t/\log t$ has density.

THEOREM 7.11. *Suppose that $\mathcal{N}$ is a g-number system for which*

$$\int_2^\infty t^{-2}|\Pi(t) - t/\log t|\,dt < \infty. \tag{7.3}$$

Then the g-integers of $\mathcal{N}$ have density.

Without loss of generality, we can replace (7.3) by

$$\int_1^\infty t^{-2}|\Pi(t) - \Pi_c(t)|\,dt < \infty \tag{7.4}$$

where

$$\Pi_c(t) = \int_1^t \frac{1-u^{-1}}{\log u}\,du, \tag{1.9 bis}$$

since

$$\int_2^\infty t^{-2}|\Pi_c(t) - t/\log t|\,dt \ll \int_2^\infty \frac{dt}{t\log^2 t} < \infty.$$

The key step in the proof is the following lemma, showing that Π can be decomposed into two parts that come within a prescribed ϵ of fulfilling the conditions of Theorem 7.8. The proof will be concluded by applying the lemma iteratively.

LEMMA 7.12. *Suppose Π satisfies condition* (7.4). *Then there exist constants B (depending on Π) and C (absolute) such that for any $\epsilon \in (0,\,1)$ there is a decomposition $\Pi = \Pi_{\epsilon 1} + \Pi_{\epsilon 2}$, where $\Pi_{\epsilon 1}\uparrow$,*

$$1-\epsilon < \Pi_{\epsilon 1}(x)/\Pi_c(x) < 1+\epsilon, \quad 1 \le x < \infty, \tag{7.5}$$

$$\int_1^\infty t^{-2}|\Pi_{\epsilon 1}(t) - \Pi_c(t)|\,dt \le B < \infty, \tag{7.6}$$

and

$$\int_4^\infty t^{-1}|d\Pi_{\epsilon 2}(t)| \le C/\epsilon^2 < \infty. \tag{7.7}$$

PROOF. Let $x_0 \ge 1$ be a point at which

$$|\Pi(x_0) - \Pi_c(x_0)| \ge \epsilon\Pi_c(x_0).$$

If no such point x_0 exists, then the choice $\Pi_{\epsilon 1} := \Pi$ and $\Pi_{\epsilon 2} := 0$ obviously suffices.

Suppose that $\Pi(x_0) \ge (1+\epsilon)\Pi_c(x_0)$. Let $I = I(x_0) = [a,\,b]$ denote the maximal interval containing x_0 such that $\Pi(x) \ge (1+\epsilon/3)\Pi_c(x)$ holds for all $x \in I$. This interval is closed at a by the right-continuity of $\Pi - (1+\epsilon/3)\Pi_c$, and it is closed at b because $\Pi(b) = (1+\epsilon/3)\Pi_c(b)$. It is bounded because of condition (7.4).

Next suppose that $\Pi(x_0) \le (1-\epsilon)\Pi_c(x_0)$. Let $J = J(x_0)$ denote the maximal interval containing x_0 such that $\Pi(x) < (1-\epsilon/3)\Pi_c(x)$ holds for all $x \in J$. Similar reasoning as for I shows that J is bounded and open. Write $J = (c,\,d)$.

We shall call points of each interval I or J "ϵ-bad," or, briefly, "bad," and those of the complementary intervals of $[1,\,\infty)$ are called "ϵ-good" or just "good." (Note that points of an interval on which $|\Pi(x) - \Pi_c(x)| > \epsilon\Pi_c(x)/3$ holds are not bad unless there exists some point x_0 of the interval at which $|\Pi(x_0) - \Pi_c(x_0)| \ge \epsilon\Pi_c(x)$.)

For $x \geq 1$, set

$$\Pi_{\epsilon 1}(x) = \begin{cases} \Pi(x), & x \text{ good}, \\ (1+\epsilon/3)\Pi_c(x), & x \in \cup I, \\ (1-\epsilon/3)\Pi_c(x), & x \in \cup J, \end{cases}$$

and

$$\Pi_{\epsilon 2}(x) = \begin{cases} 0, & x \text{ good}, \\ \Pi(x) - (1+\epsilon/3)\Pi_c(x), & x \in \cup I, \\ \Pi(x) - (1-\epsilon/3)\Pi_c(x), & x \in \cup J. \end{cases}$$

By construction, $\Pi_{\epsilon 1}$ is nondecreasing and it satisfies (7.5). Also, we have

$$|\Pi_{\epsilon 1}(x) - \Pi_c(x)| \leq |\Pi(x) - \Pi_c(x)|, \quad 1 \leq x < \infty,$$

and so $\Pi_{\epsilon 1}$ satisfies (7.6). It remains to prove (7.7); to do this, we apply an argument of a tauberian character in which we show that each bad interval is reasonably long. Separate but analogous arguments are needed for the I and J cases. We treat I in detail and sketch the argument for J.

The behavior of $d\Pi_{\epsilon 2}$ on $I = [a, b]$ is as follows.

$$\begin{aligned} 0 \leq d\Pi_{\epsilon 2}\{a\} &= \Pi(a) - (1+\epsilon/3)\Pi_c(a) \leq \Pi(a) - \Pi(a-) = d\Pi\{a\}, \\ d\Pi_{\epsilon 2} &= d\Pi - (1+\epsilon/3)d\Pi_c, \quad a < x < b, \\ d\Pi_{\epsilon 2}\{b\} &= 0. \end{aligned}$$

Our plan for estimating the integral over I (recognizing a possible jump of $\Pi_{\epsilon 2}$ at a) is first to show that

$$\int_I T^{-1}|d\Pi_{\epsilon 2}| = \int_{a-}^{b} t^{-1}|d\Pi_{\epsilon 2}(t)| \leq 3\Pi(b)/b + 3\int_I \Pi(t)t^{-2}dt, \tag{7.8}$$

and then that each of the last two terms is bounded by a constant multiple of $\epsilon^{-2}\int_I |\Pi(t) - \Pi_c(t)|t^{-2}dt$.

For $a \leq t \leq b$ we have

$$\Pi(a-) - \Pi_c(a) \leq \frac{\epsilon}{3}\Pi_c(a) \leq \frac{\epsilon}{3}\Pi_c(t) \leq \Pi(t) - \Pi_c(t).$$

It follows by integration by parts that

$$\int_I T^{-1}(d\Pi - d\Pi_c) = \int_{a-}^{b} t^{-1}(d\Pi - d\Pi_c)(t) \geq 0,$$

and thus

$$\int_I T^{-1}|d\Pi_{\epsilon 2}| \leq \int_I T^{-1}\{(1+\epsilon/3)d\Pi_c + d\Pi\} \leq \frac{7}{3}\int_I T^{-1}d\Pi.$$

If we integrate the last expression by parts, drop the negative term, and change 7/3 to 3, we get (7.8).

To estimate $\Pi(b)/b$, we first show the interval $I = [a, b]$ is rather long. Indeed, since $a \leq x_0 < b$ and $\Pi \uparrow$, we have

$$\Pi(y) \geq \Pi(x_0) \geq (1+\epsilon)\Pi_c(x_0)$$

for $y \geq x_0$. The concavity of Π_c implies that

$$\frac{\Pi_c(y) - \Pi_c(x_0)}{y - x_0} \leq \frac{\Pi_c(x_0) - 0}{x_0 - 1}.$$

It follows that

$$\Pi(y) \geq (1+\epsilon)\,\frac{x_0 - 1}{y-1}\,\Pi_c(y) > (1+\epsilon/3)\Pi_c(y)$$

for $y < 1 + (x_0 - 1)(1+\epsilon)/(1+\epsilon/3)$. Therefore if $x_0 \geq 3$, we must have

$$b \geq 1 + (x_0 - 1)\,\frac{1+\epsilon}{1+\epsilon/3} \geq x_0(1+\epsilon/3).$$

or $b \geq (1+\epsilon/3)a$. Now

$$\int_a^b |\Pi(t) - \Pi_c(t)|\,\frac{dt}{t^2} \geq \frac{\epsilon}{3}\int_{b/(1+\epsilon/3)}^b \Pi_c(t)\,\frac{dt}{t^2} \geq \frac{\epsilon^2}{9}\Pi_c\Big(\frac{b}{1+\epsilon/3}\Big)\,b^{-1} \geq \frac{\epsilon^2}{9b}\,\Pi_c(3b/4),$$

and for $b \geq 4$, the concavity of Π_c and the preceding inequalities yield

$$\begin{aligned}\frac{\Pi(b)}{b} &= \Big(1+\frac{\epsilon}{3}\Big)\frac{\Pi_c(b)}{b} \leq \frac{(1+\epsilon/3)(b-1)}{(3b/4-1)\,b}\,\Pi_c(3b/4)\\ &\leq \frac{2\Pi_c(3b/4)}{b} \leq \frac{18}{\epsilon^2}\int_a^b t^{-2}|\Pi(t) - \Pi_c(t)|\,dt.\end{aligned}$$

Also, for $a \leq t \leq b$ we have

$$\Pi(t) \leq \frac{1+\epsilon/3}{\epsilon/3}\{\Pi(t) - \Pi_c(t)\},$$

and hence

$$\int_I t^{-2}\Pi(t)\,dt \leq \frac{4}{\epsilon}\int_I t^{-2}|\Pi(t) - \Pi_c(t)|\,dt.$$

If we insert the preceding estimates into (7.8), we get

$$\int_I T^{-1}|d\Pi_{\epsilon 2}| \leq \frac{66}{\epsilon^2}\int_a^b t^{-2}|\Pi(t) - \Pi_c(t)|\,dt,$$

provided that $a \geq 4$. Arguing similarly with intervals $J = (c,\, d)$, we show this time that the interval is long by moving from x_0 to the left. We find that

$$\int_J T^{-1}|d\Pi_{\epsilon 2}| \leq \frac{24}{\epsilon^2}\int_a^b t^{-2}|\Pi(t) - \Pi_c(t)|\,dt$$

provided that $c \geq 3$. Also, $\Pi_{\epsilon 2} \equiv 0$ on $S = \complement\{(\cup I) \cup (\cup \overline{J})\}$, and hence, trivially,

$$\int_S T^{-1}|d\Pi_{\epsilon 2}| = 0 \leq \int_S t^{-2}|\Pi(t) - \Pi_c(t)|\,dt$$

Combining the three estimates for $\int T^{-1}|d\Pi_{\epsilon 2}|$, we conclude that (7.7) holds with $C = 66$. □

PROOF OF THE THEOREM. Let $\epsilon_n = 1/n$, $n = 1, 2, 3, \ldots$. Take $x_1 = 1$ and for $n = 2, 3, \ldots$ choose a strictly monotonic sequence $\{x_n\}$ such that (i) x_n is $1/(n+1)$ good and (ii) x_n is so large that

$$\int_{x_n}^\infty T^{-1}|d\Pi_{1/n,2}| < 2^{-n},$$

where $\Pi_{1/n,2}$ is as in the preceding lemma. Define functions Π_1 and Π_2 on $[1,\, \infty)$ by taking $\Pi_1(x) = \Pi(x)$ and $\Pi_2(x) = 0$ for $1 \leq x < x_2$; and for $n \geq 2$, take $\Pi_1(x) = \Pi_{1/n,1}(x)$ and $\Pi_2(x) = \Pi_{1/n,2}(x)$ for $x_n \leq x < x_{n+1}$.

By construction, we have $\Pi = \Pi_1 + \Pi_2$, $\Pi_1 \uparrow$, $\Pi_1 \sim x/\log x$ as $x \to \infty$, and

$$\int_1^\infty T^{-1}|d\Pi_2| \le \int_1^{x_2} T^{-1}|d\Pi_2| + \sum_{n=2}^\infty 2^{-n} < \infty. \tag{7.9}$$

Finally, for $s > 1$, write

$$\int_1^\infty t^{-s} d\Pi_1(t) - \log\frac{s}{s-1} = \int_1^\infty x^{-s}(d\Pi - d\Pi_c)(x) - \int_1^\infty x^{-s} d\Pi_2(x).$$

The last integral has a limit as $s \to 1+$ by (7.9). Integration by parts gives

$$\int_1^\infty x^{-s}(d\Pi - d\Pi_c)(x) = s\int_1^\infty x^{-s-1}\{\Pi(x) - \Pi_c(x)\}\,dx,$$

and this integral also tends to a limit as $s \to 1+$ by the hypothesis of the theorem and the fact that $\Pi_c(x) - x/\log x \ll x/\log^2 x$.

Now the hypotheses of Corollary 7.9 are satisfied, and hence $\mathcal{N}$ has density. □

7.5. Estimates of $N(x)$ with an error term

In the last section we showed that a g-integer system has density if an integral involving $\Pi(t) - t/\log t$ converges. Here we show how error bounds assumed for $\Pi(t)$ lead to corresponding estimates of $N(x)$.

THEOREM 7.13. *Suppose there exists a positive number A such that*

$$\int_1^x \frac{d\Pi(t)}{t} = \int_1^x \frac{(1-t^{-1})\,dt}{t\log t} + \log A + E(x) \tag{7.10}$$

where $E(x)$ is an error term.

(A) If $E(x) = O(\log^{-\alpha} x)$ for some $\alpha > 2$, then

$$N(x) = Ax + O(x\log^{2-\alpha} x).$$

(B) If $E(x) = O(\exp\{-\log^a x\})$ for some $a \in (0, 1)$, then

$$N(x) = Ax + O(x\exp\{-(\log x\,\log\log x)^{a'}\}), \quad a' = a/(1+a).$$

Let

$$\Pi_c(x) := \int_1^x \frac{1-t^{-1}}{\log t}dt,$$

$$\lambda(x) := \int_1^x T^{-1}d\Pi_c = \int_1^x \frac{1-t^{-1}}{t\log t}dt,$$

$$\nu_j(x) := \int_{1-}^x T^{-1}(d\Pi - d\Pi_c - \{\log A\}\delta_1)^{*j}, \quad j \ge 1,$$

and $\nu(x) := \nu_1(x)$. Note that $d\nu_j = d\nu^{*j}$,

$$\int_1^x |d\nu| \le 2\lambda(x) + \nu(x) + 2|\log A|,$$

and

$$\lambda(x) \le \int_1^e dt/t + \int_e^x dt/(t\log t) = 1 + \log\log x \quad \text{for } x \ge e.$$

We have (cf. (3.11))

$$dN = \exp^* d\Pi = A\,\delta_1 * (\delta_1 + dt) * \exp^*(d\Pi - d\Pi_c - \{\log A\}\delta_1),$$

so, by Lemma 3.8,

$$N(x) = A\int_{1-}^{x} (\delta_1 + dt) * \exp^*(T\,d\nu) = A\int_{1-}^{x} \frac{x}{t}\,T\exp^* d\nu = Ax\int_{1-}^{x} \exp^* d\nu.$$

Thus

$$|N(x) - Ax| \le Ax\sum_{j=1}^{\infty} |\nu_j(x)|/j!\,,$$

and we shall show the last expression to be suitably small.

The key to proving each assertion is an estimate of $|\nu_j(x)|$. For (A) one shows that there exist constants A_0 and A_1 such that for all x, $j \ge 1$,

$$|\nu_j(x)| \le jA_0(2\log\log 3x + A_1)^{j-1}\log^{-\alpha} x. \tag{7.11}$$

The proof of this inequality and its application are quite similar to that of (B) and are given in [**Bzd99**]. Here we establish the analogue of (7.11) and prove (B).

LEMMA 7.14. *Suppose that for some $b \ge 0$ and $a \in (0, 1)$*

$$|\nu(x)| \le A_0\log^b ex\,\exp\{-(\log x)^a\}, \quad 1 \le x < \infty.$$

Then there exists a constant A_1 such that for all x, $j \ge 1$

$$|\nu_j(x)| \le jA_0(2\lambda(x) + A_1)^{j-1}\log^b ex\,\exp\{-(j^{-1}\log x)^a\}. \tag{7.12}$$

The factor $\log^b ex$ will be used in Corollary 7.15 below.

PROOF. The case $j = 1$ is true. Suppose that the jth case holds and set $y = x^{1/(j+1)}$ and $z = x/y$. By the Dirichlet hyperbola method,

$$\begin{aligned}\nu_{j+1}(x) &= \iint_{st\le x} d\nu^{*j}(s)\,d\nu(t)\\ &= \int_1^y \nu_j(x/t)\,d\nu(t) + \int_1^z \nu(x/t)\,d\nu_j(t) - \nu(y)\nu_j(z)\\ &= I + II - III, \text{ say.}\end{aligned}$$

Now

$$\begin{aligned}|I| &\le \int_1^y jA_0(2\lambda(x/t) + A_1)^{j-1}\log^b(ex/t)\exp\{-(j^{-1}\log(x/t))^a\}\,|d\nu|(t)\\ &\le jA_0(2\lambda(x) + A_1)^{j-1}\log^b(ex)\exp\{-(j^{-1}\log z)^a\int_1^y |d\nu|\\ &\le jA_0(2\lambda(x) + A_1)^{j-1}\log^b(ex)\exp\{-(\{j+1\}^{-1}\log x)^a\}\\ &\qquad\times(2\lambda(x) + B + 2|\log C|),\end{aligned}$$

where $B = \sup|\nu(x)| < \infty$. Also,

$$\begin{aligned}|II| &\le \int_1^z A_0\log^b(ex/t)\exp\{-(\log x/t)^a\}|d\nu|^{*j}(t)\\ &\le A_0\log^b ex\,\exp\{-(\log x/z)^a\}\Big(\int_1^x |d\nu|\Big)^j\\ &\le A_0\log^b ex\,\exp\{-(\{j+1\}^{-1}\log x)^a\}(2\lambda(x) + B + 2|\log C|)^j,\end{aligned}$$

and

$$|III| \le BjA_0(2\lambda(x) + A_1)^{j-1}\log^b ex\,\exp\{-(\{j+1\}^{-1}\log x)^a\}.$$

Thus

$$|\nu_{j+1}(x)| \le (j+1)A_0(2\lambda(x)+A_1)^j \log^b ex \exp\{-(\{j+1\}^{-1}\log x)^a\}$$

with $A_1 = 2B + 2|\log C|$. □

PROOF OF THE THEOREM. We have

$$\frac{|N(x)-Ax|}{Ax} \le \sum_{j=1}^{\infty} |\nu_j(x)|/j! = \sum_1^K + \sum_{K+1}^{\infty},$$

say, with K to be determined, and with the common summand

$$A_0(2\lambda(x)+A_1)^{j-1}\exp\{-(j^{-1}\log x)^a\}/(j-1)!.$$

To estimate $\sum_1^K$, we first give an upper bound for the logarithm of

$$\exp\{-(j^{-1}\log x)^a\}/(j-1)!$$

over all j. Recalling the crude lower bound $j! \ge j^j e^{-j}$, it suffices to asymptotically approximate the larger and simpler function

$$F(u) := -\frac{\log^a x}{u^a} - (u-1)\log u + u$$

for large x. We have

$$u^{a+1}F'(u) = a\log^a x - u^{a+1}(\log u - 1/u).$$

For large x, there is a unique zero U, for which

$$U^{a+1}\log U - U^a = a\log^a x. \tag{7.13}$$

We have

$$F(U) = \frac{-\log^a x - U^a(U-1)\log U + U^{a+1}}{U^a} \sim \frac{-\log^a x - U^{a+1}\log U}{U^a},$$

and inserting (7.13) into the last formula yields

$$F(U) \sim \frac{-\log^a x - a\log^a x - U^a}{U^a} \sim \frac{-(a+1)\log^a x}{U^a}.$$

If we drop the smaller term in (7.13), and set $a' = a/(a+1)$, we obtain

$$U^a \sim a^{a'}(\log x)^{aa'}/(\log U)^{a'}.$$

Then taking logarithms in (7.13), we find that $\log U \sim a' \log\log x$.

The last three asymptotic formulas together give

$$F(U) \sim -(a+1)^{1/(a+1)}(\log x \, \log\log x)^{a'}$$

Now, for any fixed arbitrarily small positive number ϵ, there exists $x_0 = x_0(\epsilon)$ such that for all $x \ge x_0$ and all $j \ge 1$,

$$\frac{\exp\{-(j^{-1}\log x)^a\}}{(j-1)!} \le \exp\{(\epsilon - (a+1)^{1/(a+1)})(\log x \, \log\log x)^{a'}\}. \tag{7.14}$$

Let $C(\epsilon, x, a) = C(\epsilon)$ denote the right side of (7.14). With K still to be determined, we have

$$\sum_1^K < \sum_{j=1}^K A_0 C(\epsilon)(2\lambda(x)+A_1)^{j-1} < A_0 C(\epsilon)\cdot 2(2\lambda(x)+A_1)^{K-1},$$

provided that $2\lambda(x) + A_1 \ge 2$.

Next,

$$\sum_{K+1}^{\infty} < A_0 \sum_{j=K}^{\infty} (2\lambda(x) + A_1)^j / j! \le 2A_0(2\lambda(x) + A_1)^K / K!,$$

provided that $K \ge 2(2\lambda(x) + A_1)$.

Take $K := \lfloor (\log x)^{a'} \rfloor$, and recall that $\lambda(x) = O(\log\log x)$. For any $\epsilon' > \epsilon$, there is an $x_1 = x_1(\epsilon')$ such that

$$|N(x) - Ax|/Ax < \sum_{1}^{K} + \sum_{K+1}^{\infty} < C(\epsilon'), \quad x \ge x_1.$$

If we choose ϵ and ϵ' such that $(1+a)^{1/(1+a)} - \epsilon' > 1$, then (B) holds. □

COROLLARY 7.15. *Suppose $d\Pi \ge 0$, $0 < a < 1$, and*

$$\Pi(x) = \Pi_c(x) + O(x \exp\{-(\log x)^a\}). \tag{7.15}$$

Then, with $a' := a/(a+1)$, there exists a constant C such that

$$N(x) = Ax + O(x \exp\{-(\log x \log\log x)^{a'}\}).$$

PROOF. Integrating by parts,

$$\begin{aligned}\int_1^x T^{-1}(d\Pi - d\Pi_c) &= \{\Pi(x) - \Pi_c(x)\}/x + \int_1^x \{\Pi(t) - \Pi_c(t)\} dt/t^2 \\ &= O\{e^{-(\log x)^a}\} + \Big(\int_1^\infty - \int_x^\infty\Big)\{\Pi(t) - \Pi_c(t)\} dt/t^2 \\ &= O\{e^{-(\log x)^a}\} + \log C + O\Big(\int_x^\infty e^{-(\log t)^a} dt/t\Big).\end{aligned}$$

The last integral is asymptotic to $(\log x)^{1-a} \exp(-\log^a x)/a$, as can be seen by l'Hospital's rule. Thus (7.15) implies that there is a number $\log C$ such that

$$\int_{1-}^x T^{-1}(d\Pi - d\Pi_c - \{\log C\}\,\delta_1) = O\{(\log x)^{1-a} \exp(-\log^a x)\}.$$

The last integral is the function $\nu(x)$. We estimate $\nu_j(x)$ by (7.12), taking $b = 1-a$. It follows from the proof of the theorem that

$$|N(x) - Ax|/x < (\log x)^{1-a} C(\epsilon', x, a), \quad x \ge x_1.$$

We may absorb the $(\log x)^{1-a}$ factor into $C(\epsilon', x, a)$ by choosing an $\epsilon'' > \epsilon'$ but still small enough that $(1+a)^{1/(1+a)} - \epsilon'' > 1$. Thus, there exists x_2 such that for all $x \ge x_2$,

$$N(x) = Ax + O(x \exp\{-(\log x\ \log\log x)^{a'}\}).$$

□

7.6. Notes

Much of the material of this chapter comes from the authors' articles [**Di70a**], [**Di77**], and [**Zh88**].

§7.3. The reader is encouraged to give a direct proof of Corollary 7.5 starting from Chebychev's Identity 3.3.

The reader is invited to prove the following assertion by a small tauberian argument: *If the counting function N of a g-number system $\mathcal{N}$ satisfies*

$$\int_1^x N(t)t^{-1}dt \sim Ax$$

for some positive constant A, then $\mathcal{N}$ has density A.

§§7.3, 7.4. Several examples are given in [**Di77**] showing that Theorems 7.8 and 7.11 can fail if various of the hypotheses are not satisfied. For example, suppose φ is any positive, continuous strictly increasing function on $[1, \infty)$ for which

$$\int_1^\infty t^{-2}\varphi(t)\,dt = \infty$$

and $\varphi(x)$ is bounded pointwise by $Cx/\log x$ (a mild condition, but admittedly one not present in Theorem 7.11). Then there exists a g-prime system for which $\Pi(x) - \Pi_c(x) \ll \varphi(x)$ but

$$\limsup_{x\to\infty} x^{-1}N(x) = \infty \quad \text{and} \quad \liminf_{x\to\infty} x^{-1}N(x) = 0.$$

§7.4. We ask whether there is a simpler proof of Theorem 7.11.

§7.5. Another example of a g-number system that is somewhat close to the rational integers $\mathbb{Z}$ is given in [**AMH15**]: define a generalized Chebyshev function by $\psi(x) := \lfloor x \rfloor - 1$. A small calculation shows its zeta function to be given by

$$\zeta(s) = \exp\left\{\sum_{n=2}^\infty \frac{n^{-s}}{\log n}\right\}, \qquad \Re s > 1.$$

This system satisfies Theorem 7.13 with error term $E(x) \ll 1/\{x\log x\}$, and so formula (B) holds for $N(x)$. In the article, under assumption of the Riemann Hypothesis, O- and Ω-bounds are given for $N(x) - Ax$; the omega bound of $x^{1-o(1)}$ shows this number system is in fact not too close to $\mathbb{Z}$.

CHAPTER 8

Simple Estimates of $\pi(x)$

Individual prime are ornery; it is easier to manage a whole herd

SUMMARY. Under weak conditions, $\pi(x)$ is unbounded. An irregular prime distribution can lead to a surprising integer distribution. Simple prime estimates for integers that are regularly distributed or have logarithmic density.

8.1. Unboundedness of $\pi(x)$

In this chapter we give some simple estimates of the prime-counting function $\pi(x)$ for a g-prime system $\mathcal{P}$. *We assume throughout this chapter that $\mathcal{P}$ is a discrete set of primes.* To start, we show that, under a weak hypothesis, the g-primes, like their classical counterpart, must be infinite in number.

PROPOSITION 8.1. *Suppose that $\mathcal{N} = \mathcal{N}_{\mathcal{P}}$ is a number system whose zeta function has abscissa of convergence strictly greater than 0. Then $\mathcal{P}$ is infinite.*

PROOF. Suppose $\mathcal{N}$ were generated by only a finite number of primes. Then the Euler product for $\zeta_{\mathcal{N}}(s)$ would extend over a finite number of factors and so would converge at *each* positive value of s. By the usual proof of the equality of the Euler product and Dirichlet series for the zeta function, the Dirichlet series would converge there as well. This contradicts the hypothesis that the abscissa of convergence exceeds 0. □

An equivalent form of this proposition has the hypothesis that

$$\limsup_{x\to\infty} N(x)/x^{\alpha} > 0 \tag{8.1}$$

for some $\alpha > 0$. The two conditions can be shown equivalent using Theorem 6.9 of [**BD69**]. Here is a direct proof that (8.1) implies an infinitude of primes. (This argument introduces an estimate that will be used again below.)

PROOF. Let $n \le x$ be an integer composed entirely of the first r elements of $\mathcal{P}$, expressed with appropriate multiplicity. We have

$$p_1^{\nu_1} \cdots p_r^{\nu_r} \le x,$$

where the ν_i are nonnegative integers and clearly satisfy $\nu_i \le \lfloor \log x/\log p_i \rfloor$ for $1 \le i \le r$. Thus the number of integers that can be formed from $p_1, \ldots, p_r$ is at most

$$\left\lfloor 1 + \frac{\log x}{\log p_1} \right\rfloor \cdots \left\lfloor 1 + \frac{\log x}{\log p_r} \right\rfloor < \left(1 + \frac{\log x}{\log p_1}\right)^r. \tag{8.2}$$

If $\mathcal{P}$ contained only r primes, then we would have $N(x) \ll \log^r x$ and hence $N(x)/x^{\alpha} \to 0$ for every positive value of α, in violation of (8.1). □

8.2. Can there be as many primes as integers?

This seems a strange question: since each prime is an integer and 1 is an integer not counted as a prime, we have $N(x) \geq \pi(x) + 1$ for $x \geq 1$, and so the answer is, of course, No. In §4.5, we showed that if a g-prime set is sufficiently thin, then its g-integer counting function will have the same order of growth. Here, we give an example showing that, absent some regularity of $\mathcal{N}$ (cf. Theorem 8.5), the ratio of the g-prime and g-integer counting functions can get arbitrarily near to 1.

EXAMPLE 8.2. A SURFEIT OF PRIMES. There exists a number system $\mathcal{N} = \mathcal{N}_{\mathcal{P}}$ for which

$$\limsup_{x\to\infty} \frac{\pi_{\mathcal{P}}(x)}{N_{\mathcal{P}}(x)} = 1. \tag{8.3}$$

We create a lacunary sequence of primes with each element having high multiplicity. Let $q_1 = 2$ and $q_2, q_3, \ldots$ be an increasing sequence of positive rational integers to be chosen successively. Let $\mathcal{P}$ be the prime system formed with nq_n as a prime with multiplicity q_n for $n = 1, 2, \ldots$. For given n, set

$$m := q_1 + \cdots + q_{n-1};$$

then $p_j = nq_n$ for $m < j \leq m + q_n$.

Assuming that $q_1, \ldots, q_{n-1}$ have been selected, take q_n to be an integer so large that

$$\Big(1 + \frac{\log nq_n}{\log 2}\Big)^m q_n^{-1/2} < 1.$$

Such q_n exists, since $(a + b\log x)^m\, x^{-1/2} \to 0$ as $x \to \infty$ for fixed a, b, and m.

The integers not exceeding nq_n consist of the prime nq_n with multiplicity q_n along with the integers formed from the m primes smaller than nq_n. The number of composite integers, by (8.2), is at most

$$\Big(1 + \frac{\log nq_n}{\log 2}\Big)^m.$$

Thus

$$N(nq_n) \leq q_n + \Big(1 + \frac{\log nq_n}{\log 2}\Big)^m < q_n + q_n^{1/2}.$$

It is clear from the inequality $\pi(x)+1 \leq N(x)$ that $\limsup \pi(nq_n)/N(nq_n) \leq 1$. For the other inequality, note that

$$\pi(nq_n) = q_n + q_{n-1} + \cdots + q_1 \geq q_n$$

and hence

$$\limsup \frac{\pi(nq_n)}{N(nq_n)} \geq \limsup \frac{q_n}{q_n + q_n^{1/2}} = 1.$$

Thus (8.3) holds for $\mathcal{P}$.

REMARK 8.3. The g-number system created here has O-log density. We have

$$\sum_{n_i\leq x} \frac{1}{n_i} < \prod_{kq_k\leq x} \Big(1 - \frac{1}{kq_k}\Big)^{-q_k} = \exp\Big\{\sum_{kq_k\leq x} -q_k \log\Big(1 - \frac{1}{kq_k}\Big)\Big\}$$

$$= \exp\Big\{\sum_{kq_k\leq x} \Big(\frac{q_k}{kq_k} + O\Big(\frac{q_k}{kq_k^2}\Big)\Big)\Big\} \ll \exp\Big\{\sum_{kq_k\leq x} \frac{1}{k}\Big\},$$

since $\sum_k q_k/(kq_k)^2$ is bounded.

By the construction,

$$\big(\log q_k\big)^{q_{k-1}} < \Big(1 + \frac{\log kq_k}{\log 2}\Big)^{q_1+\cdots+q_{k-1}} < q_k^{1/2}$$

or

$$q_{k-1} \log\log q_k < (1/2)\log q_k.$$

If k is large enough that $\log\log q_k \geq 1/2$, we have

$$q_k > \exp q_{k-1}, \tag{8.4}$$

showing that the q sequence grows very rapidly.

Suppose that $kq_k \leq x$. With two applications of (8.4), we get

$$\exp\exp q_{k-2} < \exp q_{k-1} < q_k < x.$$

Since $k-2 < q_{k-2}$, we have $k-2 < \log\log x$, and hence $k-1 \leq \log x$.

Returning to the first set of inequalities, we obtain

$$\sum_{kq_k \leq x} \frac{1}{k} \leq \sum_{k \leq 1+\log x} \frac{1}{k} \leq \log\log x + C_1,$$

and so

$$\sum_{n_i \leq x} \frac{1}{n_i} < C_2 \exp\{\log\log x + C_1\} \ll \log x.$$

8.3. $\pi(x)$ estimates via regular growth

The following combinatorial lemma is a g-number version of the Moebius inversion formula. For it, we recall the notion of regular growth from §4.6 and introduce here an item of specialized notation. Given a discrete g-number system $\mathcal{N} = \mathcal{N}_{\mathcal{P}}$ and a positive integer r, let $N_{\mathcal{P}}^{(r)}(x) = N^{(r)}(x)$ denote the number of elements of $\mathcal{N} \cap [1, x]$ having no prime factor among the initial r g-primes. More formally, let $p_1 \leq p_2 \leq \ldots$ be the primes of $\mathcal{P}$ and $N^{(r)}(x)$ the function on $[1, \infty)$ that counts the number of sequences $\{\nu_1, \nu_2, \ldots\}$ of nonnegative integers for which

$$\nu_1 = \cdots = \nu_r = 0 \ \text{ and } \ \prod p_i^{\nu_i} \leq x.$$

LEMMA 8.4. *Let $N^{(r)}(x)$ be the counting function defined above. Then*

$$N^{(r)}(x) = \sum_{\delta_1=0}^{1} \cdots \sum_{\delta_r=0}^{1} (-1)^{\delta_1+\cdots+\delta_r} N(xp_1^{-\delta_1}\cdots p_r^{-\delta_r}). \tag{8.5}$$

Further, if $N(x)$ has regular growth with exponent α, then

$$N^{(r)}(x) \sim N(x) \prod_{i=1}^{r} (1 - p_i^{-\alpha}), \qquad x \to \infty. \tag{8.6}$$

PROOF. We have

$$p_1^{\mu_1} p_2^{\mu_2} \cdots \leq xp_1^{-\delta_1} \cdots p_r^{-\delta_r}$$

if and only if

$$p_1^{\mu_1+\delta_1} \cdots p_r^{\mu_r+\delta_r} p_{r+1}^{\mu_{r+1}} p_{r+2}^{\mu_{r+2}} \cdots \leq x.$$

Thus

$$N(xp_1^{-\delta_1} \cdots p_r^{-\delta_r}) = \sum_{p_1^{\nu_1} p_2^{\nu_2} \cdots \leq x}^{*} 1,$$

where $\sum^*$ denotes the restriction $\nu_1 \geq \delta_1, \ldots, \nu_r \geq \delta_r$.

Let $M^{(r)}(x)$ denote the right side of (8.5). We have

$$M^{(r)}(x) = \sum_{\delta_1=0}^{1} \cdots \sum_{\delta_r=0}^{1} (-1)^{\delta_1+\cdots+\delta_r} \sum_{p_1^{\nu_1} p_2^{\nu_2} \cdots \leq x}^{*} 1,$$

with the same ν restriction. Changing the summation order, we get

$$M^{(r)}(x) = \sum_{p_1^{\nu_1} p_2^{\nu_2} \cdots \leq x} \sum_{\delta_1=0}^{\min(\nu_1,1)} \cdots \sum_{\delta_r=0}^{\min(\nu_r,1)} (-1)^{\delta_1+\cdots+\delta_r}.$$

Since

$$\sum_{\delta=0}^{\min(\nu,1)} (-1)^{\delta} = \begin{cases} 0 \text{ if } \nu \geq 1, \\ 1 \text{ if } \nu = 0, \end{cases}$$

we have

$$M^{(r)}(x) = \sum_{\substack{p_1^{\nu_1} p_2^{\nu_2} \cdots \leq x \\ \nu_1 = \cdots = \nu_r = 0}} 1 = N^{(r)}(x).$$

If $N(x)$ has regular growth with exponent α, then, as $x \to \infty$,

$$\begin{aligned} \frac{N^{(r)}(x)}{N(x)} &= \sum_{\delta_1=0}^{1} \cdots \sum_{\delta_r=0}^{1} (-1)^{\delta_1+\cdots+\delta_r} \frac{N(x p_1^{-\delta_1} \cdots p_r^{-\delta_r})}{N(x)} \\ &\sim \sum_{\delta_1=0}^{1} \cdots \sum_{\delta_r=0}^{1} (-1)^{\delta_1+\cdots+\delta_r} (p_1^{-\delta_1} \cdots p_r^{-\delta_r})^{\alpha} = \prod_{i=1}^{r} (1 - p_i^{-\alpha}). \qquad \square \end{aligned}$$

We apply this lemma to estimate $\pi(x)$.

THEOREM 8.5. *Let $\mathcal{P}$ be a g-prime system whose integer counting function $N = N_{\mathcal{P}}$ has regular growth with exponent of growth α. For any positive integer r we have*

$$\limsup_{x \to \infty} \frac{\pi(x)}{N(x)} \leq \prod_{i=1}^{r} (1 - p_i^{-\alpha}).$$

In particular, if $\sum 1/p_i{}^{\alpha}$ diverges, then $\pi(x) = o(N(x))$.

PROOF. Let $N^{(r)}$ be the counting function of the last lemma. We have

$$N^{(r)}(x) = \sum_{p_{r+1}^{\nu_{r+1}} p_{r+1}^{\nu_{r+1}} \cdots \leq x} 1 \geq \sum_{\substack{p_i \leq x \\ i > r}} 1 = \pi(x) - r.$$

Combining this inequality with (8.6), we find that

$$\limsup_{x \to \infty} \frac{\pi(x)}{N(x)} \leq \limsup_{x \to \infty} \frac{r + N^{(r)}(x)}{N(x)} = \prod_{i=1}^{r} (1 - p_i^{-\alpha}).$$

Now

$$\prod_{1}^{r} (1 - p_i^{-\alpha}) \leq \exp\Big\{ - \sum_{1}^{r} p_i{}^{-\alpha} \Big\};$$

if the last sum diverges as $r \to \infty$, then the product diverges to 0, and we get $\pi(x) = o(N(x))$. $\square$

8.4. Lower bounds for $\sum 1/p_i$ via lower log-density

We showed in Theorem 4.7 that a g-number system has positive or infinite lower log density if and only if

$$\int_1^x u^{-1} d\Pi(u) \geq \log\log x + O(1).$$

Here we extend one implication of that result to treat $\sum 1/p_i$.

LEMMA 8.6. *We have*

$$\Pi(x) - \pi(x) \leq \Pi(x^{1/2}), \quad x \geq 1.$$

PROOF. By (1.10),

$$\begin{aligned} &\Pi(x) - \pi(x) \\ &= \left\{\frac{1}{2}\pi(x^{\frac{1}{2}}) + \frac{1}{3}\pi(x^{\frac{1}{3}})\right\} + \left\{\frac{1}{4}\pi(x^{\frac{1}{4}}) + \frac{1}{5}\pi(x^{\frac{1}{5}})\right\} + \left\{\frac{1}{6}\pi(x^{\frac{1}{6}}) + \frac{1}{7}\pi(x^{\frac{1}{7}})\right\} + \dots \\ &\leq \pi(x^{1/2}) + (1/2)\,\pi(x^{1/4}) + (1/3)\,\pi(x^{1/6}) + \dots = \Pi(x^{1/2}). \end{aligned}$$

□

COROLLARY 8.7. *Suppose that $\mathcal{N}$ is a g-number system generated by discrete g-primes $p_1 \leq p_2 \leq \dots$ and $\mathcal{N}$ has positive lower logarithmic density. Then the g-primes of $\mathcal{N}$ satisfy*

$$\sum_{p_i \leq x} \frac{1}{p_i} \geq \{1 + o(1)\} \log\log x.$$

If, in addition,

$$\int_1^\infty u^{-2} d\Pi(u) < \infty, \tag{8.7}$$

then

$$\sum_{p_i \leq x} \frac{1}{p_i} \geq \log\log x + O(1).$$

PROOF. We have $\Pi(x) - \pi(x) \leq \Pi(x^{1/2})$ by the last lemma. Combining this inequality with Lemma 2.1, we find that

$$\int_1^x u^{-1}\{d\Pi(u) - d\pi(u)\} \leq \int_1^x u^{-1} d\Pi(u^{1/2}) \leq \int_1^x u^{-2} d\Pi(u).$$

If (8.7) holds, then the second conclusion follows at once from Theorem 4.7. Otherwise, for large K, write

$$\int_1^x u^{-1} d\pi(u) \geq \int_1^x \{u^{-1} - u^{-2}\} d\Pi(u) > \left(1 - \frac{1}{K}\right) \int_K^x u^{-1} d\Pi(u).$$

Since K is arbitrary, we have

$$\sum_{p_i \leq x} \frac{1}{p_i} \geq \{1 + o(1)\} \log\log x.$$

□

8.5. Notes

§§8.2, 8.3. Much of the material of these sections comes from [**BD69**].

CHAPTER 9

Chebyshev Bounds – Elementary Theory

Are ballpark estimates OK off the field?

SUMMARY. A review of Chebyshev's method. Conditions for a g-prime counting function to have the expected order of growth. An example of failure. (In Chapter 11 Chebyshev bounds are established under somewhat weaker hypotheses by analytic methods.)

9.1. Introduction

Before the PNT was established for the rational integers $\mathbb{N}$, the true order of the prime-counting function $\pi(x)$ was established for the first time by P. L. Chebyshev. He showed by elementary methods (see Chapter 14 for a discussion of this term) that there exist two numbers $\alpha > 0$ and $\beta < \infty$ such that

$$\liminf_{x\to\infty} \frac{\pi(x)}{x/\log x} \geq \alpha, \quad \limsup_{x\to\infty} \frac{\pi(x)}{x/\log x} \leq \beta. \tag{9.1}$$

We call the preceding inequalities lower and upper Chebyshev bounds. The PNT asserts that (9.1) holds with $\alpha = \beta = 1$.

In this chapter we apply elementary methods to study the g-number analogues of these relations. As in the classical case, we call the inequalities *Chebyshev bounds.* We give several conditions for a g-integer counting function $N(x)$ that insure the validity of one or the other of the bounds (9.1). As a warm-up and also for its historical interest, we give a version of Chebyshev's argument.

We express our estimates in terms of Chebyshev's function

$$\psi(x) := \sum_{n\leq x} \Lambda(n),$$

with $\Lambda(n) = \log p$ if $n = p^\alpha$ and $\Lambda(n) = 0$ otherwise. We establish inequalities

$$\liminf_{x\to\infty} \frac{\psi(x)}{x} \geq \alpha, \quad \limsup_{x\to\infty} \frac{\psi(x)}{x} \leq \beta\,; \tag{9.2}$$

the form (9.1) with the same values of α and β is obtained by summation by parts.

9.2. Chebyshev bounds for natural primes

The starting point of Chebyshev's investigation is the formula

$$\log n = \sum_{p^k|n} \Lambda(p^k).$$

This can be expressed in (multiplicative) convolution terms as

$$L1 = \Lambda * 1 \tag{9.3}$$

where $L(n) = \log n$ and $1(n) = 1$ for all $n \geq 1$. (In §3.1 we considered the measure version of this relation.) For $k = 1, 2, \ldots$ let functions e_k be defined by

$$e_k(n) = \delta_{k,n} = \begin{cases} 1, & n = k, \\ 0, & n \neq k. \end{cases}$$

If we convolve each side of (9.3) by the Moebius function $\mu(n)$ and sum the resulting expression, we obtain

$$\sum_{n\leq x} L1 * \mu(n) = \sum_{n\leq x} \Lambda * 1 * \mu(n) = \sum_{n\leq x} \Lambda * e_1(n) = \psi(x), \tag{9.4}$$

since $\mu * 1 = e_1$ and e_1 is the identity element of multiplicative convolution.

This formula is difficult to use directly, because rather little is known at the outset about the factor μ on the left hand side. A good strategy is to introduce in place of $\mu(n)$ an arithmetic function $f(n)$ which serves as an "approximate convolution inverse" of $1(n)$. Specifically, f will be chosen to have the following properties:

$$\sum_{n=1}^{\infty} f(n)/n = 0, \tag{9.5}$$

$$\sum_{n\leq x} |f(n)| = O(x \log^{-2-\epsilon} x) \quad \text{with some} \quad \epsilon > 0, \tag{9.6}$$

and one of the one-sided conditions

$$\sum_{n\leq x} f * 1(n) \leq C, \quad x \geq 1, \tag{9.7}$$

or

$$\sum_{n\leq x} f * 1(n) \geq \begin{cases} c_1, & 1 \leq x < B, \\ 0, & x \geq B, \end{cases} \tag{9.8}$$

or

$$\sum_{n\leq x} f * 1(n) \geq \begin{cases} 0, & 1 \leq x < B, \\ c_2, & x \geq B, \end{cases} \tag{9.9}$$

for some real constants $C < \infty$, $c_1, c_2 > 0$, and $B > 1$. It is known from classical prime number theory that $f = \mu$ satisfies (9.5) but not (9.6). Also, μ satisfies (9.7), (9.8), and (9.9) with $C = c_1 = c_2 = 1$ and $B > 1$. Property (9.6) makes f more tractable than μ.

PROPOSITION 9.1. *Let f satisfy* (9.5) *and* (9.6) *and set*

$$\alpha_f := -\sum_{1}^{\infty} f(n) n^{-1} \log n.$$

Then

$$\psi(x) \geq \frac{\alpha_f x}{C} + O(x \log^{-\epsilon} x) \quad \textit{if } (9.7) \textit{ holds} \tag{9.10}$$

and

$$\psi(x) \leq \begin{cases} \dfrac{\alpha_f x}{c_1(1 - B^{-1})} + O(x \log^{-\epsilon} x) & \textit{if } (9.8) \textit{ holds}, \\ \dfrac{\alpha_f B x}{c_2} + O(x \log^{-\epsilon} x) & \textit{if } (9.9) \textit{ holds}. \end{cases} \tag{9.11}$$

PROOF. If we convolve each side of (9.3) with f and sum, we obtain

$$\sum_{n\le x} L1 * f(n) = \sum_{n\le x} \Lambda * (1 * f)(n)$$

and hence

$$\sum_{n\le x}\Big(\sum_{m\le x/n} \log m\Big) f(n) = \sum_{n\le x}\Big(\sum_{m\le x/n} 1 * f(m)\Big)\Lambda(n). \tag{9.12}$$

Using the simple estimate

$$\sum_{m\le y} \log m = y\log y - y + O(\log y)$$

and (9.5) and (9.6), we evaluate the left hand side of (9.12) as

$$\begin{aligned}
&\sum_{n\le x}\Big(\frac{x}{n}\log\frac{x}{n} - \frac{x}{n} + O(\log x)\Big) f(n)\\
&= (x\log x - x)\sum_{n\le x}\frac{f(n)}{n} - x\sum_{n\le x}\frac{f(n)\log n}{n} + O\Big(\log x \sum_{n\le x}|f(n)|\Big)\\
&= -(x\log x - x)\sum_{n>x}\frac{f(n)}{n} + \alpha_f x + x\sum_{n>x}\frac{f(n)\log n}{n} + O(x\log^{-\epsilon-1} x).
\end{aligned}$$

Then we evaluate the two sums on the right hand side of the last equality by partial summation and use property (9.6). We have

$$\sum_{n>x}\frac{f(n)}{n} = \int_x^\infty y^{-1}\, dS(y),$$

where $S(y) := \sum_{n\le y} f(n)$, and we find that

$$\sum_{n>x}\frac{f(n)}{n} = O(\log^{-1-\epsilon} x).$$

Similarly, we find that

$$\sum_{n>x} f(n)\, n^{-1}\log n = \int_x^\infty y^{-1}\log y\, dS(y) = O(\log^{-\epsilon} x).$$

Therefore the left hand side of (9.12) equals $\alpha_f x + O(x\log^{-\epsilon} x)$.

Now we evaluate the right hand side of (9.12). In case (9.7) holds, we have

$$\sum_{n\le x}\Big(\sum_{m\le x/n} 1 * f(m)\Big)\Lambda(n) \le C\sum_{n\le x}\Lambda(n) = C\psi(x),$$

and the desired lower bound for $\psi(x)$ follows. If (9.8) holds, we use the inequality

$$\sum_{m\le x/n} 1 * f(m) \ge \begin{cases} c_1, & \text{for } 1\le x/n < B,\\ 0, & \text{for } x/n \ge B.\end{cases}$$

In this case, we have

$$\sum_{n\le x}\Big(\sum_{m\le x/n} 1 * f(m)\Big)\Lambda(n) \ge \sum_{x/B<n\le x} c_1\Lambda(n) = c_1\{\psi(x) - \psi(x/B)\},$$

and

$$\psi(x) - \psi(x/B) \le \alpha_f x/c_1 + O(x \log^{-\epsilon} x)$$

follows. Now we replace x in the last inequality by $x/B, x/B^2, \cdots, x/B^{K-1}$ in turn, where $x/B^K < 2 \le x/B^{K-1}$, and sum the resulting inequalities. This gives

$$\psi(x) \le \frac{\alpha_f x}{c_1} \sum_{k \le K-1} \frac{1}{B^k} + O\left(\sum_{k \le K-1} \frac{x}{B^k} \log^{-\epsilon} \frac{x}{B^k} \right).$$

The error term has order

$$\begin{aligned} \sum_{k \le K-1} \frac{x}{B^k} \log^{-\epsilon} \frac{x}{B^k} &\le \sum_{k \le \log(x^{1/2})/\log B} \frac{x}{B^k} \log^{-\epsilon} \sqrt{x} + \sum_{k > \log(x^{1/2})/\log B} \frac{x}{B^k} \\ &\ll x \log^{-\epsilon} x, \end{aligned}$$

so the desired upper bound for ψ holds.

Finally, assuming (9.9), we apply the inequality

$$\sum_{m \le x/n} 1 * f(m) \ge \begin{cases} 0, & 1 \le x/n < B, \\ c_2, & x/n \ge B, \end{cases}$$

in the right hand side of (9.12) and obtain

$$\alpha_f x + O(x \log^{-\epsilon} x) \ge c_2 \sum_{n \le x/B} \Lambda(n) = c_2 \psi(x/B).$$

Thus the upper bound for ψ follows here too. □

EXAMPLE 9.2. Take $f = e_1 - 2e_2$. Then

$$\sum_{n \le x} f * 1(n) = \left\lfloor \frac{x}{1} \right\rfloor - 2 \left\lfloor \frac{x}{2} \right\rfloor = \begin{cases} 0, & \lfloor x \rfloor \text{ even}, \\ 1, & \lfloor x \rfloor \text{ odd}. \end{cases}$$

Thus we may take $c_1 = C = 1$, $B = 2$ in (9.7) and (9.8) and $\alpha_f = \log 2$ and obtain

$$x \log 2 + o(x) \le \psi(x) \le 2x \log x + o(x).$$

EXAMPLE 9.3. (Chebyshev) Take $f = e_1 - e_2 - e_3 - e_5 + e_{30}$. Then

$$\sum_{n \le x} f * 1(n) = \left\lfloor \frac{x}{1} \right\rfloor - \left\lfloor \frac{x}{2} \right\rfloor - \left\lfloor \frac{x}{3} \right\rfloor - \left\lfloor \frac{x}{5} \right\rfloor + \left\lfloor \frac{x}{30} \right\rfloor,$$

a periodic function of x with period 30 that assumes only the values 0 and 1. Also, the value 1 is assumed for $1 \le x < 6$. Thus properties (9.7) and (9.8) are satisfied with $c_1 = C = 1$ and $B = 6$. Furthermore,

$$\sum_1^\infty \frac{f(n)}{n} = 0, \quad \sum_1^\infty |f(n)| = O(1),$$

and

$$\alpha_f = \frac{\log 2}{2} + \frac{\log 3}{3} + \frac{\log 5}{5} - \frac{\log 30}{30} = .92129\ldots.$$

Thus we find Chebyshev's estimates

$$.92129\cdots + o(1) \le \psi(x)/x \le 1.10555\cdots + o(1).$$

Summation by parts yields (9.1) with $\alpha = .92129\ldots$ and $\beta = 1.10555\ldots.$

9.3. An auxiliary function

In the next section, we shall establish Chebyshev bounds for g-primes under the assumption that

$$E^*(x) := \sup_{y \geq x} |N(y) - Ay|/y$$

satisfies

$$\int_1^\infty E^*(x)\, x^{-1} dx < \infty. \tag{9.13}$$

The function $E^*(x)$ has several useful properties: it is a positive valued, nonincreasing upper bound for $|N(x) - Ax|/x$ satisfying an integral condition. However, it would be desirable for our analysis to have a bounding function with the further property of not decreasing too rapidly. The goal of this section is to construct such a function to use as a replacement for E^*.

LEMMA 9.4. *Assume $E^*(x)$ satisfies* (9.13). *Then there exists a positive valued function $Q(x)$ having the following four properties for $1 \leq x < \infty$:*

$$Q(x) \geq E^*(x) \tag{9.14}$$

$$Q(x) \text{ is nonincreasing} \tag{9.15}$$

$$Q(x) \leq 4Q(x^2) \tag{9.16}$$

$$\int_1^\infty Q(x)\, x^{-1}\, dx < \infty\,. \tag{9.17}$$

REMARKS 9.5. (i) For $\gamma \in (1, 2]$, the function $\log^{-\gamma} ex$ has properties (9.15), (9.16), (9.17) of $Q(x)$; readers willing to settle for Chebyshev inequalities proved under the stronger hypothesis $E(x) \ll \log^{-\gamma} ex$ can skip the construction of $Q(x)$ below and simply use $\log^{-\gamma} ex$ in its place in the next section.

(ii) The condition (9.16) that we have introduced is reasonably sharp: if it were replaced by $Q(x) \leq 2Q(x^2)$, say, with $Q(x)$ assumed positive on some interval $(1, 1+\epsilon)$, then the inequality

$$\int_X^{X^2} \frac{Q(x)}{x}\, dx \leq 2 \int_X^{X^2} \frac{Q(x^2)}{x}\, dx = \int_{X^2}^{X^4} \frac{Q(u)}{u}\, du$$

shows that (9.17) could not hold.

PROOF. Define $Q(x)$ recursively by setting $Q(x) := E^*(1)$ for $1 \leq x \leq 2$ and

$$Q(x) := \max\{E^*(2^{2^m}), 4^{-1}Q(2^{2^m})\}$$

for $2^{2^m} < x \leq 2^{2^{m+1}}$, $m = 0, 1, 2, \ldots$. We shall verify that $Q(x)$ satisfies conditions (9.14) – (9.17).

- $Q(x) \geq E^*(x)$. This is clear on $[1, 2]$, since $Q(x) = E^*(1) \geq E^*(x)$. For $x \in (2^{2^m}, 2^{2^{m+1}}]$, we have

$$Q(x) = \max\{E^*(2^{2^m}), 4^{-1}Q(2^{2^m})\} \geq E^*(2^{2^m}) \geq E^*(x),$$

since $E^*(x)$ is nonincreasing.

• $Q(x)$ is nonincreasing. We show that $Q(2^{2^{m+1}}) \le Q(2^{2^m})$, $m = 0, 1, 2 \ldots$. Since Q is constant on intervals $(2^{2^m}, 2^{2^{m+1}}]$ this will show the monotonicity. By definition, $Q(1) = Q(2) = E^*(1)$. Thus,

$$Q(4) := \max\{E^*(2), 4^{-1}Q(2)\} \le \max\{E^*(1), 4^{-1}E^*(1)\} = Q(2).$$

Similarly, for $m \ge 1$, by (9.14), $E^*(2^{2^m}) \le Q(2^{2^m})$, and hence

$$Q(2^{2^{m+1}}) = \max\{E^*(2^{2^m}), 4^{-1}Q(2^{2^m})\} \le Q(2^{2^m}).$$

• $Q(x) \le 4Q(x^2)$. For $1 \le x \le 2$, we have $1 \le x^2 \le 2^2$. If $1 \le x^2 \le 2$, $Q(x) = Q(x^2)$ and the result holds. If $2 < x^2 \le 2^2$, then

$$Q(x^2) = \max\{E^*(2), 4^{-1}Q(2)\} \ge 4^{-1}Q(2) = 4^{-1}Q(x).$$

For $2^{2^{m-1}} < x \le 2^{2^m}$, $m \ge 1$, we have $2^{2^m} < x^2 \le 2^{2^{m+1}}$ and, similarly,

$$Q(x^2) \ge 4^{-1}Q(2^{2^m}) = 4^{-1}Q(x).$$

• $\int_1^\infty Q(x)\,x^{-1}\,dx < \infty$. Since $Q(x)$ is constant on intervals, we have

$$\begin{aligned}\int_{2^{2^m}}^{2^{2^{m+1}}} Q(x)\,x^{-1}\,dx &= Q(2^{2^{m+1}})\log 2^{2^m}\\ &= \max\{E^*(2^{2^m}), 4^{-1}Q(2^{2^m})\}\,2^m \log 2.\end{aligned}$$

Suppose that, for some positive integers μ, ν,

$$4^{-1}Q(2^{2^m}) \ge E^*(2^{2^m}) \tag{9.18}$$

holds for $\mu \le m < \mu + \nu$. In this case, we have inductively for $1 \le \lambda \le \nu$,

$$Q(2^{2^{\mu+\lambda}}) = 4^{-\lambda}Q(2^{2^\mu}),$$

and thus

$$\int_{2^{2^\mu}}^{2^{2^{\mu+\nu}}} Q(x)\,x^{-1}\,dx = Q(2^{2^\mu})\bigl(\log 2^{2^{\mu-1}}\bigr)\sum_{\lambda=1}^{\nu} 2^{-\lambda}. \tag{9.19}$$

Now we distinguish two cases, according to occurrences of (9.18).

Case I. Suppose (9.18) holds for *all* $m \ge \lambda$ for some $\lambda \in \mathbb{N}$. By the preceding formula,

$$\int_{2^{2^\lambda}}^{\infty} Q(x)\,x^{-1}\,dx = Q(2^{2^\lambda})\log 2^{2^{\lambda-1}} < \infty,$$

and thus (9.17) holds here.

Case II. Now suppose there exists an infinite sequence

$$m_1 < m_2 < \cdots < m_k < m_{k+1} < \cdots$$

such that

$$E^*(2^{2^m}) > 4^{-1}Q(2^{2^m}) \tag{9.20}$$

holds for each m_k and the reverse inequality, (9.18), holds for each $m \in (m_k, m_{k+1})$. In this case, we show that

$$\sum_{m=m_k}^{m_{k+1}-1} \int_{2^{2^m}}^{2^{2^{m+1}}} Q(x)\, x^{-1}\, dx \le 4E^*(2^{2^{m_k}}) \log\left(2^{2^{m_k-1}}\right). \tag{9.21}$$

First, if $m_{k+1} = m_k + 1$, the left hand side of (9.21) equals

$$\begin{aligned}\int_{2^{2^{m_k}}}^{2^{2^{m_k+1}}} Q(x)\, x^{-1}\, dx &= E^*(2^{2^{m_k}}) \int_{2^{2^{m_k}}}^{2^{2^{m_k+1}}} x^{-1}\, dx \\ &= E^*(2^{2^{m_k}}) \log\left(2^{2^{m_k}}\right) = 2E^*(2^{2^{m_k}}) \log\left(2^{2^{m_k-1}}\right),\end{aligned}$$

which establishes (9.21).

Next, suppose $m_{k+1} \ge m_k + 2$. The preceding calculation gives

$$\int_{2^{2^{m_k}}}^{2^{2^{m_k+1}}} Q(x)\, x^{-1}\, dx = 2E^*(2^{2^{m_k}}) \log\left(2^{2^{m_k-1}}\right).$$

For the remaining intervals of (9.21) we can apply (9.19) to obtain

$$\int_{2^{2^{m_k+1}}}^{2^{2^{m_{k+1}}}} Q(x)\, x^{-1}\, dx < Q\left(2^{2^{m_k+1}}\right) \log\left(2^{2^{m_k}}\right) = E^*(2^{2^{m_k}}) 2 \log\left(2^{2^{m_k-1}}\right),$$

since

$$Q\left(2^{2^{m_k+1}}\right) = \max\{E^*(2^{2^{m_k}}),\, 4^{-1} Q(2^{2^{m_k}})\} = E^*(2^{2^{m_k}}).$$

Together, the last two integral expressions give (9.21).

Thus we have

$$\begin{aligned}\sum_{m=m_1}^{\infty} \int_{2^{2^m}}^{2^{2^{m+1}}} Q(x)\, x^{-1}\, dx &= \sum_{k=1}^{\infty} \sum_{m=m_k}^{m_{k+1}-1} \int_{2^{2^m}}^{2^{2^{m+1}}} Q(x)\, x^{-1}\, dx \\ &\le 4 \sum_{k=1}^{\infty} E^*(2^{2^{m_k}}) \log\left(2^{2^{m_k-1}}\right) \le 4 \sum_{m=1}^{\infty} E^*(2^{2^m}) \log\left(2^{2^{m-1}}\right).\end{aligned}$$

To show the last sum is finite, we note that

$$\int_{2^{2^{m-1}}}^{2^{2^m}} E^*(x) x^{-1}\, dx \ge E^*(2^{2^m}) \int_{2^{2^{m-1}}}^{2^{2^m}} x^{-1}\, dx = E^*(2^{2^m}) \log\left(2^{2^{m-1}}\right),$$

since $E^*(x)$ is nonincreasing. Combining the inequalities, we find

$$\begin{aligned}\int_{2^{2^{m_1}}}^{\infty} Q(x)\, x^{-1}\, dx &\le 4 \sum_{m=1}^{\infty} E^*(2^{2^m}) \log\left(2^{2^{m-1}}\right) \le 4 \sum_{m=1}^{\infty} \int_{2^{2^{m-1}}}^{2^{2^m}} E^*(x) x^{-1}\, dx \\ &= 4 \int_2^{\infty} E^*(x) x^{-1}\, dx < \infty\end{aligned}$$

by (9.13). Thus $\int_1^\infty Q(x)\, x^{-1}\, dx < \infty$. □

LEMMA 9.6. *Assume* (9.15) *and* (9.17) *hold. Then* $Q(x) = o(\log^{-1} ex)$.

PROOF. Given $\epsilon > 0$, we have $\int_x^\infty Q(t) t^{-1}\, dt < \epsilon/2$ for all $x \ge X = X(\epsilon)$. Thus, for $x \ge X^2$,

$$Q(x) \log x \le 2 \int_{\sqrt{x}}^{x} Q(t) t^{-1}\, dt < \epsilon.$$

□

The following lemma shows a kind of "stability" of $Q(x)$ under multiplicative convolution.

LEMMA 9.7. *Assume that* $Q(x)$ *satisfies* (9.15), (9.16), *and* (9.17). *Then we have* $\int_1^x t^{-1}Q(x/t)Q(t)\,dt \le c_1 Q(x)$ *for some positive constant* c_1.

PROOF. We have

$$\int_1^x t^{-1}Q(x/t)Q(t)\,dt = 2\int_1^{\sqrt{x}} t^{-1}Q(x/t)Q(t)\,dt$$
$$\le 2Q(\sqrt{x})\int_1^{\sqrt{x}} Q(t)t^{-1}dt \le c_1 Q(x)$$

since $Q(x/t) \le Q(\sqrt{x}) \le 4Q(x)$ for $1 \le t \le \sqrt{x}$ and $\int_1^\infty Q(t)t^{-1}dt < \infty$. □

9.4. Chebyshev bounds for g-primes

In this section, we give finite upper- and positive lower Chebyshev bounds for g-primes. (We shall revisit this topic in Chapter 11, when we have developed further analytic tools.) We first show a result in the opposite direction: that the lower bound cannot be too big nor the upper bound too small.

PROPOSITION 9.8. *Let* α *and* β *be numbers satisfying* (9.1). *If a g-number system* $\mathcal{N}$ *has finite O-log density, then* $\alpha \le 1$; *if* $\mathcal{N}$ *has positive lower log density, then* $\beta \ge 1$. *In particular, if* $N(x) \ll x$, *then* $\alpha \le 1$ *and if* $N(x) \gg x$, *then* $\beta \ge 1$.

PROOF. First assume that $\mathcal{N}$ has finite O-log density. By Proposition 4.2, $(s-1)\zeta(s)$ is bounded above as $s \to 1+$. Thus

$$\limsup_{s\to 1+} \log\{(s-1)\zeta(s)/s\} = \limsup_{s\to 1+} s\int_1^\infty x^{-s-1}\{\Pi(x) - \Pi_c(x)\}\,dx < \infty,$$

with the continuous "prime counting function" of (1.9),

$$\Pi_c(x) := \int_1^x \frac{1-u^{-1}}{\log u}\,du \sim \frac{x}{\log x}\,.$$

Suppose $\alpha > 1$, i.e.

$$\liminf_{x\to\infty} \frac{\pi(x)}{x/\log x} > 1+\epsilon$$

for some $\epsilon > 0$. Then

$$\Pi(x) - \Pi_c(x) > \epsilon\Pi_c(x), \quad x \ge X,$$

for some number $X \ge 1$, and

$$\log\{(s-1)\zeta(s)/s\} > O(1) + \epsilon\int_X^\infty x^{-s-1}\Pi_c(x)\,dx \to \infty$$

as $s \to 1+$, contradicting the boundedness of the function.

The case of $\mathcal{N}$ having positive lower log density is handled analogously. □

Now we turn to the more significant Chebyshev bounds $\alpha > 0$ and $\beta < \infty$, using an approximate convolution inverse relation analogous to (9.9). Again, we write the counting function $N(x)$ of the g-integers of a system $\mathcal{N}$ as

(9.22) $$N(x) = Ax + xE(x), \quad \text{for} \quad x \ge 1,$$

with A a positive constant, and set

$$E^*(x) := \sup_{y \geq x} |E(y)|.$$

THEOREM 9.9. *Suppose* (9.22) *holds with*

$$\int_1^\infty x^{-1} E^*(x)\, dx < \infty.$$

Then there exist numbers $\alpha > 0$ *and* $\beta < \infty$ *satisfying Chebyshev's bounds* (9.2).

COROLLARY 9.10. *If*

$$N(x) = Ax + O(x \log^{-\gamma} ex) \tag{9.23}$$

is satisfied with some constant $\gamma > 1$, *then the Chebyshev bounds* (9.2) *hold.*

PROOF OF THE COROLLARY. Say $|E(x)| \leq c \log^{-\gamma} ex$ for some $c > 0$. Then $E^*(x) \leq c \log^{-\gamma} ex$. It follows that

$$\int_1^\infty x^{-1} E^*(x)\, dx \leq c \int_1^\infty x^{-1} \log^{-\gamma} ex < \infty,$$

and the claim of the corollary follows from the theorem. □

We begin proving the theorem by establishing a convolution identity that will be used in the two succeeding lemmas.

LEMMA 9.11. *For* $x \geq 1$ *and* $\eta > 0$, *we have*

$$\int_{1-}^x (\delta_1 + dt) * (\delta_1 - \eta t^{-\eta}\, dt) = x^{1-\eta}. \tag{9.24}$$

PROOF. The left side of (9.24) is

$$\int_{1-}^x (x/t)(\delta_1 - \eta t^{-\eta}\, dt) = x + x t^{-\eta}\Big|_1^x = x^{1-\eta}.$$ □

REMARK 9.12. If $\eta = 1$, then the right side of (9.24) is 1. The associated measures satisfy

$$(\delta_1 + dt) * (\delta_1 - t^{-1}\, dt) = \delta_1,$$

i.e., $\delta_1 - t^{-1}\, dt$ is the convolution inverse of $\delta_1 + dt$. (The usual group theory argument for the uniqueness of inverses justifies the word "the.")

Now we introduce the measure $(\delta_1 - \epsilon t^{-\epsilon}\, dt) * Q(t)\, dt$, with a suitable $\epsilon > 0$, to prove Theorem 9.9. We shall show that this measure acts as an "approximate convolution inverse of dN" – it is an analogue of the function f in (9.9), as the following lemma shows.

LEMMA 9.13. *Assume that* $Q(x)$ *satisfies the four properties of Lemma* 9.4. *Then, for a sufficiently small (fixed) number* $\epsilon > 0$, *we have*

$$u_\epsilon(x) := \int_{1-}^x dN * (\delta_1 - \epsilon t^{-\epsilon} dt) * Q(t)\, dt \geq 0$$

for all $x \geq 1$; *also,* $u_\epsilon(x) \to \infty$ *as* $x \to \infty$.

PROOF. Note that

$$\int_{1-}^{x} dN * t^{-\epsilon}\,dt = \int_{1}^{x} N(x/t)t^{-\epsilon}\,dt \le N(x)\int_{1}^{x} dt.$$

Thus, if $1 \le x \le 1/\epsilon$, then

$$\int_{1-}^{x} dN * (\delta_1 - \epsilon t^{-\epsilon}\,dt) \ge N(x) - \epsilon N(x)(x-1) \ge 0.$$

The lemma is certainly true for $1 \le x \le 1/\epsilon$, since the third convolution factor in the u_ϵ integral is everywhere nonnegative.

If $x > 1/\epsilon$, we utilize all the convolution factors. With A the density of the g-number system (cf. (9.22)), we shall subtract from dN and then add back the measure $A(\delta_1 + dt)$. Write

$$\begin{aligned} u_\epsilon(x) = A &\int_{1-}^{x} (\delta_1 + dt) * (\delta_1 - \epsilon t^{-\epsilon}\,dt) * Q(t)\,dt \\ &+ \int_{1-}^{x} (dN - A\delta_1 - Adt) * (\delta_1 - \epsilon t^{-\epsilon}\,dt) * Q(t)\,dt =: I_1 + I_2, \end{aligned}$$

say. We shall show that $I_1 > 0$ on $(1, \infty)$, that $I_1 \to \infty$ as $x \to \infty$, and that $|I_2|$ is smaller than I_1. Using (9.24) on the first two factors in I_1, we find

$$I_1 = Ax^{1-\epsilon}\int_{1}^{x} t^{-1+\epsilon}Q(t)\,dt = Ax\int_{1}^{x} u^{-\epsilon-1}Q(x/u)\,du.$$

Thus $I_1 > 0$, and $I_1 \gg x^{1-\epsilon} \to \infty$ as $x \to \infty$.

To estimate I_2, note first that

$$|N(x) - Ax| = x|E(x)| \le xE^*(x) \le xQ(x), \quad x \ge 1.$$

This inequality and Lemma 9.7 give

$$\begin{aligned} \left|\int_{1-}^{x} (dN - A\delta_1 - Adt) * Q(t)\,dt\right| &= \left|\int_{1}^{x} (N(x/t) - Ax/t)Q(t)\,dt\right| \\ &\le \int_{1}^{x} (x/t)Q(x/t)Q(t)\,dt \le c_1 xQ(x). \end{aligned}$$

It follows that

$$|I_2| \le c_1\int_{1-}^{x} (x/t)Q(x/t)(\delta_1 + \epsilon t^{-\epsilon}\,dt) = c_1 xQ(x) + c_1\epsilon x\int_{1}^{x} t^{-\epsilon-1}Q(x/t)\,dt.$$

Now choose $\epsilon > 0$ so small that $\log(1/\epsilon) > 4c_1/A$; then $1/\epsilon > 4c_1/A$ holds as well. Thus the last term in the $|I_2|$ estimate is smaller than $I_1/4$. Also, for $x \ge 1/\epsilon$, we have

$$\begin{aligned} \frac{1}{2}I_1 = \frac{1}{2}Ax\int_{1}^{x} t^{-\epsilon-1}Q(x/t)\,dt &\ge \frac{1}{2}AxQ(x)\int_{1}^{1/\epsilon} t^{-\epsilon-1}\,dt \\ &\ge \frac{1}{2}AxQ(x)\epsilon^{\epsilon}\log\frac{1}{\epsilon} > c_1 xQ(x), \end{aligned}$$

since $\epsilon^\epsilon \ge \exp(-e^{-1}) > 1/2$. Therefore, $I_2 < (3/4)I_1$, and $u_\epsilon(x) \ge I_1 - |I_2| > I_1/4$. Thus $u_\epsilon(x)$ is always positive and tends to infinity with x. □

We shall use the following approximation to estimate $\psi(x)$.

LEMMA 9.14. *Suppose that* $N(x) = Ax + o(x\log^{-1} ex)$. *Then, for given* $\eta > 0$,

$$\int_{1-}^{x} LdN * \big(\delta_1 - \eta t^{-\eta}\,dt\big) = Ax/\eta + o(x).$$

PROOF. Write the integral in the iterated form

$$\int_{1-}^{x} \Big\{ \int_{1}^{x/t} LdN \Big\} (\delta_1 - \eta t^{-\eta}\,dt).$$

Applying integration by parts and the hypothesis, we find

$$\int_{1}^{u} LdN = N(u)\log u - \int_{1}^{u} N(t)t^{-1}\,dt = Au\log u - Au + o(u).$$

Thus the integral of the lemma can be written as

$$\begin{aligned}
&\int_{1-}^{x} \Big\{ A\frac{x}{t}\log\frac{x}{t} - A\frac{x}{t} + o\Big(\frac{x}{t}\Big) \Big\} (\delta_1 - \eta t^{-\eta}\,dt) \\
&\quad = A(\log x - 1)\int_{1-}^{x} \frac{x}{t}\,(\delta_1 - \eta t^{-\eta}\,dt) \\
&\quad\quad - Ax\int_{1-}^{x} t^{-1}(\log t)\,(\delta_1 - \eta t^{-\eta}\,dt) + \int_{1-}^{x} o\Big(\frac{x}{t}\Big)(\delta_1 + \eta t^{-\eta}\,dt) \\
&\quad =: I_1 + I_2 + I_3, \text{ say.}
\end{aligned}$$

By Lemma 9.11, $I_1 = A(\log x - 1)x^{1-\eta} = o(x)$. Integration by parts shows

$$I_2 = A\eta x \int_{1}^{x} (\log t)\,t^{-1-\eta}\,dt = Ax/\eta + o(x),$$

and direct calculation gives $I_3 = o(x)$. □

LEMMA 9.15. *Assume that* $Q(x)$ *satisfies the four properties of Lemma* 9.4. *Then, for fixed* $\epsilon > 0$,

$$\int_{1-}^{x} LdN * (\delta_1 - \epsilon t^{-\epsilon}\,dt) * Q(t)\,dt = O(x).$$

PROOF. Lemma 9.6 implies that the hypothesis of Lemma 9.14 is satisfied; by this and (9.17), the integral on the left hand side equals

$$\int_{1-}^{x} O(x/t)Q(t)\,dt = O\Big(x\int_{1-}^{x} t^{-1}Q(t)\,dt\Big) = O(x). \qquad \square$$

We are now ready to establish Theorem 9.9. The fundamental relation is Chebyshev's identity $d\psi * dN = LdN$, which was established in Chapter 3. We convolve both side of the identity by an appropriate measure for each of the bounds.

PROOF OF THE CHEBYSHEV UPPER BOUND. Here we start with the identity

$$\int_{1-}^{x} dN * (\delta_1 - \epsilon t^{-\epsilon}\,dt) * Q(t)\,dt * d\psi = \int_{1-}^{x} LdN * (\delta_1 - \epsilon t^{-\epsilon}\,dt) * Q(t)\,dt. \tag{9.25}$$

The left side is expressible as

$$\int_{1-}^{x} du_\epsilon * d\psi = \int_{1}^{x} u_\epsilon(x/t)\,d\psi(t) \ge \int_{1}^{x/B} d\psi(t),$$

since $u_\epsilon(x)$, the function introduced in Lemma 9.13, is everywhere nonnegative and is at least 1 for all x exceeding some number B.

The right side of (9.25) is $O(x)$ by Lemma 9.15. Combining the estimates of (9.25), we obtain $\psi(x/B) = O(x)$, whence

$$\psi(x) = O(x). \qquad \square$$

PROOF OF THE CHEBYSHEV LOWER BOUND. This time we take $\delta_1 - t^{-1}\,dt$ as an approximate inverse of dN. Starting again from the Chebyshev identity, we have

$$\int_{1-}^{x} dN * (\delta_1 - t^{-1}\,dt) * d\psi = \int_{1-}^{x} L dN * (\delta_1 - t^{-1}\,dt).$$

The right side of the last equation is $Ax + o(x)$ by Lemma 9.14 with $\eta = 1$.

For the left side, set

$$v(u) := \int_{1-}^{u} dN * (\delta_1 - t^{-1}\,dt),$$

and restate the starting identity as

$$\int_{1}^{x} v(x/t)\,d\psi(t) = Ax + o(x). \tag{9.26}$$

Now write $v(u) = I + II$ with

$$I = I(u) := A \int_{1-}^{u} \frac{u}{t}\,(\delta_1 - t^{-1}\,dt),$$

$$II = II(u) := \int_{1-}^{u} \{N(u/t) - Au/t\}\,(\delta_1 - t^{-1}\,dt).$$

We have $I = A$ by (9.24) with $\eta = 1$. Since $|N(w) - Aw| \le wQ(w)$, we have

$$|II| \le \int_{1-}^{u} (u/t)Q(u/t)\,(\delta_1 + t^{-1}\,dt)$$

$$= uQ(u) + \int_{1}^{u} Q(w)\,dw = O\Big(1 + \int_{1}^{u} Q(w)\,dw\Big),$$

the last because $\int_1^u Q(t)\,dt \ge (u-1)Q(u) \ge (u/2)Q(u)$ for $u \ge 2$. Combining the estimates, we see that

$$v(u) = O\Big(1 + \int_{1}^{u} Q(w)\,dw\Big).$$

Insert into (9.26) both the preceding bound and $\psi(t) \le ct$. We find

$$\int_{1}^{x} v(x/t)\,d\psi(t) \le K \int_{1}^{x} \Big(1 + \int_{1}^{x/t} Q(u)\,du\Big)\,d\psi(t)$$

$$\le K\psi(x) + K \int_{1}^{x} d\psi * Q(t)\,dt = K\psi(x) + K \int_{1}^{x} \psi(x/t)Q(t)\,dt$$

$$\le K\Big\{\psi(x)\Big(1 + \int_{1}^{B} Q(t)\,dt\Big) + cx \int_{B}^{x} t^{-1}Q(t)\,dt\Big\}$$

with some constant $K > 0$ and any number $B > 1$. For B sufficiently large, we have $cK \int_B^\infty t^{-1}Q(t)\,dt \le \frac{1}{3}A$. Fixing such a B and recalling (9.26), we find

$$K\psi(x)\Big(1 + \int_{1}^{B} Q(t)\,dt\Big) + \frac{1}{3}Ax \ge \int_{1}^{x} v(x/t)\,d\psi(t) = Ax + o(x).$$

Thus, for x sufficiently large,

$$K\psi(x)\Big(1+\int_1^B Q(t)\,dt\Big) \geq \frac{1}{3}Ax,$$

i.e. $\psi(x) \gg x$. This completes the proof of Theorem 9.9. □

9.5. A failure of Chebyshev bounds

In this section, we show that Chebyshev bounds need not hold for g-number systems satisfying condition (9.23) with $\gamma < 1$. Thus, apart from the case $\gamma = 1$, Corollary 9.10 is sharp. In Chapter 11, we consider this limiting case.

THEOREM 9.16 (R. S. Hall). *Let $\alpha \in [0,1]$, $\beta \in [1,\infty]$ and $\gamma \in [0,1)$ be given. There exists a g-prime system for which*

(1) $N(x) = Ax + O(x\log^{-\gamma} x)$;
(2) $\liminf_{x\to\infty} \pi(x)(\log x)/x = \alpha$;
(3) $\limsup_{x\to\infty} \pi(x)(\log x)/x = \beta$.

PROOF. We assume that $\gamma \in (0,1)$, for a construction with this restriction clearly satisfies condition (1) with $\gamma = 0$.

Let $\pi_0(x)$ be the counting function of the rational primes. For each $n \in \mathbb{N}$, define the interval $I_n := (a_n, b_n]$ with end points

$$a_n := \exp\{2^{n/(1-\gamma)}\} \quad \text{and} \quad b_n := a_n 2^{n/4}.$$

The intervals $\{I_n\}$ are nonoverlapping, for a small calculation shows that $b_n < a_{n+1}$.

Let C_n be the set consisting of the ν_n largest rational primes in I_n, with

$$\nu_n := \lfloor (1-\alpha)\{\pi_0(b_n) - \pi_0(a_n)\}\rfloor.$$

(C_n consists of all primes in I_n if $\alpha = 0$, and $C_n = \emptyset$ if $\alpha = 1$.) If $\beta < \infty$, let D_n be a set consisting of any $\lfloor(\beta-1)\{\pi_0(b_n) - \pi_0(a_n)\}\rfloor$ distinct real numbers in the interval $(b_n - 1, b_n]$. (If $\beta = 1$, then $D_n = \emptyset$.) If $\beta = \infty$, let D_n be a set consisting of any $\lfloor 2^{n/8}\{\pi_0(b_n) - \pi_0(a_n)\}\rfloor$ distinct real numbers in $(b_n - 1, b_n]$. Let

$$C = \bigcup_{n\ \mathrm{odd}} C_n, \quad D = \bigcup_{n\ \mathrm{even}} D_n.$$

Define $\mathcal{P} := \{p_j\}_{j=1}^\infty$ as the nondecreasing sequence of elements of $(R \setminus C) \cup D$, where R is the set of rational primes, and take $\mathcal{N} := \{n_k\}_{k=1}^\infty$ to be the sequence of g-integers generated by $\mathcal{P}$, with $n_1 = 1$ and appropriate multiplicity for any integers that occur repeatedly.

We next show that the contribution of the primes of $C \cup D$ is of modest size. Since $(\log x)^\gamma/x$ is decreasing for $x > e^\gamma$,

$$\{\pi_0(b_n) - \pi_0(a_n)\}\frac{\log^\gamma(b_n-1)}{b_n-1} \leq \{\pi_0(b_n) - \pi_0(a_n)\}\frac{\log^\gamma a_n}{a_n}.$$

By Chebyshev's bounds for rational primes, the last quantity is of order

$$\Big(\frac{b_n}{\log b_n}\Big)\frac{\log^\gamma a_n}{a_n} = \frac{2^{n/4}\log^\gamma a_n}{\log a_n + (n/4)\log 2} \leq \frac{2^{n/4}}{\log^{1-\gamma} a_n} = \frac{1}{2^{3n/4}}.$$

Then, for some $K = K(\beta)$, we have

$$\begin{aligned}\sum_{p\in C}\frac{\log^\gamma p}{p}+\sum_{p\in D}\frac{\log^\gamma q}{q} &\le \sum_{n\text{ odd}}\{\pi_0(b_n)-\pi_0(a_n)\}\frac{\log^\gamma a_n}{a_n}\\ &\quad + K\sum_{n\text{ even}}2^{n/8}\{\pi_0(b_n)-\pi_0(a_n)\}\frac{\log^\gamma(b_n-1)}{b_n-1}\\ &= O\Big(\sum_{n=1}^{\infty}\frac{1}{2^{n/2}}\Big)=O(1).\end{aligned}$$

It follows that

$$\prod_{p\in C}(1+(\log^\gamma p)/p)\prod_{q\in D}(1-(\log^\gamma q)/q)^{-1}$$

converges.

Let $\{k_j\}_{j=1}^{\infty}$ be the sequence of products $p^n q^m$, $n, m = 0, 1, 2, \ldots$ of powers p^n of primes $p \in C$ and powers q^m of $q \in D$, arranged in ascending order. Also, we define a sequence $\{c_j\}_{j=1}^{\infty}$ by

$$\sum_{j=1}^{\infty}\frac{c_j}{{k_j}^s}=\prod_{p\in C}\Big(1-\frac{1}{p^s}\Big)\prod_{q\in D}\Big(1-\frac{1}{q^s}\Big)^{-1},$$

where $\Re s \ge 1$. Then we have

$$\sum_{j=1}^{\infty}\frac{|c_j|\log^\gamma k_j}{k_j}\le\prod_{p\in C}\Big(1+\frac{\log^\gamma p}{p}\Big)\prod_{q\in D}\Big(1-\frac{\log^\gamma q}{q}\Big)^{-1} \tag{9.27}$$

by repeated use of the inequality

$$\log^\gamma(xy)=(\log x+\log y)^\gamma\le\big(\log^\gamma x\big)\big(\log^\gamma y\big),$$

valid for $x, y \ge e^2$. (Note that each $k_j \ge a_1 \ge e^2$.)

Since the g-number system $\mathcal{N}$ was generated by g-primes $\mathcal{P} = (R\setminus C)\cup D$, the zeta function of $\mathcal{N}$ has the representation

$$\zeta(s):=\sum_{k=1}^{\infty}\frac{1}{n_k^s}=\zeta_0(s)\prod_{p\in C}\Big(1-\frac{1}{p^s}\Big)\prod_{q\in D}\Big(1-\frac{1}{q^s}\Big)^{-1},\quad \sigma>1,$$

where $\zeta_0(s)$ is the Riemann zeta function. It follows that

$$\sum_{k=1}^{\infty}\frac{1}{{n_k}^s}=\sum_{m=1}^{\infty}\frac{1}{m^s}\cdot\sum_{j=1}^{\infty}\frac{c_j}{{k_j}^s},$$

and hence $N(x)$, the counting function of g-integers in $\mathcal{N}$, is given by

$$N(x)=\sum_{n_k\le x}1=\sum_{k_j\le x}c_j\Big\lfloor\frac{x}{k_j}\Big\rfloor=x\sum_{k_j\le x}\frac{c_j}{k_j}+O\Big(\sum_{k_j\le x}|c_j|\Big).$$

From (9.27) and the convergence of the infinite product on the right, the infinite series $\sum_{j=1}^{\infty}c_j/k_j$ converges absolutely and, by the definition of c_j,

$$\sum_{j=1}^{\infty}c_j/k_j=\prod_{p\in C}\Big(1-\frac{1}{p}\Big)\prod_{q\in D}\Big(1-\frac{1}{q}\Big)^{-1}=:A>0.$$

Hence we have

$$N(x) = Ax + O\Big(x \sum_{k_j > x} \frac{|c_j|}{k_j}\Big) + O\Big(\sum_{k_j \le x} |c_j|\Big).$$

Also, we have

$$\log^\gamma x \sum_{k_j > x} \frac{|c_j|}{k_j} \le \sum_{k_j > x} \frac{|c_j| \log^\gamma k_j}{k_j} = o(1)$$

as $x \to \infty$ and

$$\sum_{k_j \le x} |c_j| \le 1 + \frac{x}{\log^\gamma x} \sum_{k_j \le x} \frac{|c_j| \log^\gamma k_j}{k_j} = O\left(\frac{x}{\log^\gamma x}\right).$$

Thus

$$N(x) = Ax + O(x/\log^\gamma x).$$

Hence Property (1) holds.

We next prove Property (2). Since $\pi_0(x) \sim x/\log x$ by the PNT for $\mathbb{N}$, we have

$$\lim_{n\to\infty} \frac{\pi_0(a_n)}{\pi_0(b_n)} = 0 \quad \text{and} \quad \lim_{n\to\infty} \frac{\pi_0(b_{n-1})}{\pi_0(a_n)} 2^{n/8} \le \lim_{n\to\infty} \frac{\pi_0(a_{n-1}2^{(n-1)/4})}{\pi_0(a_{n-1}^2)} 2^{n/8} = 0.$$

From the inequality

$$\pi(x) \ge \pi_0(x) - \sum_{b_n \le x,\, n \text{ odd}} \lfloor (1-\alpha)\{\pi_0(b_n) - \pi_0(a_n)\rfloor \ge \pi_0(x) - (1-\alpha)\pi_0(x),$$

we see that

$$\liminf_{x\to\infty} \frac{\pi(x)\log x}{x} = \liminf_{x\to\infty} \frac{\pi(x)}{\pi_0(x)} \ge \alpha.$$

In the other direction,

$$\pi(a_n) \le \pi_0(a_n) + K \sum_{k<n,\, k \text{ even}} 2^{k/8}\{\pi_0(b_k) - \pi_0(a_k)\} \le \pi_0(a_n) + K2^{n/8}\pi_0(b_{n-1}).$$

It follows that $\lim_{n\to\infty} \pi(a_n)/\pi_0(b_n) = 0$. Also, by construction and the preceding inequalities, for n odd,

$$\pi(b_n) \le \pi(a_n) + \alpha\{\pi_0(b_n) - \pi_0(a_n)\} + 1 = (\alpha + o(1))\pi_0(b_n).$$

(Note that this inequality is valid as well for $\alpha = 0$, for in this case we have $\pi(b_n) \le \pi(a_n) + 1 = o(\pi_0(b_n))$.) Now, Property (2) holds, since

$$\liminf_{x\to\infty} \frac{\pi(x)}{\pi_0(x)} \le \liminf_{n\to\infty,\, n \text{ odd}} \frac{\pi(b_n)}{\pi_0(b_n)} = \alpha.$$

Finally, we turn to Property (3). If $\beta < \infty$, then

$$\pi(x) \le \pi_0(x) + \sum_{b_n \le x,\, n \text{ even}} (\beta - 1)\{\pi_0(b_n) - \pi_0(a_n)\} \le \beta\pi_0(x).$$

Similarly, it follows that

$$\begin{aligned}\limsup_{x\to\infty} \frac{\pi(x)}{\pi_0(x)} &\ge \limsup_{n\to\infty,\, n \text{ even}} \frac{\pi(b_n)}{\pi_0(b_n)} \\ &\ge \limsup_{n\to\infty,\, n \text{ even}} \frac{\pi(a_n) + \beta\{\pi_0(b_n) - \pi_0(a_n)\} - 1}{\pi_0(b_n)} = \beta.\end{aligned}$$

Thus, $\limsup_{x\to\infty} \pi(x)(\log x)/x = \beta$ if β is finite.

If $\beta = \infty$, then

$$\begin{aligned}\limsup_{x\to\infty} \frac{\pi(x)}{\pi_0(x)} &\geq \limsup_{n\to\infty,\, n\,\text{even}} \frac{\pi(b_n)}{\pi_0(b_n)} \\ &\geq \limsup_{n\to\infty,\, n\,\text{even}} \frac{\pi(a_n) + (2^{n/8}+1)\{\pi_0(b_n) - \pi_0(a_n)\} - 1}{\pi_0(b_n)} = \infty.\end{aligned}$$

This establishes Property (3) of Theorem 9.16 . □

9.6. Notes

§9.2. The results of Proposition 9.1 and the "approximate convolution inverse" come mainly from [**Di73b**].

§9.4. Theorem 9.9 is proved in [**Zh87a**]; Corollary 9.10 was established earlier in [**Di73a**]. Can Chebyshev bounds be established under the weaker assumption $N(x) = Ax + O(x/\log x)$? This seems unlikely to us. In Theorem 11.1 such bounds are established using (an average form of) this assumption along with the additional condition (11.1).

§9.5. Theorem 9.16 is based on the article [**Ha73**].

CHAPTER 10

Wiener-Ikehara Tauberian Theorems

How does one unscramble an egg?

SUMMARY. Versions of the Wiener-Ikehara Theorem, including necessary and sufficient conditions and one-sided estimates.

10.1. Introduction

In the last chapter, we deduced Chebyshev bounds under various hypotheses on the g-integer counting function $N(x)$ via elementary methods. We can establish such bounds under weaker hypotheses using analysis. The general method is to introduce a generating function and extract number-theoretic conclusions from it by a tauberian ("unaveraging") argument. We have met arguments of this type already in Chapter 5.

A particularly useful tauberian method is the theorem of Wiener and Ikehara (W-I), which lies at the heart of a well-known proof of the PNT for the rational primes; the only other ingredient in a proof of the PNT via W-I is the nonvanishing of $\zeta(1+it)$. In this chapter, we establish a general form of W-I which applies in cases – such as ours – where the generating function does not have good analytic behavior on its abscissa of convergence.

In addition, we give versions of the theorem to treat upper and lower estimates, which are useful for studying Chebyshev-type bounds for g-primes. These results will be applied in the following several chapters. Other variants of the W-I theorem are listed in the references. Also, we give converse forms of our theorems.

Unlike the usual Laplace inversion formula, the W-I method uses integration along a vertical line segment in $\mathbb{C}$, rather than along an entire line. Simple truncation would give rise to convergence issues in the subsequent analysis; the clever device of Wiener to overcome this problem is to introduce a function that vanishes at the end points of the interval.

10.2. Wiener-Ikehara Theorems

The following notation and conventions will be in force throughout this chapter. Let $F(x)$ be a nondecreasing, nonnegative function on $[0,\infty)$ such that

$$\lim_{\sigma\to 1+}(\sigma-1)\int_0^\infty F(x)e^{-\sigma x}\,dx = L, \tag{10.1}$$

for some constant $L \geq 0$, and for $s=\sigma+it$ with $\sigma>1$, set

$$G(s) := \int_0^\infty F(x)e^{-sx}\,dx - \frac{L}{s-1}\,. \tag{10.2}$$

THEOREM 10.1 (W-I Upper Bound). *If there exist positive constants* λ, y_0, *and* K *such that*

$$\limsup_{\sigma\to 1+}\left|\int_{-2\lambda}^{2\lambda}\frac{1}{2}\Big(1-\frac{|t|}{2\lambda}\Big)e^{ity}G(\sigma+it)\,dt\right|\le K \tag{10.3}$$

for each $y \ge y_0$, *then* $F(x)e^{-x} \le C$ *holds for some constant* $C > 0$.

Conversely, if $F(x)e^{-x} \le C$ *for some constant* $C > 0$ *then, for every* $\lambda > 0$, (10.3) *holds with* $K = C\pi$ *uniformly for* $y \in \mathbb{R}$.

REMARK 10.2. If there exist positive constants λ_0, y_0, and K such that (10.3) holds with $\lambda = \lambda_0$ for all $y \ge y_0$ then this relation holds (with some other constant K) for all $\lambda > 0$ and all y.

COROLLARY 10.3. *If there exists* $\lambda > 0$ *such that*

$$\limsup_{\sigma\to 1+}\int_{-2\lambda}^{2\lambda}|G(\sigma+it)|\,dt<\infty$$

then $F(x)e^{-x} \le C$ *for some constant* $C > 0$.

We see from (10.2) that $G(\sigma)$ is real for real $\sigma > 1$. By the reflection principle, $G(\bar{s}) = \overline{G(s)}$ for complex s, and thus

$$\int_{-2\lambda}^{2\lambda}\frac{1}{2}\Big(1-\frac{|t|}{2\lambda}\Big)e^{ity}G(\sigma+it)\,dt$$

is real for real y.

THEOREM 10.4 (W-I Lower Bound). *If* $F(x)e^{-x} \ge c$ *for all* $x \ge x_0 \ge 0$ *and some constant* $c > 0$, *then there exists a constant* $\gamma > 0$ *such that, for every* $\lambda > 0$,

$$\liminf_{\sigma\to 1+}\int_{-2\lambda}^{2\lambda}\frac{1}{2}\Big(1-\frac{|t|}{2\lambda}\Big)e^{ity}G(\sigma+it)\,dt\ge(-L+\gamma)\pi \tag{10.4}$$

uniformly for $y \ge y_0(\lambda)$.

Conversely, assume there exist positive constants λ, γ, *and* y_0 *such that* (10.4) *holds for all* $y \ge y_0$ *and that* $F(x)e^{-x} \le C$ *for some constant* $C > 0$. *Then* $F(x)e^{-x} \ge c$ *for all* $x \ge x_0$, *with some constant* $c > 0$.

THEOREM 10.5 (Wiener-Ikehara). *Let* $F(x)$ *be a nondecreasing, nonnegative function on* $[0,\infty)$ *which satisfies* (10.1) *and* (10.2). *Then*

$$\lim_{x\to\infty}F(x)e^{-x}=L$$

if and only if there exists a constant $\lambda_0 \ge 0$ *such that*

$$\lim_{\sigma\to 1+}\int_{-2\lambda}^{2\lambda}\frac{1}{2}\Big(1-\frac{|t|}{2\lambda}\Big)e^{ity}G(\sigma+it)\,dt\quad\textit{exists} \tag{10.5}$$

for every $\lambda > \lambda_0$ *and every* $y \ge y_0(\lambda)$, *and such that*

$$\lim_{y\to\infty}\lim_{\sigma\to 1+}\int_{-2\lambda}^{2\lambda}\frac{1}{2}\Big(1-\frac{|t|}{2\lambda}\Big)e^{ity}G(\sigma+it)\,dt=0 \tag{10.6}$$

for every $\lambda > \lambda_0$.

COROLLARY 10.6. *If there exists a constant* $\lambda_0 \geq 0$ *such that, for every* $\lambda > \lambda_0$,

$$\lim_{\sigma,\,\sigma' \to 1+} \int_{-2\lambda}^{2\lambda} \frac{1}{2}\Big(1 - \frac{|t|}{2\lambda}\Big) e^{ity} \{G(\sigma + it) - G(\sigma' + it)\}\, dt = 0 \tag{10.7}$$

uniformly for $y \geq y_0(\lambda)$, *then* $\lim_{x\to\infty} F(x)e^{-x} = L$.

In particular, if

$$\lim_{\sigma,\,\sigma' \to 1+} \int_{-2\lambda}^{2\lambda} |G(\sigma + it) - G(\sigma' + it)|\, dt = 0 \tag{10.8}$$

for every $\lambda > 0$, *then* $\lim_{x\to\infty} F(x)e^{-x} = L$.

REMARK 10.7. Our theorems apply for a Laplace transform having abscissa of convergence $\sigma_c = 1$. For other abscissa values, analogous theorems hold; alternatively, as noted in §4.1, one could normalize the function F to make $\sigma_c = 1$.

REMARK 10.8. In Theorem 10.5 we must assume that (10.6) is satisfied for *all* $\lambda > \lambda_0$; if this relation holds only for $0 < \lambda \leq \lambda_1$, then we can conclude only that $\liminf F(x)e^{-x} \leq \limsup F(x)e^{-x} < \infty$. An example with $\liminf F(x)e^{-x} < \limsup F(x)e^{-x}$ is given in §10.5.

10.3. The Fejér kernel

Our proofs of Theorems 10.1, 10.4, and 10.5 depend on analytic properties of the Fejér kernel, which is defined on $\mathbb{R}$ for every $\lambda > 0$ by

$$k_\lambda(x) := \int_{-2\lambda}^{2\lambda} \frac{1}{2}\Big(1 - \frac{|t|}{2\lambda}\Big) e^{ixt}\, dt. \tag{10.9}$$

For completeness, we state and prove some familiar properties of k_λ.

LEMMA 10.9. *Let* $x \in \mathbb{R}$, $\lambda > 0$, *and* $\delta > 0$. *Then*

$$k_\lambda(x) = \lambda\Big(\frac{\sin \lambda x}{\lambda x}\Big)^2, \tag{10.10}$$

$$0 \leq k_\lambda(x) \leq \min\{\lambda,\, \lambda^{-1}x^{-2}\}, \tag{10.11}$$

$$\int_{-\infty}^{\infty} k_\lambda(x)\, dx = \pi, \tag{10.12}$$

$$\int_{|u|\geq\delta} k_\lambda(u)\, du < \frac{2}{\lambda\delta}\,. \tag{10.13}$$

PROOF. A small manipulation gives

$$k_\lambda(x) = \int_0^{2\lambda} \Big(1 - \frac{t}{2\lambda}\Big) \cos(xt)\, dt,$$

and hence $k_\lambda(0) = \lambda$. For $x \neq 0$, integrating by parts yields

$$k_\lambda(x) = \frac{1}{2\lambda x} \int_0^{2\lambda} \sin(xt)\, dt = \frac{1 - \cos 2\lambda x}{2\lambda x^2} = \frac{(\sin \lambda x)^2}{\lambda x^2}\,,$$

which establishes (10.10). Then (10.11) follows using the inequalities

$$0 \leq u^{-2} \sin^2 u \leq \min\{1,\, u^{-2}\}.$$

The integral in (10.13) converges and satisfies the stated inequality by the second estimate of (10.11). The integral in (10.12) converges by another use of (10.11). A change of variable shows that this integral equals $\int_{-\infty}^{\infty} v^{-2}\sin^2 v\,dv$ for any value of $\lambda > 0$. Then, integration by parts and some familiar relations yield

$$\int_{-\infty}^{\infty} \frac{\sin^2 v}{v^2}\,dv = 2\int_0^{\infty} \frac{\sin 2v}{v}\,dv = 2\int_0^{\infty} \frac{\sin u}{u}\,du = \pi. \qquad \square$$

The next lemma estimates L^1-norms of

$$k_\lambda^{(\nu)}(x) = \int_{-2\lambda}^{2\lambda} \frac{1}{2}\Big(1 - \frac{|t|}{2\lambda}\Big)(it)^\nu e^{ixt}\,dt, \tag{10.14}$$

the ν-th derivative of k_λ. We shall use these results to establish Chebyshev bounds in the next chapter.

LEMMA 10.10. *Let $0 < \lambda \le 1/2$. Then, for $\nu = 1, 2, 3, \ldots,$*

$$|k_\lambda^{(\nu)}(x)| \le \frac{3(2\lambda)^{\nu-1}}{1+x^2}, \qquad x \in \mathbb{R}, \tag{10.15}$$

$$\int_{-\infty}^{\infty} |k_\lambda^{(\nu)}(x)|\,dx \le \frac{8(2\lambda)^\nu}{\nu+1}. \tag{10.16}$$

PROOF. We begin with an absolute bound for $k_\lambda^{(\nu)}(x)$, to be applied for $|x|$ small. From (10.14), we have the simple inequality

$$|k_\lambda^{(\nu)}(x)| \le \int_0^{2\lambda} \Big(1 - \frac{t}{2\lambda}\Big) t^\nu\,dt = \frac{(2\lambda)^{\nu+1}}{(\nu+1)(\nu+2)}, \qquad x \in \mathbb{R}. \tag{10.17}$$

For application to larger $|x|$, we show that

$$|k_\lambda'(x)| \le \frac{8}{3x^2}, \qquad \lambda|x| \ge 6/5, \tag{10.18}$$

$$|k_\lambda^{(\nu)}(x)| \le \frac{8(2\lambda)^{\nu-1}}{3x^2}, \qquad \nu \ge 2, \quad x \ne 0. \tag{10.19}$$

For $\nu = 1$, we have

$$\begin{aligned} |k_\lambda'(x)| &= \Big|\frac{2\sin\lambda x\cos\lambda x}{x^2} - \frac{2\sin^2\lambda x}{\lambda x^3}\Big| \\ &\le \frac{|\sin 2\lambda x|}{x^2} + \frac{2\sin^2\lambda x}{\lambda|x|^3} \le \frac{1}{x^2} + \frac{2}{\lambda|x|\,x^2} \le \frac{8}{3x^2}, \qquad \lambda|x| \ge 6/5. \end{aligned}$$

For $\nu \ge 2$, start with the expression

$$k_\lambda(x) = \int_0^{2\lambda} \Big(1 - \frac{t}{2\lambda}\Big)\cos xt\,dt$$

and make ν differentiations. We find

$$k_\lambda^{(\nu)}(x) = \int_0^{2\lambda} \Big(1 - \frac{t}{2\lambda}\Big) t^\nu T(xt)\,dt,$$

where $T = \pm\sin$ or $\pm\cos$, depending on ν (mod 4). Integrating by parts twice, with $T_1 := \int T = \mp\cos$ or $\pm\sin$ and $T_2 := \int T_1 = \mp\sin$ or $\mp\cos$, we find

$$k_\lambda^{(\nu)}(x) = \frac{T_2(2\lambda x)}{x^2}(2\lambda)^{\nu-1} + \int_0^{2\lambda} \frac{T_2(xt)}{x^2}\nu t^{\nu-2}\Big\{\nu - 1 - \frac{(\nu+1)t}{2\lambda}\Big\}\,dt,$$

and so

$$|k_\lambda^{(\nu)}(x)| \le \frac{(2\lambda)^{\nu-1}}{x^2} + \int_0^{2\lambda} \frac{\nu t^{\nu-2}}{x^2} \left|\nu - 1 - \frac{(\nu+1)t}{2\lambda}\right| dt.$$

To treat the last integral, set $t^* = 2\lambda(\nu-1)/(\nu+1)$ and note that

$$\nu - 1 - \frac{(\nu+1)t}{2\lambda} \begin{cases} > 0, & \text{if } 0 \le t < t^*, \\ < 0, & \text{if } t^* < t \le 2\lambda. \end{cases}$$

Therefore

$$\begin{aligned} |k_\lambda^{(\nu)}(x)| &\le \frac{(2\lambda)^{\nu-1}}{x^2} + \frac{1}{x^2}\int_0^{t*} \nu t^{\nu-2}\Big\{\nu - 1 - \frac{(\nu+1)t}{2\lambda}\Big\}\, dt \\ &\quad - \frac{1}{x^2}\int_{t*}^{2\lambda} \nu t^{\nu-2}\Big\{\nu - 1 - \frac{(\nu+1)t}{2\lambda}\Big\}\, dt \\ &= 2(2\lambda)^{\nu-1}x^{-2}\Big\{1 + \Big(\frac{\nu-1}{\nu+1}\Big)^{\nu-1}\Big\}. \end{aligned}$$

Now we note that $\{(\nu-1)/(\nu+1)\}^{\nu-1}$ is decreasing for $\nu \ge 2$ and hence

$$\{(\nu-1)/(\nu+1)\}^{\nu-1} \le 1/3 \quad \text{for} \quad \nu = 2, 3, \ldots.$$

It follows that (10.19) holds for $x \ne 0$ and all integers $\nu \ge 2$.

To show (10.15), we treat separately the case $\nu = 1$. Recall that $0 < \lambda \le 1/2$. If $\lambda|x| \le 2$, we have

$$2\lambda^2(1+x^2) \le 1/2 + 2(2)^2 < 9,$$

whence

$$\frac{(2\lambda)^2}{6} < \frac{3}{1+x^2}.$$

On the other hand, if $\lambda|x| > 2$, then $|x| > 4$ and $8 + 8x^2 < 9x^2$, so

$$\frac{8}{3x^2} < \frac{3}{1+x^2}.$$

For $\nu \ge 2$, we establish (10.15) by an analogous but simpler argument, treating separately the cases $|x| \le 3$ and $x > 3$.

While we could estimate $\int |k_\lambda^{(\nu)}(x)|\, dx$ by integrating (10.15), a sharper bound comes from applying (10.17) and (10.19) (resp. (10.18) for $\nu = 1$ and $|x| \ge 2/\lambda$) once again. We have

$$\begin{aligned} \int_{-\infty}^{\infty} |k_\lambda^{(\nu)}(x)|\, dx &\le \frac{2(2\lambda)^{\nu+1}X}{(\nu+1)(\nu+2)} + 2\int_X^\infty \frac{8(2\lambda)^{\nu-1}}{3x^2}\, dx \\ &= 8(2\lambda)^{\nu-1}\Big(\frac{\lambda^2 X}{(\nu+1)(\nu+2)} + \frac{2}{3X}\Big) \le \frac{8(2\lambda)^\nu}{\nu+1} \end{aligned}$$

by taking

$$X = \frac{1}{\lambda}\sqrt{\frac{2(\nu+1)(\nu+2)}{3}}. \qquad \square$$

10.4. Proof of the Wiener-Ikehara Theorems

To simplify the notation, let

$$\Delta_\lambda(t) := \frac{1}{2}\Big(1 - \frac{|t|}{2\lambda}\Big)^+.$$

PROOF OF THEOREM 10.1. We begin our argument by rewriting (10.2) as

$$G(\sigma + it) = \int_0^\infty e^{-(\sigma+it)x}\{F(x) - Le^x\}\,dx, \qquad \sigma > 1,$$

multiplying by $\Delta_\lambda(t)e^{ity}$, and integrating over $-2\lambda < t < 2\lambda$. If then we change the order integration in the resulting double integral, we find

$$\begin{aligned}(10.20)\qquad &\int_{-2\lambda}^{2\lambda} \Delta_\lambda(t)e^{ity}G(\sigma + it)\,dt \\ &\quad = \int_0^\infty e^{-\sigma x}F(x)\,k_\lambda(y - x)\,dx - L\int_0^\infty e^{-(\sigma-1)x}k_\lambda(y - x)\,dx.\end{aligned}$$

Suppose $F(x)e^{-x} \le C$ holds for some constant $C > 0$. In this case, by (10.1),

$$(10.21)\quad L = \lim_{\sigma\to 1+}(\sigma - 1)\int_0^\infty F(x)e^{-\sigma x}\,dx \le C\lim_{\sigma\to 1+}(\sigma - 1)\int_0^\infty e^{-(\sigma-1)x}\,dx = C.$$

Then the left hand side of (10.20) is at most

$$C\int_0^\infty e^{-(\sigma-1)x}k_\lambda(y - x)\,dx \le C\pi,$$

and it is at least

$$-L\int_0^\infty e^{-(\sigma-1)x}k_\lambda(y - x)\,dx \ge -L\pi \ge -C\pi$$

for all $\lambda > 0$ and $y \in \mathbb{R}$ by (10.12) and (10.21). Thus (10.3) holds for all y.

Conversely, suppose (10.3) is satisfied for each $y \ge y_0$. In this case, the modulus of the left hand side of (10.20) is at most $K + 1$ for $y \ge y_0$ and $1 < \sigma < 1 + \delta(y)$, for some $\delta(y) > 0$. Hence, by (10.12),

$$\int_0^\infty e^{-\sigma x}F(x)\,k_\lambda(y - x)\,dx \le K + 1 + L\pi.$$

Let B denote the constant on the right hand side. By the monotone convergence theorem,

$$\int_0^\infty e^{-x}F(x)\,k_\lambda(y - x)\,dx \le B$$

for all $y \ge y_0$; thus

$$B \ge \int_y^{y+1} e^{-x}F(x)\,k_\lambda(y - x)\,dx \ge e^{-y-1}F(y)\int_{-1}^0 k_\lambda(u)\,du$$

for $y \ge y_0$. It follows that

$$F(y)e^{-y} \le eB\Big(\int_{-1}^0 k_\lambda(u)\,du\Big)^{-1}$$

for $y \ge y_0$, and therefore $F(x)e^{-x} \le C$ holds with a suitable constant $C > 0$. $\square$

PROOF OF THEOREM 10.4. Suppose that $F(x)e^{-x} \geq c$, for all $x \geq x_0$ with some constant $c > 0$. We have first, $L \geq c$, by an analogue of (10.21). Next, by (10.20) and the $F(x)e^{-x}$ condition, we have

$$\begin{aligned}
&\int_{-2\lambda}^{2\lambda} \Delta_\lambda(t)e^{ity}G(\sigma+it)\,dt \\
&\geq (c-L)\int_0^\infty e^{-(\sigma-1)x}k_\lambda(y-x)\,dx + \int_0^{x_0} e^{-(\sigma-1)x}(F(x)e^{-x}-c)k_\lambda(y-x)\,dx \\
&\geq (c-L)\int_0^\infty e^{-(\sigma-1)x}k_\lambda(y-x)\,dx - R\int_0^{x_0} e^{-(\sigma-1)x}k_\lambda(y-x)\,dx
\end{aligned}$$

for R a positive constant and every $\lambda > 0$. Hence, by (10.11) and (10.12),

$$\begin{aligned}
&\liminf_{\sigma\to 1+} \int_{-2\lambda}^{2\lambda} \Delta_\lambda(t)e^{ity}G(\sigma+it)\,dt \\
&\qquad \geq (c-L)\int_{-\infty}^{y} k_\lambda(u)\,du - R\int_{y-x_0}^{y} k_\lambda(u)\,du \geq (c-L)\pi - \frac{R}{\lambda(y-x_0)}
\end{aligned}$$

for $y > x_0$. Let $\gamma = c/2$ and fix $y_0 = y_0(\lambda)$ ($> x_0$) satisfying

$$\frac{c\pi}{2} > \frac{R}{\lambda(y_0-x_0)}.$$

Then (10.4) holds uniformly for $y \geq y_0$.

Conversely, if there exist positive constants λ, γ, and y_0 such that (10.4) holds for all $y \geq y_0$ then, by (10.20),

$$\int_0^\infty e^{-\sigma x}F(x)\,k_\lambda(y-x)\,dx - L\int_0^\infty e^{-(\sigma-1)x}k_\lambda(y-x)\,dx \geq (-L+\gamma/2)\pi$$

for every $y \geq y_0$ and $1 < \sigma < 1+\eta(y)$ with $\eta(y) > 0$. Hence, by the monotone convergence theorem,

$$\int_0^\infty e^{-x}F(x)\,k_\lambda(y-x)\,dx \geq L\int_0^\infty k_\lambda(y-x)\,dx + (-L+\gamma/2)\pi \tag{10.22}$$

for $y \geq y_0$. If we assume also that $F(x)e^{-x} \leq C$ for some constant $C > 0$ then the left hand side is bounded above by

$$C\int_{|y-x|\geq\delta} k_\lambda(y-x)\,dx + \int_{y-\delta}^{y+\delta} e^{-x}F(x)\,k_\lambda(y-x)\,dx$$

for $y \geq \delta > 0$. Applying (10.12) and (10.13), the left hand side of (10.22) is bounded above further by

$$\frac{2C}{\lambda\delta} + \pi e^{2\delta}e^{-(y+\delta)}F(y+\delta)$$

for any $\delta > 0$ and $y \geq \max\{\delta, y_0\}$. Choosing δ sufficiently large so that $2C/(\lambda\delta) < \gamma\pi/4$, i.e., taking $\delta > 8C/(\gamma\pi\lambda)$ and fixing it, (10.22) yields

$$\pi e^{2\delta}\liminf_{y\to\infty} e^{-(y+\delta)}F(y+\delta) \geq L\pi + \left(-L+\frac{\gamma}{2}\right)\pi - \frac{2C}{\lambda\delta} > \frac{\gamma\pi}{4},$$

and hence

$$\liminf_{x\to\infty} e^{-x}F(x) > \frac{\gamma e^{-2\delta}}{4} > 0. \qquad \square$$

PROOF OF THEOREM 10.5. If $\lim_{x\to\infty} F(x)e^{-x} = L$ then $F(x)e^{-x} = L + r(x)$, where $r(x) \to 0$ as $x \to \infty$. Hence the right hand side of (10.20) equals

$$\int_0^\infty e^{-(\sigma-1)x} r(x) k_\lambda(y-x)\, dx.$$

Since $\int_0^\infty |r(x)| k_\lambda(y-x)\, dx < \infty$, then, by the dominated convergence theorem, the right hand side of (10.20) has a limit as $\sigma \to 1+$. Hence, so does the left hand side of (10.20), and we have

$$\lim_{\sigma\to 1+} \int_{-2\lambda}^{2\lambda} \Delta_\lambda(t) e^{ity} G(\sigma+it)\, dt = \int_0^\infty r(x) k_\lambda(y-x)\, dx,$$

which shows (10.5). Also, since $r(x) \to 0$ as $x \to \infty$ and the mass of $k_\lambda(y-x)$ is concentrated in the region where x is near y, a small argument shows that the last integral goes to 0 as $y \to \infty$. This establishes (10.6) for every $\lambda > 0$.

Conversely, assume that the limit (10.5) exists for all $\lambda > \lambda_0$ and $y \geq y_0(\lambda)$, and that (10.6) is satisfied for every $\lambda > \lambda_0$. Then, by (10.20) again,

$$\int_0^\infty e^{-x} F(x)\, k_\lambda(y-x)\, dx - L \int_0^\infty k_\lambda(y-x)\, dx = \lim_{\sigma\to 1+} \int_{-2\lambda}^{2\lambda} \Delta_\lambda(t) e^{ity} G(\sigma+it)\, dt$$

for $\lambda > \lambda_0$ and $y \geq y_0(\lambda)$. From (10.6),

$$\int_0^\infty e^{-x} F(x)\, k_\lambda(y-x)\, dx = L \int_0^\infty k_\lambda(y-x)\, dx + o_\lambda(1),$$

where $o_\lambda(1) \to 0$ as $y \to \infty$. The last integral tends to π as $y \to \infty$ by (10.11). Thus

$$\int_0^\infty e^{-x} F(x)\, k_\lambda(y-x)\, dx = L\pi + o_\lambda(1) \tag{10.23}$$

for $\lambda > \lambda_0$.

For $\delta_0 > 0$ to be specified later, we have, by (10.23),

$$\int_{y-\delta_0}^{y+\delta_0} e^{-x} F(x)\, k_\lambda(y-x)\, dx \leq L\pi + o_\lambda(1).$$

Note that the left hand side is at least

$$e^{-(y+\delta_0)} F(y-\delta_0) \int_{y-\delta_0}^{y+\delta_0} k_\lambda(y-x)\, dx.$$

By (10.12) and (10.13),

$$\int_{y-\delta_0}^{y+\delta_0} k_\lambda(y-x)\, dx = \pi - \int_{|u|\geq\delta_0} k_\lambda(u)\, du \geq \pi\Big(1 - \frac{2}{\pi\lambda\delta_0}\Big) > 0$$

for $\lambda \geq \lambda_1$ with λ_1 sufficiently large. Therefore we obtain

$$e^{-(y-\delta_0)} F(y-\delta_0) e^{-2\delta_0} \pi\Big(1 - \frac{2}{\pi\lambda\delta_0}\Big) \leq L\pi + o_\lambda(1)$$

for $\lambda \geq \lambda_1$. Given $\epsilon > 0$, we first choose $\delta_0 > 0$ satisfying $e^{2\delta_0} < 1 + \epsilon/2$, and next choose λ_1 so large that

$$\lambda_1 \geq \lambda_0, \quad \lambda_1\delta_0 \geq 2, \quad \text{and} \quad \frac{e^{2\delta_0}}{1 - 2/(\pi\lambda_1\delta_0)} < 1 + \epsilon.$$

Then we obtain

$$e^{-(y-\delta_0)}F(y-\delta_0) \le \frac{e^{2\delta_0}}{1-2/(\pi\lambda\delta_0)}(L+o_\lambda(1)) \le L(1+\epsilon)+o_\lambda(1)$$

for $\lambda \ge \lambda_1\,(> \lambda_0)$. It follows that

$$\limsup_{y\to\infty} e^{-y}F(y) \le L(1+\epsilon),$$

and since ϵ can be taken arbitrarily small,

$$\limsup_{y\to\infty} e^{-y}F(y) \le L. \tag{10.24}$$

Furthermore, by (10.24), the left hand side of (10.23) is at most

$$K\int_{|y-x|\ge\delta} k_\lambda(y-x)\,dx + \int_{y-\delta}^{y+\delta} e^{-x}F(x)\,k_\lambda(y-x)\,dx$$
$$\le \frac{2K}{\lambda\delta} + e^{-y+\delta}F(y+\delta)\int_{y-\delta}^{y+\delta} k_\lambda(y-x)\,dx$$

for some constant $K>0$ and every $\delta>0$. Hence

$$\pi e^{-y+\delta}F(y+\delta) + \frac{2K}{\lambda\delta} \ge L\pi + o_\lambda(1)$$

for $\lambda > \lambda_0$ and $y \ge y_0(\lambda)$, or

$$e^{2\delta}e^{-y-\delta}F(y+\delta) \ge L + o_\lambda(1) - \frac{2K}{\pi\lambda\delta}.$$

It follows that

$$\liminf_{y\to\infty} e^{-y}F(y) \ge (L-2K/(\pi\lambda\delta))e^{-2\delta}$$

for every $\lambda > \lambda_0$. Therefore, letting $\lambda\to\infty$, we obtain

$$\liminf_{y\to\infty} e^{-y}F(y) \ge Le^{-2\delta}$$

for every $\delta > 0$. Now letting $\delta \to 0+$, we get

$$\liminf_{y\to\infty} e^{-y}F(y) \ge L.$$

This bound and (10.24) complete the proof of Theorem 10.5. □

Proof of Corollary 10.6. If, for every $\lambda > \lambda_0$, (10.7) holds uniformly for $y \ge y_0(\lambda)$, then by Cauchy's convergence criterion, (10.5) holds for every $\lambda > \lambda_0$ and every $y \ge y_0(\lambda)$. Let $\ell(y) = \ell(y,\lambda)$ denote the limit function in (10.5). Then

$$\int_{-2\lambda}^{2\lambda} \Delta_\lambda(t)e^{ity}G(\sigma+it)\,dt$$

converges to $\ell(y)$ as $\sigma\to 1+$ uniformly for $y \ge y_0(\lambda)$. Hence, given $\epsilon > 0$,

$$\Big|\int_{-2\lambda}^{2\lambda} \Delta_\lambda(t)e^{ity}G(\sigma+it)\,dt - \ell(y)\Big| < \epsilon \tag{10.25}$$

for $1<\sigma<1+\delta(\lambda)$ and $y \ge y_0(\lambda)$. Fixing $\sigma=\sigma_0<1+\delta(\lambda)$ and noting that $G(\sigma_0+it)$ is a continuous function of t, we have by the Riemann-Lebesgue lemma

$$\lim_{y\to\infty}\int_{-2\lambda}^{2\lambda} \Delta_\lambda(t)e^{ity}G(\sigma_0+it)\,dt = 0.$$

It follows from (10.25) that

$$\limsup_{y\to\infty} \ell(y) \le \lim_{y\to\infty} \int_{-2\lambda}^{2\lambda} \Delta_\lambda(t) e^{ity} G(\sigma_0 + it)\, dt + \epsilon = \epsilon.$$

Similarly, $\liminf_{y\to\infty} \ell(y) \ge -\epsilon$, whence $\lim_{y\to\infty} \ell(y) = 0$, i.e., (10.6) is satisfied for every $\lambda > \lambda_0$. Thus $\lim_{x\to\infty} e^{-x} F(x) = L$ by Theorem 10.5. □

10.5. A W-I oscillatory example

Let $\lambda_1 > 0$ and $F(x)$ be defined on $[0, \infty)$ by

$$F(x) := \int_0^x (1 - \cos\{2\lambda_1 u\})\, e^u\, du.$$

We show that $F(x)e^{-x}$ does not have a limit at infinity and the Laplace transform of F has poles at nonreal points of $\{\sigma = 1\}$.

First, direct integration shows

$$F(x) = e^x - \frac{e^x \sin(2\lambda_1 x + \theta)}{\sqrt{1 + 4\lambda_1^2}} - \frac{4\lambda_1^2}{1 + 4\lambda_1^2},$$

whence

$$\liminf_{x\to\infty} F(x)e^{-x} = 1 - \frac{1}{\sqrt{1+4\lambda_1^2}}, \qquad \limsup_{x\to\infty} F(x)e^{-x} = 1 + \frac{1}{\sqrt{1+4\lambda_1^2}}.$$

The Laplace transform of F can be written as

$$\int_0^\infty e^{-sx} F(x)\, dx = \frac{1}{s}\int_0^\infty e^{-(s-1)x}(1 - \cos 2\lambda_1 x)\, dx = \frac{1}{s-1} + G(s),$$

for $\sigma > 1$, where

$$G(s) := -\frac{1}{s} - \frac{1}{2s}\left(\frac{1}{s - 1 - 2i\lambda_1} + \frac{1}{s - 1 + 2i\lambda_1}\right).$$

It is easy to see that the condition (10.1) is satisfied here.

To examine the W-I expression associated with F, let

$$I(\lambda, \sigma, y) := \int_{-2\lambda}^{2\lambda} \Delta_\lambda(t) e^{ity} G(\sigma + it)\, dt, \quad \lambda > 0,\ \sigma > 1.$$

We shall show that $\lim_{\sigma\to 1+} I(\lambda, \sigma, y)$ exists for all $\lambda > 0$ and $y > 0$ and

$$\lim_{\sigma\to 1+} I(\lambda, \sigma, y) = \begin{cases} o_\lambda(1), & 0 < \lambda \le \lambda_1, \\ -\dfrac{\pi}{2}\left(1 - \dfrac{\lambda_1}{\lambda}\right)\dfrac{\sin(2\lambda_1 y + \theta)}{\sqrt{1+4\lambda_1^2}} + o_\lambda(1), & \lambda > \lambda_1, \end{cases} \tag{10.26}$$

where $o_\lambda(1) \to 0$ as $y \to \infty$. Therefore (10.5) holds for all $\lambda > 0$ and $y > 0$, but (10.6) holds only for $0 < \lambda \le \lambda_1$, and this limit does not exist for *any* $\lambda > \lambda_1$.

Using the identity

$$\frac{1}{s(s-a)} = \frac{1}{a}\left(\frac{1}{s-a} - \frac{1}{s}\right)$$

and a small manipulation, we find

$$G(s) = -\frac{4\lambda_1^2}{1+4\lambda_1^2}\frac{1}{s} - \frac{1}{2(1+2i\lambda_1)}\frac{1}{s-1-2i\lambda_1} - \frac{1}{2(1-2i\lambda_1)}\frac{1}{s-1+2i\lambda_1}.$$

Then

$$I(\lambda,\sigma,y) = I_1(\lambda,\sigma,y) + I_2(\lambda,\sigma,y) + I_3(\lambda,\sigma,y),$$

where

$$I_1(\lambda,\sigma,y) := -\frac{4\lambda_1^2}{1+4\lambda_1^2}\int_{-2\lambda}^{2\lambda}\frac{\Delta_\lambda(t)e^{ity}\,dt}{\sigma+it},$$

$$I_2(\lambda,\sigma,y) := -\frac{1}{2(1+2i\lambda_1)}\int_{-2\lambda}^{2\lambda}\frac{\Delta_\lambda(t)e^{ity}\,dt}{\sigma-1+i(t-2\lambda_1)},$$

and

$$I_3(\lambda,\sigma,y) := -\frac{1}{2(1-2i\lambda_1)}\int_{-2\lambda}^{2\lambda}\frac{\Delta_\lambda(t)e^{ity}\,dt}{\sigma-1+i(t+2\lambda_1)}.$$

By the continuity of the integrand, $\lim_{\sigma\to 1+} I_1(\lambda,\sigma,y)$ exists for all $y \geq 0$ and $\lambda > 0$, and, by the Riemann-Lebesgue lemma, this function tends to 0 as $y \to \infty$. Similarly, $\lim_{\sigma\to 1+} I_{2,3}(\lambda,\sigma,y)$ each exist for $0 < \lambda < \lambda_1$, $y \geq 0$ and each tends to 0 as $y \to \infty$ for $0 < \lambda < \lambda_1$.

It remains to study $\lim_{\sigma\to 1+} I_{2,3}(\lambda,\sigma,y)$ for $\lambda \geq \lambda_1$, where the integrands are not so well-behaved. First, for $\lambda > \lambda_1$, we evaluate $\lim_{\sigma\to 1+} I_2(\lambda,\sigma,y)$ by splitting the integration interval $[-2\lambda, 2\lambda]$ into parts

$$[-2\lambda,\, 2\lambda_1-\epsilon] \cup [2\lambda_1-\epsilon,\, 2\lambda_1+\epsilon] \cup [2\lambda_1+\epsilon,\, 2\lambda]$$

for some ϵ satisfying $0 < \epsilon < \min\{2\lambda - 2\lambda_1, 2\lambda_1\}$. The integrand of $I_2(\lambda,\sigma,y)$ over the first and third subintervals are continuous and, as before, have limits as $\sigma \to 1+$ for $y \geq 0$ and these limits tend to zero as $y \to \infty$.

By changing the integration variable, the remaining integral becomes

$$\int_{2\lambda_1-\epsilon}^{2\lambda_1+\epsilon}\frac{\Delta_\lambda(t)e^{ity}\,dt}{\sigma-1+i(t-2\lambda_1)} = \frac{e^{2i\lambda_1 y}}{2}\left(1-\frac{\lambda_1}{\lambda}\right)\int_{-\epsilon}^{\epsilon}\frac{e^{iuy}\,du}{\sigma-1+iu} + \frac{e^{2i\lambda_1 y}}{4\lambda}\int_{-\epsilon}^{\epsilon}\frac{ue^{iuy}\,du}{\sigma-1+iu}. \tag{10.27}$$

Let A and B denote the first and second integrals respectively on the right side of (10.27). By the bounded convergence theorem, B has the limit

$$\int_{-\epsilon}^{\epsilon}\frac{e^{iuy}}{i}\,du = -\frac{e^{i\epsilon y} - e^{-i\epsilon y}}{y}$$

as $\sigma \to 1+$ for $y > 0$, and this tends to zero as $y \to \infty$.

Also,

$$\begin{aligned} A &= \int_0^{\epsilon}\left(\frac{e^{iuy}}{\sigma-1+iu} + \frac{e^{-iuy}}{\sigma-1-iu}\right)du \\ &= 2\int_0^{\epsilon}\frac{(\sigma-1)\cos uy}{(\sigma-1)^2+u^2}\,du + 2\int_0^{\epsilon}\frac{u\sin uy}{(\sigma-1)^2+u^2}\,du := 2A_1 + 2A_2. \end{aligned} \tag{10.28}$$

A further change of integration variable makes A_1 into

$$\int_0^{\epsilon/(\sigma-1)}\frac{\cos(\{\sigma-1\}vy)}{1+v^2}\,dv,$$

and hence, as $\sigma \to 1+$, it has a limit

$$\int_0^{\infty}\frac{dv}{1+v^2} = \frac{\pi}{2}$$

by the dominated convergence theorem. Then, by the bounded convergence theorem again, A_2 has a limit

$$\int_0^{\epsilon} \frac{\sin uy}{u}\, du = \int_0^{y\epsilon} \frac{\sin v}{v}\, dv = \frac{\pi}{2} + o(1)$$

as $\sigma \to 1+$. Therefore, $\lim_{\sigma\to 1+} I_2(\lambda, \sigma, y)$ exists for $\lambda > \lambda_1$ and $y > 0$ and

$$\lim_{\sigma\to 1+} I_2(\lambda, \sigma, y) = -\frac{e^{2i\lambda_1 y}}{2(1+2i\lambda_1)}\left(1 - \frac{\lambda_1}{\lambda}\right)(\pi + o(1)).$$

Similarly, to evaluate $\lim_{\sigma\to 1+} I_3(\lambda, \sigma, y)$, we split the integration interval into parts

$$[-2\lambda,\, -2\lambda_1 - \epsilon] \cup [-2\lambda_1 - \epsilon,\, -2\lambda_1 + \epsilon] \cup [-2\lambda_1 + \epsilon,\, 2\lambda].$$

We can show that

$$\lim_{\sigma\to 1+} I_3(\lambda, \sigma, y) = -\frac{e^{-2i\lambda_1 y}}{2(1-2i\lambda_1)}\left(1 - \frac{\lambda_1}{\lambda}\right)(\pi + o(1)).$$

Therefore, $\lim_{\sigma\to 1+} I(\lambda, \sigma, y)$ exists for all $\lambda > \lambda_1$ and $y > 0$ and

$$\lim_{\sigma\to 1+} I(\lambda, \sigma, y) = -\frac{\pi}{2}\left(1 - \frac{\lambda_1}{\lambda}\right)\frac{\sin(2\lambda_1 y + \theta)}{\sqrt{1+4\lambda_1^2}} + o_\lambda(1).$$

Finally, for $\lambda = \lambda_1$, to evaluate $\lim_{\sigma\to 1+} I_2(\lambda_1, \sigma, y)$, we split the integration interval $[-2\lambda_1, 2\lambda_1]$ into $[-2\lambda_1, 2\lambda_1 - \epsilon] \cup [2\lambda_1 - \epsilon, 2\lambda_1]$ with ϵ satisfying $0 < \epsilon < 2\lambda_1$. The integral over the first subinterval has a limit as $\sigma \to 1+$ for all $y \geq 0$, and this limit tends to zero as $y \to \infty$. By changing the integration variable, the integral over the rest of the interval becomes

$$\int_{-\epsilon}^{0} \frac{1}{2}\left(1 - \frac{2\lambda_1 + u}{2\lambda_1}\right)\frac{e^{2i\lambda_1 y + iuy}\, du}{\sigma - 1 + iu} = -\frac{e^{2i\lambda_1 y}}{4\lambda_1}\int_{-\epsilon}^{0} \frac{ue^{iuy}\, du}{\sigma - 1 + iu}.$$

By the bounded convergence theorem, the last integral has a limit

$$\int_{-\epsilon}^{0} \frac{e^{iuy}}{i}\, du = -\frac{1 - e^{-i\epsilon y}}{y}$$

as $\sigma \to 1+$, and this limit also tends to zero as $y \to \infty$. Hence $\lim_{\sigma\to 1+} I_2$ exists and is $o(1)$ as $y \to \infty$. Similarly, $\lim_{\sigma\to 1+} I_3(\lambda_1, \sigma, y)$ exists and is $o(1)$ as $y \to \infty$. Therefore, $\lim_{\sigma\to 1+} I(\lambda_1, \sigma, y)$ exists and is $o(1)$ as $y \to \infty$. This completes the proof of (10.26).

10.6. Notes

§10.2. The Wiener-Ikehara tauberian theorem is discussed in some detail in [**Ko04**], [**Ha49**], and [**Te95**].

The proof of the W-I theorem we give is based on one of S. Bochner (see [**Ch68**], Notes).

Converse forms of the W-I theorem, including upper and lower bound estimates, are given in [**Zh14**].

A form of W-I using the hypothesis (10.8) is given in [**BD69**]. See also the Notes for Chapter 13.

§10.5. The example of this section is given in [**Be37**] and [**BD69**].

CHAPTER 11

Chebyshev Bounds – Analytic Methods

Of course you can do better – Your mother

SUMMARY. Analytic methods are applied to establish upper and lower bounds for Chebyshev's function $\psi(x)$ under hypotheses weaker than those of Chapter 9. In the next chapter, we show the optimality of these results.

11.1. Introduction

Let $N(x) = Ax + xE(x)$ $(A > 0)$ denote the integer counting function of a g-number system $\mathcal{N}$. It was conjectured by the first author that Chebyshev bounds for $\psi(x)$ could be established under the assumption

$$\int_1^\infty x^{-1}|E(x)|\,dx < \infty. \tag{11.1}$$

Later, the second author established the Chebyshev bounds given in Theorem 9.9 using the stronger hypothesis

$$\int_1^\infty x^{-1} \sup_{y\geq x}|E(y)|\,dx < \infty.$$

The condition (11.1) was shown to be too weak by an example of J.-P. Kahane. Subsequently, Chebyshev bounds were established by J. Vindas and the authors under various auxiliary conditions on $E(x)$ (see Notes). Here we show the bounds under an average hypothesis on $E(x)$. In the next chapter, we shall show that these results are optimal in the class of systems $\mathcal{N}$ which satisfy (11.1) and a pointwise bound on $E(x)\log x$.

THEOREM 11.1. *Suppose that the counting function $N(x)$ of integers of a g-number system $\mathcal{N}$ satisfies both* (11.1) *and*

$$\int_1^x E(u)\log u\,du \ll x, \quad x \geq 1. \tag{11.2}$$

Then the Chebyshev bounds $x \ll \psi(x) \ll x$ hold.

Our proof is based on the application of W-I theorems to the Mellin transform $\widehat{\psi}(s)$. Key ingredients in the argument are estimates of L^1 norms of derivatives of the Fejér kernel given in Lemma 10.10 and the treatment of $I_3(y)$ (given in §11.4) by a concrete version of Wiener's theorem on division algebras.

11.2. Wiener-Ikehara setup

We begin with a few preliminaries. By (11.1) and Proposition 7.7, we have $N(x) \sim Ax$. The weak estimate

$$\psi(x) := \sum_{p^\nu \le x} \log p \le N(x) \log x \sim x \log x$$

follows, since every power p^ν of a g-prime p is a g-integer. This bound insures convergence of the Mellin transform

$$\widehat{\psi}(s) := \int_1^\infty x^{-s} d\psi(x) = s \int_1^\infty x^{-s-1} \psi(x)\, dx = -\frac{\zeta'(s)}{\zeta(s)}$$

in the half plane $\{s = \sigma + it, \sigma > 1\}$. Here, as usual, $\zeta(s)$ is the zeta function associated with the g-number system $\mathcal{N}$. We make a change of variables in the last integral to bring it to our Laplace form for the W-I theorems, writing

$$-\frac{\zeta'(s)}{s\,\zeta(s)} = \int_0^\infty e^{-su} \psi(e^u)\, du.$$

We claim that this transform has a right hand residue 1 at $s = 1$. Indeed, since $N(x) \sim Ax$, as $\sigma \to 1+$,

$$\zeta(\sigma) = \sigma \int_1^\infty x^{-\sigma-1} N(x)\, dx \sim \frac{A}{\sigma - 1},$$

$$\zeta'(\sigma) = \int_1^\infty x^{-\sigma-1} N(x)\, dx - \sigma \int_1^\infty x^{-\sigma-1} (\log x) N(x)\, dx \sim \frac{-A}{(\sigma-1)^2};$$

thus

$$\lim_{\sigma \to 1+} (\sigma - 1) \int_0^\infty e^{-\sigma u} \psi(e^u)\, du = 1,$$

and condition (10.1) is satisfied with $L = 1$.

Following (10.2), for $\sigma > 1$, write

$$\begin{aligned} G(s) &:= \int_0^\infty e^{-su} \psi(e^u)\, du - \frac{1}{s-1} = \int_0^\infty e^{-su} \{\psi(e^u) - e^u\}\, du \\ &= -\frac{\zeta'(s)}{s\,\zeta(s)} - \frac{1}{s-1} = -\frac{1}{s} - \frac{1}{s}\frac{d}{ds} \log\{(s-1)\zeta(s)\}. \end{aligned} \tag{11.3}$$

Let λ be a positive number, to be chosen later. (Spoiler alert: unlike the proof of the PNT, here λ will be bounded.) Following the W-I method of the preceding chapter, multiply through (11.3) by e^{ity} and the smoothing factor

$$\Delta_\lambda(t) := \frac{1}{2}\left(1 - \frac{|t|}{2\lambda}\right)^+.$$

Then integrate both sides of the resulting formula over $-2\lambda < t < 2\lambda$ and exchange the integration order (which is valid by absolute convergence). For $\sigma > 1$, we find

$$\begin{aligned} \int_{-2\lambda}^{2\lambda} \Delta_\lambda(t) e^{ity} G(\sigma + it)\, dt &= \int_0^\infty e^{-\sigma u} \{\psi(e^u) - e^u\}\, k_\lambda(y - u)\, du \\ &= -\int_0^\infty e^{-\sigma u} k_\lambda(y-u)\, du - I_\sigma(y), \end{aligned} \tag{11.4}$$

where

$$k_\lambda(x) := \int_{-2\lambda}^{2\lambda} \frac{1}{2}\left(1 - \frac{|t|}{2\lambda}\right) e^{itx}\, dt$$

is the Fejér kernel for $\mathbb{R}$ and (still with $\sigma > 1$),

$$I_\sigma(y) := \int_{-2\lambda}^{2\lambda} \Delta_\lambda(t) \frac{e^{ity}}{s} \frac{d}{ds} \log\{(s-1)\zeta(s)\}\, dt.$$

Now let $\sigma \to 1+$ in (11.4). Since $k_\lambda \geq 0$, each of the integrals

$$\int_0^\infty e^{-\sigma u}\psi(e^u) k_\lambda(y-u)\, du, \quad \int_0^\infty e^{-\sigma u} e^u k_\lambda(y-u)\, du, \quad \int_0^\infty e^{-\sigma u} k_\lambda(y-u)\, du$$

has a limit by the monotone convergence theorem. Also, since $\int_0^\infty k_\lambda(y-u)\, du < \pi$, by (10.12), the last two integrals have finite limits for each y. (It remains to be shown that the limit of the first integral also is finite.) Hence $I_\sigma(y)$ has a limit as well. It follows that

$$\lim_{\sigma \to 1+} \int_{-2\lambda}^{2\lambda} \Delta_\lambda(t) e^{ity} G(\sigma + it)\, dt = -\int_0^\infty e^{-u} k_\lambda(y-u)\, du - \lim_{\sigma \to 1+} I_\sigma(y).$$

The integral on the right side of the last formula can be rewritten as

$$\int_{-\infty}^{y/2} e^{v-y} k_\lambda(v)\, dv + \int_{y/2}^{y} e^{v-y} k_\lambda(v)\, dv < \pi e^{-y/2} + \int_{y/2}^{y} k_\lambda(v)\, dv \to 0$$

as $y \to \infty$. Thus we see that

$$\lim_{\sigma \to 1+} \int_{-2\lambda}^{2\lambda} \Delta_\lambda(t) e^{ity} G(\sigma + it)\, dt = o(1) - \lim_{\sigma \to 1+} I_\sigma(y) \quad \text{as} \quad y \to \infty. \tag{11.5}$$

Our proof of the Chebyshev bounds will depend on the size of $\lim_{\sigma \to 1+} I_\sigma(y)$*: if it is bounded, then* $\psi(x) \ll x$ *by Theorem* 10.1*; if it is sufficiently small, then* $\psi(x) \gg x$ *by Theorem* 10.4. The analysis of $I_\sigma(y)$ is more delicate than that of the classical W-I proof of the PNT, mainly because, here, $(d/ds)\log\{(s-1)\zeta(s)\}$ need not have a continuous extension to the closed half plane $\sigma \geq 1$.

11.3. A first decomposition

It is convenient to express $I_\sigma(y)$ in terms of a more tractable function. We begin by examining $(s-1)\zeta(s)$ near $s = 1$. By (11.1),

$$H := \int_1^\infty x^{-1}|E(x)|\, dx = \int_0^\infty |E(e^u)|\, du < \infty. \tag{11.6}$$

Also, we set

$$g(s) := \frac{1}{A}\int_1^\infty x^{-s} E(x)\, dx = \frac{\zeta(s)}{As} - \frac{1}{s-1}. \tag{11.7}$$

The integral for g converges uniformly for $\sigma \geq 1$, and defines a continuous function there. Since

$$(s-1)\zeta(s) = As + (s-1)s\int_1^\infty x^{-s}E(x)\, dx = A(s + (s-1)sg(s)), \tag{11.8}$$

$(s-1)\zeta(s)$ has a continuous extension to the half plane $\sigma \geq 1$, and

$$\frac{(s-1)\zeta(s)}{A} \to 1 \quad \text{as} \quad s \to 1, \ \sigma \geq 1.$$

Thus there exists a number $\eta_1 > 0$ such that

$$\left|\frac{(s-1)\zeta(s)}{A} - 1\right| \le \frac{1}{2} \quad \text{for } |s-1| \le \eta_1,\ \sigma \ge 1.$$

For s in this semidisc, we can write

$$\log\{(s-1)\zeta(s)\} = \log A + \log\{1 + (s-1)(1 + sg(s))\},$$

and, differentiating, obtain

$$\frac{d}{ds}\log\{(s-1)\zeta(s)\} = \frac{1 + (2s-1)g(s) + s(s-1)g'(s)}{1 + (s-1)(1 + sg(s))}. \tag{11.9}$$

It follows, for $0 < \lambda \le \eta_1/4$ and $1 < \sigma \le 1 + \eta_1/2$, that

$$\begin{aligned} I_\sigma(y) &= \int_{-2\lambda}^{2\lambda} \Delta_\lambda(t)\frac{e^{ity}}{s}\frac{1 + (2s-1)g(s)}{1 + (s-1)(1 + sg(s))}\,dt \\ &\quad + \int_{-2\lambda}^{2\lambda} \Delta_\lambda(t)\frac{e^{ity}(s-1)g'(s)}{1 + (s-1)(1 + sg(s))}\,dt \\ &:= I_{1,\sigma}(y) + I_{2,\sigma}(y). \end{aligned} \tag{11.10}$$

The integrand of $I_{1,\sigma}(y)$ is continuous on the closed semidisc, and so

$$\lim_{\sigma\to 1+} I_{1,\sigma}(y) = \int_{-2\lambda}^{2\lambda} \Delta_\lambda(t)\frac{e^{ity}}{1+it}\frac{1 + (1+2it)g(1+it)}{1 + it\{1 + (1+it)g(1+it)\}}\,dt$$

exists and is finite for each y. By the Riemann-Lesbegue lemma,

$$\lim_{\sigma\to 1+} I_{1,\sigma}(y) = o(1) \tag{11.11}$$

as $y \to \infty$. Since $I_\sigma(y)$ has a limit and $I_{1,\sigma}(y)$ has a finite limit as $\sigma \to 1+$, it follows that $\lim_{\sigma\to 1+} I_{2,\sigma}(y)$ exists for each y. It remains to study the last limit.

11.4. Further decomposition of $I_{2,\sigma}(y)$

It is a nuisance to deal with many factors in the integrand of $I_{2,\sigma}(y)$ moving at once as $\sigma \to 1+$, so we write

$$\frac{s-1}{1 + (s-1)\{1 + sg(s)\}} = \frac{it}{(1+it)\{1 + itg(s)\}} + (\sigma - 1)R(s), \tag{11.12}$$

where we have (after a bit of algebra)

$$R(s) := \frac{1 - it(s-1)g(s)}{s(1+it)\,(1 + itg(s))\,(1 + (s-1)g(s))}.$$

The definition of $g(s)$ and (11.6) give $|g(s)| \le H/A$ and hence

$$|itg(s)| \le 1/2 \quad \text{for } |s-1| \le \eta_2,\ \sigma \ge 1$$

for some constant $\eta_2 > 0$. If we further require that $0 < \eta_2 \le \min\{\eta_1, 1\}$, then $R(s)$ is well-defined in the semidisc

$$D := \{s : \sigma \ge 1,\ |s-1| \le \eta_2\}.$$

In this section, we shall approximate $I_{2,\sigma}(y)$ using (11.12) and applying a geometrical series device of Wiener to treat $1/\{1 + itg(s)\}$.

Inserting (11.12) into the integrand of $I_{2,\sigma}(y)$ yields, for $0 < \lambda \leq \eta_2/4$ and $1 < \sigma \leq 1 + \eta_2/2$,

$$
\begin{aligned}
I_{2,\sigma}(y) &= \int_{-2\lambda}^{2\lambda} \Delta_\lambda(t) \frac{e^{ity} g'(s)\, it}{(1+it)(1+itg(s))} dt \\
&\quad + \int_{-2\lambda}^{2\lambda} \Delta_\lambda(t) e^{ity} (\sigma - 1) g'(s) R(s)\, dt \qquad (11.13)\\
&:= I_{3,\sigma}(y) + I_{4,\sigma}(y).
\end{aligned}
$$

The function $R(s)$ is continuous on the compact semidisc D and bounded there by some constant R, say. It follows that

$$
|I_{4,\sigma}(y)| \leq R \int_{-2\lambda}^{2\lambda} \Delta_\lambda(t)(\sigma - 1)|g'(s)|\, dt, \quad y \in \mathbb{R}.
$$

Recalling the definition of $g(s)$, we have

$$
(\sigma - 1)|g'(s)| \leq \frac{1}{A} \int_1^\infty x^{-\sigma}(\sigma - 1)|E(x)| \log x\, dx.
$$

Note that

$$
x^{-\sigma}(\sigma - 1)|E(x)| \log x = \frac{(\sigma - 1)\log x}{\exp\{(\sigma - 1)\log x\}} x^{-1}|E(x)| \leq x^{-1}|E(x)|,
$$

which is integrable by the hypothesis (11.1), and

$$
x^{-(\sigma-1)}(\sigma - 1)\log x \to 0
$$

at each point $x \geq 1$ as $\sigma \to 1+$. By the dominated convergence theorem, the last integral tends to 0 as $\sigma \to 1+$. Therefore

$$
(\sigma - 1)|g'(s)| \to 0 \quad \text{as} \quad \sigma \to 1+
$$

uniformly for all $t \in \mathbb{R}$, and so

$$
\lim_{\sigma \to 1+} I_{4,\sigma}(y) = 0. \tag{11.14}
$$

It remains to study

$$
I_3(y) := \lim_{\sigma \to 1+} I_{3,\sigma}(y)\,;
$$

this limit exists since $I_{2,\sigma}(y)$ and $I_{4,\sigma}(y)$ have limits as $\sigma \to 1+$. Since $|itg(s)| \leq 1/2$ for $s \in D$, we can write

$$
\frac{it}{(1+it)(1+itg(s))} = \frac{it}{1+it} \sum_{\nu \geq 0} (-1)^\nu (it)^\nu g(s)^\nu.
$$

Thus

$$
I_{3,\sigma}(y) = \sum_{\nu \geq 0} (-1)^\nu J_{\nu,\sigma}(y),
$$

where

$$
J_{\nu,\sigma}(y) = \int_{-2\lambda}^{2\lambda} \Delta_\lambda(t) e^{ity} g'(s) \frac{(it)^{\nu+1} g(s)^\nu}{1+it}\, dt. \tag{11.15}
$$

We shall evaluate $J_{\nu,\sigma}$ by expressing it as an additive convolution of L^1 functions.

We have

$$
k_\lambda^{(\nu+1)}(y) = \int_{-2\lambda}^{2\lambda} \Delta_\lambda(t)(it)^{\nu+1} e^{ity}\, dt,
$$

the $\nu + 1$st derivative of the Fejér kernel, and for $\sigma \geq 1$, define

$$G_\sigma(u) := A^{-1} e^{-(\sigma-1)u} E(e^u), \quad G_\sigma^D(u) := -uG_\sigma(u), \quad Z(u) := e^{-u},$$

the last three functions with support in $[0, \infty)$. We commit a small abuse of notation by writing here the *Fourier* transforms of the functions as

$$\widehat{G_\sigma}(t) = \frac{1}{A} \int_0^\infty e^{-itu} e^{-(\sigma-1)u} E(e^u)\, du = g(s), \quad \sigma \geq 1,$$

$$\widehat{G_\sigma^D}(t) = -\frac{1}{A} \int_0^\infty e^{-itu} e^{-(\sigma-1)u} uE(e^u)\, du = g'(s), \quad \sigma > 1,$$

and

$$\widehat{Z}(t) = \int_0^\infty e^{-itu} e^{-u}\, du = \frac{1}{1+it}.$$

Below and in the next section, we make another exception: *we write '$*$' for additive convolution.* By direct calculation or familiar Fourier relations, we obtain the representation

$$J_{\nu,\sigma}(y) = (k_\lambda^{(\nu+1)} * G_\sigma^D * G_\sigma^{*\nu} * Z)(y), \quad \sigma > 1. \tag{11.16}$$

11.5. Chebyshev bounds

Using the last formula and the bounds (10.16) for L^1-norms of $k_\lambda^{(\nu)}(x)$, we shall derive the inequality (11.19) (below) for $I_3(y)$. This in turn, will yield an upper bound for $\psi(x)/x$. Afterward, with a further restriction on λ, we sharpen the estimate of $I_3(y)$ to establish a lower bound. Here, we introduce for the first time the hypothesis (11.2): there is a number $B > 0$ such that

$$\left| x^{-1} \int_1^x E(u) \log u\, du \right| \leq B, \quad 1 \leq x < \infty. \tag{11.17}$$

Our first task is to study $J_{\nu,\sigma}(y)$ ($\nu = 0, 1, 2, \ldots$) and estimate limits as $\sigma \to 1+$. To overcome convergence problems of (11.16) at $\sigma = 1+$, we combine G_σ^D with Z: for $\sigma \geq 1$ let

$$m_\sigma(u) := -\frac{1}{A} \int_0^u e^{-(u-v)} e^{-(\sigma-1)v} vE(e^v)\, dv.$$

Then $m_\sigma(u) = (Z * G_\sigma^D)(u)$, and, since $e^{-(\sigma-1)v}$ is decreasing in v, we find from (11.17) by integration by parts that

$$|m_\sigma(u)| \leq B/A \ \text{ for } \sigma \geq 1,\ u \geq 0. \tag{11.18}$$

By the dominated convergence theorem, $m_1(u) = \lim_{\sigma \to 1+} m_\sigma(u)$. Also, by (11.6), $\int_0^\infty |G_1(u)|\, du = H/A$. Induction on ν shows that $\int_0^\infty |G_1|^{*\nu}(u)\, du = (H/A)^\nu$; hence $\int_0^\infty |G_1^{*\nu}(u)|\, du \leq (H/A)^\nu$. It follows that

$$(G_\sigma^{*\nu} * Z * G_\sigma^D)(u) = (G_\sigma^{*\nu} * m_\sigma)(u) \to (G_1^{*\nu} * m_1)(u)$$

as $\sigma \to 1+$, again by the dominated convergence theorem, and, with (11.18),

$$|(G_\sigma^{*\nu} * Z * G_\sigma^D)(u)|,\ |(G_1^{*\nu} * m_1)(u)| \leq \frac{BH^\nu}{A^{\nu+1}}.$$

Combining the preceding bounds for $(G_\sigma^{*\nu} * Z * G_\sigma^D)(u)$ with (10.16), we see that

$$J_{\nu,\sigma}(y) = \int_{-\infty}^y (G_\sigma^D * G_\sigma^{*\nu} * Z)(y-t)\, k_\lambda^{(\nu+1)}(t)\, dt$$

is absolutely convergent. By one last application of the dominated convergence theorem, we conclude that $J_\nu(y) = \lim_{\sigma\to 1+} J_{\nu,\sigma}(y)$ exists. Moreover, with (10.16) again, we have

$$|J_\nu(y)| \le \frac{4B}{H}\left(\frac{2\lambda H}{A}\right)^{\nu+1} \quad \text{for} \quad \nu \ge 0,$$

where $0 < \lambda \le \eta_2/4$.

It now follows that

$$I_3(y) = \lim_{\sigma\to 1+} \sum_{\nu\ge 0} (-1)^\nu J_{\nu,\sigma}(y) = \sum_{\nu\ge 0} (-1)^\nu J_\nu(y)$$

holds for $0 < \lambda \le \min\{\eta_2,\ A/H\}/4$. Furthermore, under this λ restriction,

$$|I_3(y)| < \frac{8B\lambda/A}{1 - 2\lambda H/A} \le \frac{16B\lambda}{A} \tag{11.19}$$

uniformly for all y.

Combining (11.5) with the several I_ν estimates yields

$$\lim_{\sigma\to 1+} \int_{-2\lambda}^{2\lambda} \Delta_\lambda(t) e^{ity} G(\sigma + it)\, dt = o(1) - I_3(y) \quad \text{as} \quad y \to \infty \tag{11.20}$$

(equivalently,

$$\int_0^\infty e^{-u}\psi(e^u) k_\lambda(y-u)\, du = \pi - I_3(y) + o(1)).$$

Since $|I_3(y)|$ is bounded, we have an upper Chebyshev bound by the first part of Theorem 10.1.

Finally, we deduce a lower bound for $\psi(x)/x$ from (11.20) via the second part of Theorem 10.4. For this, we need $\limsup_{y\to\infty} |I_3(y)| < \pi$. Now (11.19) shows $I_3(y) \ll \lambda$, so by choosing λ sufficiently small, e.g.

$$0 < \lambda \le \min\{\eta_2/4,\ A/(4H),\ \pi A/(32B)\},$$

we find a number $c > 0$ such that

$$\liminf_{\sigma\to 1+} \int_{-2\lambda}^{2\lambda} \Delta_\lambda(t) e^{ity} G(\sigma + it)\, dt \ge (-1 + c)\pi$$

for all sufficiently large y (equivalently

$$\int_0^\infty e^{-u}\psi(e^u) k_\lambda(y-u)\, du \ge c, \quad y \to \infty).$$

From the preceding inequality and the fact that $\psi(x)/x$ is bounded above, we deduce the desired lower bound.

11.6. Notes

§11.1 Zhang's article appears in [**Zh87a**]; Kahane's in [**Ka98**]; Vindas established a Chebyshev upper bound assuming (11.1) and the additional hypothesis $E(x) = o(1/\log x)$ in [**Vn12**]. The authors [**DZ13a**] and Vindas [**Vn13**] independently established upper and lower bounds with the auxiliary hypothesis $E(x) = O(1/\log x)$, and the same people further weakened the $E(x)$ hypothesis to the average condition given in this chapter.

CHAPTER 12

Optimality of a Chebyshev Bound

Too much of a good thing can be wonderful – Mae West

SUMMARY. An example is given of a g-number system satisfying the integral condition of Theorem 11.1 but having $\{N(x)-Ax\}(\log x)/x$ tend to infinity, possibly arbitrarily slowly, for which the Chebyshev O-bounds fail.

12.1. Introduction

In the last chapter, we proved that if the counting function of integers of a Beurling g-number system satisfies the conditions $N(x) = Ax + xE(x)$,

$$\int_1^\infty x^{-1}|E(x)|\,dx < \infty, \tag{11.1 bis}$$

and an average of the pointwise bound

$$E(x) = O(1/\log x), \tag{12.1}$$

then the associated g-prime system satisfies Chebyshev bounds. In this chapter we show the optimality of (12.1) in the class of g-number systems satisfying (11.1) by giving an example in which both (12.1) and the Chebyshev bounds fail.

THEOREM 12.1. *Let $f(x)$ be any positive-valued, increasing unbounded function on $[1, \infty)$. Then there exists a g-number system $\mathcal{N}_B$ such that*

(1) *The associated zeta function $\zeta_B(s)$ is analytic on the open half plane $\{\sigma > 1\}$, $(s-1)\zeta_B(s)$ has a continuous extension to $\{\sigma \geq 1\}$, and $it\zeta_B(1+it) \neq 0$ for all real t;*

(2) *The g-integer counting function $N_B(x)$ satisfies* (11.1) *and also*

$$E(x) = O(f(x)/\log x); \tag{12.2}$$

(3) *The counting function $\pi_B(x)$ of the g-primes satisfies*

$$\limsup_{x\to\infty} \frac{\pi_B(x)}{x/\log x} = \infty \quad \text{and} \quad \liminf_{x\to\infty} \frac{\pi_B(x)}{x/\log x} = 0.$$

In other words, if (12.1) is replaced by (12.2) for an unbounded function f, no matter how slowly growing, then there exists a g-number system also satisfying (11.1) for which the Chebyshev bounds fail.

12.2. The g-prime system $\mathcal{P}_B$

The proof proceeds by construction of an example. We start by creating from f another function which grows at least as slowly as f, but with some more useful analytical properties.

LEMMA 12.2. *Given a positive-valued, increasing unbounded function $f(x)$ on $[1, \infty)$, there exists a function $k(x)$ also defined on $[1, \infty)$ such that*

(1) $k(x) \geq 1$ *for* $x \geq 1$ *and* $k(x) \ll f(x)$;
(2) $k(x)$ *is increasing and* $k(x) \to \infty$ *as* $x \to \infty$;
(3) $k(x)$ *is differentiable and* $(\log x)/k(x)$ *is increasing on* $(1, \infty)$.

PROOF. First, set

$$f_1(x) := \min\{f(x), \log\log(e^e x)\}, \quad x \geq 1.$$

Thus $0 < f_1(x) \leq f(x)$ for $x \geq 1$. Moreover, $f_1(x)$ is increasing and $f_1(x) \to \infty$ as $x \to \infty$. Next, set

$$f_2(x) := x^{-1} \int_1^x f_1(t)\, dt, \quad x \geq 1.$$

We have, for $x \geq 1$,

$$0 \leq f_2(x) \leq \frac{x-1}{x} f_1(x) < f_1(x).$$

Also, $f_2(x)$ is continuous ($xf_2(x)$ is an integral!) and it increases, since for $\delta > 0$,

$$\begin{aligned} f_2(x+\delta) &\geq \frac{1}{x+\delta}\left(\int_1^x f_1(t)\, dt + f_1(x)\,\delta\right) \\ &\geq \frac{1}{x+\delta}\left(\int_1^x f_1(t)\, dt + \frac{\delta}{x}\int_1^x f_1(t)\, dt\right) = f_2(x). \end{aligned}$$

Moreover, $f_2(x) \to \infty$ as $x \to \infty$, for

$$f_2(x) > \frac{1}{x}\int_{x/2}^x f_1(t)\, dt \geq \frac{1}{2} f_1(x/2) \to \infty.$$

Then set

$$f_3(x) := 1 + x^{-1}\int_1^x f_2(t)\, dt.$$

Now,

$$1 \leq f_3(x) \leq 1 + f_2(x) \leq 1 + f_1(x) \leq 1 + f(x), \quad x \geq 1.$$

Also, as before, $f_3(x)$ is increasing and $f_3(x) \to \infty$ as $x \to \infty$. Moreover, $f_3(x)$ is differentiable on $(1, \infty)$, since $f_2(x)$ is continuous there.

Finally, we define

$$k(x) = f_3(\log\log(e^e x)), \quad x \geq 1.$$

Thus we have

$$1 \leq k(x) \leq 1 + f(\log\log(e^e x)), \quad x \geq 1,$$

and, from the definition of $f_1(x)$,

$$k(x) \ll \log\log\log\log x.$$

Also, $k(x)$ is increasing and $k(x) \to \infty$ as $x \to \infty$. Moreover,

$$\left(\frac{\log x}{k(x)}\right)' = \frac{1}{xk(x)}\left(1 - \frac{f_3'(\log\log(e^e x))}{f_3(\log\log(e^e x))}\frac{\log x}{\log(e^e x)}\right).$$

Note that $f_3(x) \geq 1$ and that

$$0 \leq f_3'(x) = \frac{f_2(x)}{x} - \frac{\int_1^x f_2(t)\, dt}{x^2} < \frac{f_2(x)}{x} < \frac{\log\log(e^e x)}{x}$$

for $x > 1$. Therefore, for $x > 1$,

$$\frac{f_3'(\log\log(e^e x))}{f_3(\log\log(e^e x))}\,\frac{\log x}{\log(e^e x)} < \frac{\log\log\{e^e \log\log(e^e x)\}}{\log\log(e^e x)} < 1,$$

and hence

$$\left(\frac{\log x}{k(x)}\right)' \geq \frac{1}{xk(x)}\left(1 - \frac{\log\log\{e^e \log\log(e^e x)\}}{\log\log(e^e x)}\right) > 0. \qquad \square$$

We next determine a sparse sequence of real numbers to use in the construction. First, since $k(x)$ is increasing and $k(x) \to \infty$ as $x \to \infty$, there exists a sequence $\{c_n\}$, $n = 1, 2, \ldots$, such that

$$\sum_{n\geq 1} 1/\sqrt{k(c_n)} < \infty.$$

Next, define another sequence $\{A_n\}$, $n = 1, 2, \ldots$, recursively by $A_1 = e$ and $A_{n+1} = \max\{e^{A_n}, c_{n+1}\}$. Note that the sequence $\{\log A_n\}$ grows faster than exponentially. Hence we have

$$\sum_{n\geq 1} \frac{\log k(n)}{k(A_n)} < \infty \tag{12.3}$$

since $k(x)$ is increasing and

$$\sqrt{k(A_n)} \geq \frac{1}{2}\log k(A_n) \geq \frac{1}{2}\log k(n).$$

We now construct the g-primes $\mathcal{P}_B$ of the theorem. Let n_0 be a positive integer large enough to satisfy each of the several conditions below. Also, here and in the rest of the proof, p denotes a *rational* prime, $\mathcal{P}$ the set of all such, and $\pi(x)$ the counting function of the rational primes. We take

$$\mathcal{P}_B = \Big(\mathcal{P} \setminus \bigcup_{n\geq n_0} \{p \in (A_n, \sqrt{k(n)}A_n]\}\Big)$$
$$\bigcup_{n\geq n_0} \{A_n \text{ with multiplicity } \lfloor A_n \log k(n)/(2\log A_n)\rfloor\}.$$

In words, $\mathcal{P}_B$ consists of an initial string of rational primes, then g-primes appear at A_{n_0} with high multiplicity (a "pulse"), followed by a suitably long interval having no g-primes (a "dead zone"), after which comes a very much longer interval of rational primes, then more g-primes at A_{n_0+1}, and the cycle repeats without end.

Note that the intervals $(A_n, \sqrt{k(n)}A_n]$, $n \geq n_0$ are pairwise non-overlapping, since $k(n) \ll \log\log(e^e n)$ and $A_{n+1} \geq (n+1)A_n$. Also, we shall see that the multiplicity of the primes at each A_n has been balanced with the length of the dead zone that follows to achieve a positive density of the g-integers. To condense the notation, we generally write A_n^* in place of $\sqrt{k(n)}A_n$.

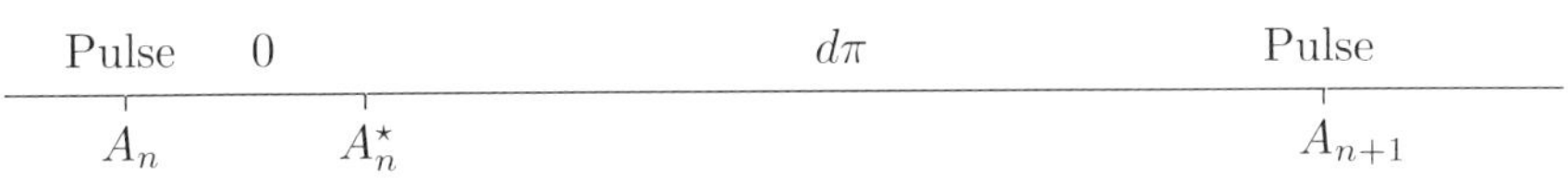

Figure 1. $d\pi_B$ on one interval

We shall show that the resulting g-integer system $\mathcal{N}_B$ satisfies the conditions of the theorem, beginning with the failure of the Chebyshev bounds.

12.3. Chebyshev bounds and the zeta function $\zeta_B(s)$

LEMMA 12.3. *Property (3) of the theorem is satisfied.*

PROOF. First, $\pi_B(A_n)$ is too large. Indeed, as $n \to \infty$,

$$\frac{\pi_B(A_n)}{A_n/\log A_n} \geq \frac{\lfloor A_n \log k(n)/(2\log A_n)\rfloor}{A_n/\log A_n} \to \infty.$$

Next, $\pi_B(A_n^*)$ is too small, where $A_n^* = \sqrt{k(n)}A_n$ lies at the end of a dead zone. We first show, by induction, that $\pi_B(A_n-) \leq \pi(A_n-)$. Indeed, this is trivially true (with equality) for $n = n_0$. Note that the number of rational primes inhabiting a dead zone $(A_n, A_n^*]$ is

$$\pi(\sqrt{k(n)}A_n) - \pi(A_n) \sim \frac{A_n\sqrt{k(n)}}{\log A_n + (1/2)\log k(n)} > \frac{A_n \log k(n)}{2\log A_n}$$

for $n \geq n_0$. Hence, from the definition of $\mathcal{P}_B$,

$$\begin{aligned}\pi_B(A_{n+1}-) &= (\pi(A_{n+1}-) - \pi(\sqrt{k(n)}A_n)) + (\pi_B(A_n) - \pi_B(A_n-)) + \pi_B(A_n-)\\ &\leq (\pi(A_{n+1}-) - \pi(\sqrt{k(n)}A_n)) + \frac{A_n\log k(n)}{2\log A_n} + \pi_B(A_n-)\\ &< \pi(A_{n+1}-).\end{aligned}$$

Thus $\pi_B(A_n-) \leq \pi(A_n-)$ holds for all (sufficiently large) n, and so

$$\begin{aligned}\frac{\pi_B(A_n^*)}{A_n^*/\log A_n^*} = \frac{\pi_B(A_n)}{A_n^*/\log A_n^*} &\leq \frac{A_n\log k(n)/(2\log A_n) + \pi(A_n)}{\sqrt{k(n)}A_n/\log(\sqrt{k(n)}A_n)}\\ &\leq (\log k(n))/\sqrt{k(n)} \to 0 \text{ as } n \to \infty.\end{aligned}$$

□

For further analysis, we introduce an auxiliary g-number system. Set

$$d\pi_0 := d(\pi_B - \pi)_v,$$

the variation of $d(\pi_B - \pi)$. Notice that

$$d\pi_0 = 0 \text{ on } [1, A_{n_0}) \cup \bigcup_{n\geq n_0}(A_n^*, A_{n+1}),$$

and for $n \geq n_0$,

$$\begin{aligned}d\pi_0\{A_n^*\} &= d\pi\{A_n^*\}\\ d\pi_0\{A_n\} &= \lfloor A_n \log k(n)/(2\log A_n)\rfloor,\end{aligned}$$

the last regardless of whether or not A_n is a rational prime.

By analogy with relations from Chapter 3, set

$$d\Pi_0(x) := \sum_{k\geq 1}\frac{1}{k}\, d\pi_0(x^{1/k})$$

and

$$N_0(x) := 1 + \sum_{n\geq 1}\frac{1}{n!}\int_1^x d\Pi_0^{*n}, \quad x \geq 1,$$

where $d\Pi_0^{*n}$ denotes the n-fold multiplicative convolution of $d\Pi_0$ with itself.

Also, we need an estimate of a sum of reciprocals of rational primes.

LEMMA 12.4.

$$\sum_{A_m<p\le\sqrt{k(m)}A_m} p^{-1} = \frac{\log k(m)}{2\log A_m} - \frac{\log^2 k(m)}{8\log^2 A_m} + O\Big(\frac{\log k(m)}{\log^2 A_m}\Big). \tag{12.4}$$

PROOF. In integral form, the left hand side equals

$$\int_{A_m}^{A_m^*} t^{-1}d\pi(t) = \int_{A_m}^{A_m^*} \frac{dt}{t\log t} + \int_{A_m}^{A_m^*} \frac{1}{t}\Big(d\pi(t) - \frac{dt}{\log t}\Big) := I_1 + I_2,$$

say. We have

$$\begin{aligned} I_1 &= \log\left(1 + \frac{\log k(m)}{2\log A_m}\right) \\ &= \frac{\log k(m)}{2\log A_m} - \frac{\log^2 k(m)}{8\log^2 A_m} + O\left(\frac{\log^3 k(m)}{\log^3 A_m}\right). \end{aligned}$$

For I_2, recall the classical PNT error bound

$$R(x) := \int_2^x \Big(d\pi(t) - \frac{dt}{\log t}\Big) \ll \frac{x}{\log^2 x}. \tag{12.5}$$

Using integration by parts, we find

$$\begin{aligned} I_2 &= \frac{R(\sqrt{k(m)}A_m)}{\sqrt{k(m)}A_m} - \frac{R(A_m)}{A_m} + \int_{A_m}^{\sqrt{k(m)}A_m} \frac{R(t)}{t^2}dt \\ &\ll \frac{1}{\log^2 A_m} + \frac{1}{\log^2 A_m}\int_{A_m}^{\sqrt{k(m)}A_m} \frac{dt}{t} = O\Big(\frac{\log k(m)}{\log^2 A_m}\Big). \end{aligned}$$

Hence (12.4) follows. □

We next show that the contribution of the auxiliary system is limited.

LEMMA 12.5.

$$\int_1^\infty x^{-1}\Big(1 + \frac{\log x}{k(x)}\Big)\, d\Pi_0(x) < \infty.$$

PROOF. We have $1 \ll (\log x)/k(x)$ for $x \ge 2$, since $(\log x)/k(x)$ is increasing. Thus it suffices to show

$$\int_1^\infty x^{-1}\frac{\log x}{k(x)}\, d\Pi_0(x) < \infty. \tag{12.6}$$

We begin by noting that

$$\begin{aligned} &\int_1^\infty x^{-1}\frac{\log x}{k(x)}\, d\pi_0(x) \\ &\quad = \sum_{n\ge n_0}\left(A_n^{-1}\frac{\log A_n}{k(A_n)}\left\lfloor\frac{A_n\log k(n)}{2\log A_n}\right\rfloor + \sum_{A_n<p\le A_n^*}\frac{\log p}{p\,k(p)}\right). \end{aligned}$$

Then, by the monotonicity of $(\log x)/k(x)$ and the last lemma,

$$\sum_{A_n<p\le A_n^*}\frac{\log p}{p\,k(p)} \le \frac{\log(A_n^*)}{k(A_n^*)}\sum_{A_n<p\le A_n^*} p^{-1} \ll \frac{\log(\sqrt{k(n)}A_n)}{k(A_n)}\frac{\log k(n)}{\log A_n}.$$

Next, $\log\left(\sqrt{k(n)}A_n\right) \ll \log A_n$. It follows that

$$\int_1^\infty x^{-1}\frac{\log x}{k(x)}\,d\pi_0(x) \ll \sum_{n\geq n_0}\frac{\log k(n)}{k(A_n)} < \infty,$$

the last by (12.3). Finally, the left hand side of (12.6) equals

$$\sum_{\ell\geq 1}\frac{1}{\ell}\int_1^\infty x^{-1}\frac{\log x}{k(x)}\,d\pi_0(x^{1/\ell}) = \sum_{\ell\geq 1}\frac{1}{\ell}\int_1^\infty u^{-\ell}\frac{\ell\log u}{k(u^\ell)}\,d\pi_0(u)$$

$$\leq \frac{1}{1-A_{n_0}^{-1}}\int_1^\infty u^{-1}\frac{\log u}{k(u)}\,d\pi_0(u) < \infty. \qquad \square$$

As usual, define the zeta function for $\mathcal{N}_B$ by $\zeta_B(s) := \int_{1-}^\infty u^{-s}dN_B(u)$; we now show that this function satisfies Property (1) of Theorem 12.1.

LEMMA 12.6. *$\zeta_B(s)$ is analytic on the open half plane $\{\sigma > 1\}$ and $(s-1)\zeta_B(s)$ has a continuous extension to $\{\sigma \geq 1\}$. Moreover, $it\zeta_B(1+it) \neq 0$ for all real t.*

PROOF. We have, by formulas from Chapter 3,

$$\zeta_B(s) = \exp\left\{\int_1^\infty x^{-s}d\Pi_B(x)\right\} = \zeta(s)\exp\left\{\int_1^\infty x^{-s}d(\Pi_B - \Pi)(x)\right\},$$

where $\zeta(s)$ is the Riemann zeta function, $\pi(x)$ the counting function of the rational primes, and $\Pi(x) = \sum_{\ell\geq 1}\ell^{-1}\pi(x^{1/\ell})$. Note that

$$d(\Pi_B - \Pi)_v(x) = \left(\sum_{\ell\geq 1}\frac{1}{\ell}\,d(\pi_B - \pi)(x^{1/\ell})\right)_v \leq \sum_{\ell\geq 1}\frac{1}{\ell}\,d(\pi_B-\pi)_v(x^{1/\ell}) = d\Pi_0(x),$$

by the triangle inequality. Also, $(\log x)/k(x) \gg 1$ for $x \geq 2$, as we noted above. Therefore, by Lemma 12.5, the last integral converges absolutely for $\sigma \geq 1$. Hence $\zeta_B(s)$ is analytic on $\{\sigma > 1\}$ and, by familiar properties of the Riemann zeta function,

$$(s-1)\zeta_B(s) = (s-1)\zeta(s)\exp\left\{\int_1^\infty x^{-s}d(\Pi_B - \Pi)(x)\right\}$$

has a continuous extension to $\{\sigma \geq 1\}$ and, for all real t,

$$it\zeta_B(1+it) = it\zeta(1+it)\exp\left\{\int_1^\infty x^{-(1+it)}d(\Pi_B - \Pi)(x)\right\} \neq 0. \qquad \square$$

12.4. The counting function $N_B(x)$

To conclude the proof of Theorem 12.1, we establish Property (2) via estimates of $N_B(x)$. We begin by showing that the auxiliary function $N_0(x)$ is suitably small.

LEMMA 12.7.

$$\int_1^\infty x^{-1}\frac{\log x}{k(x)}\,dN_0(x)$$

converges.

PROOF. Recall that $k(x)$ is increasing. Hence we have

$$1 + \frac{\log(x_1x_2\cdots x_n)}{k(x_1x_2\cdots x_n)} \leq \left(1 + \frac{\log x_1}{k(x_1)}\right)\cdots\left(1 + \frac{\log x_n}{k(x_n)}\right)$$

for $x_i \geq A_{n_0}$, $i = 1, 2, \ldots, n$. Then

$$\begin{aligned}\int_1^\infty x^{-1}\Big(1 + \frac{\log x}{k(x)}\Big)\, d\Pi_0^{*n}(x) &\leq \int_1^\infty x^{-1}\Big\{\Big(1 + \frac{\log x}{k(x)}\Big)\, d\Pi_0(x)\Big\}^{*n} \\ &= \int_1^\infty \Big\{x^{-1}\Big(1 + \frac{\log x}{k(x)}\Big)\, d\Pi_0(x)\Big\}^{*n} \\ &= \Big\{\int_1^\infty x^{-1}\Big(1 + \frac{\log x}{k(x)}\Big)\, d\Pi_0(x)\Big\}^n .\end{aligned}$$

Therefore, by Lemma 12.5,

$$\int_1^\infty x^{-1}\left(1 + \frac{\log x}{k(x)}\right) dN_0(x) \leq \exp\Big\{\int_1^\infty x^{-1}\Big(1 + \frac{\log x}{k(x)}\Big)\, d\Pi_0(x)\Big\} < \infty. \qquad \square$$

By the fundamental relation between dN and $d\Pi$ (resp. dN_B and $d\Pi_B$) and the homomorphic property of the exponential mapping, we have

$$dN_B = \exp\{d\Pi_B\} = \exp\{d\Pi + d(\Pi_B - \Pi)\} = dN * \exp\{d(\Pi_B - \Pi)\}.$$

Hence, the counting function $N_B(x)$ of g-integers has the representation

$$\begin{aligned}N_B(x) &= \int_{1-}^x N\left(\frac{x}{t}\right) \exp\{d(\Pi_B - \Pi)\}(t) \\ &= N(x) + \int_1^x N\left(\frac{x}{t}\right) \sum_{n\geq 1} \frac{1}{n!}\, d(\Pi_B - \Pi)^{*n}(t) \\ &= x + \theta(x) + x \sum_{n\geq 1} \frac{1}{n!} \int_1^x t^{-1} d(\Pi_B - \Pi)^{*n}(t) \\ &\quad + \sum_{n\geq 1} \frac{1}{n!} \int_1^x \theta\left(\frac{x}{t}\right) d(\Pi_B - \Pi)^{*n}(t),\end{aligned} \tag{12.7}$$

where $N(x)$ is the counting function of rational integers and $\theta(x) = N(x) - x$.

Let

$$c_1 := \int_1^\infty t^{-1} d(\Pi_B - \Pi)(t);$$

the integral is absolutely convergent by Lemma 12.5. Also, as we saw in the proof of Lemma 12.7,

$$\int_1^\infty t^{-1} d(\Pi_B - \Pi)^{*n}(t)$$

is absolutely convergent, and it equals

$$\left(\int_1^\infty t^{-1} d(\Pi_B - \Pi)(t)\right)^n = c_1^n.$$

Therefore, by adding and subtracting terms $\int_x^\infty t^{-1}\, d(\Pi_B - \Pi)^{*n}(t)$, we can write (12.7) as

$$N_B(x) = Ax + xE(x), \tag{12.8}$$

where

$$A = 1 + \sum_{n\geq 1} \frac{1}{n!} \int_1^\infty t^{-1}\, d(\Pi_B - \Pi)^{*n}(t) = e^{c_1}$$

and

$$E(x) := x^{-1}\theta(x) - E_1(x) + E_2(x) \tag{12.9}$$

with

$$E_1(x) := \sum_{n\geq 1} \frac{1}{n!} \int_x^\infty t^{-1} d(\Pi_B - \Pi)^{*n}(t) \tag{12.10}$$

and

$$E_2(x) := x^{-1} \sum_{n\geq 1} \frac{1}{n!} \int_1^x \theta\left(\frac{x}{t}\right) d(\Pi_B - \Pi)^{*n}(t).$$

Also, let

$$\zeta_0(s) := \int_{1-}^\infty x^{-s} dN_0(x).$$

The integral converges absolutely for $\sigma \geq 1$ by Lemma 12.7, so $\zeta_0(s)$ is analytic on $\{\sigma > 1\}$ and continuous on $\{\sigma \geq 1\}$.

LEMMA 12.8.

$$N_0(x)/x \ll \frac{k(x)}{\log x}, \quad x > 1, \tag{12.11}$$

and hence

$$|E_2(x)| \leq \frac{N_0(x)}{x} \ll \frac{k(x)}{\log x}. \tag{12.12}$$

Also,

$$\int_1^\infty x^{-1} |E_2(x)| \, dx < \infty. \tag{12.13}$$

PROOF. By Lemma 12.7,

$$\int_{A_{n_0}}^x y^{-1} \frac{\log y}{k(y)} \, dN_0(y) < \int_1^\infty y^{-1} \frac{\log y}{k(y)} \, dN_0(y) < \infty.$$

The left hand side equals, by integration by parts,

$$x^{-1} \frac{\log x}{k(x)} N_0(x) - A_{n_0}^{-1} \frac{\log A_{n_0}}{k(A_{n_0})} N_0(A_{n_0})$$
$$+ \int_{A_{n_0}}^x N_0(y)\, y^{-2} \Big(\frac{\log y - 1}{k(y)} + \frac{y\, k'(y) \log y}{k^2(y)} \Big) dy.$$

Recall that $k'(x) \geq 0$ and note that $\log A_{n_0} > 1$. Hence we have

$$\int_{A_{n_0}}^x y^{-1} \frac{\log y}{k(y)} \, dN_0(y) \geq x^{-1} \frac{\log x}{k(x)} N_0(x) - A_{n_0}^{-1} \frac{\log A_{n_0}}{k(A_{n_0})} N_0(A_{n_0}).$$

Also, the second term on the right hand side is a constant; thus (12.11) follows. Then,

$$|E_2(x)| \leq x^{-1} \sum_{n\geq 1} \frac{1}{n!} \int_1^x d\Pi_0^{*n}(t) \leq \frac{N_0(x)}{x}$$

and (12.12) follows. Moreover, by Lemma 12.7 again,

$$\int_1^\infty x^{-s} \frac{N_0(x)}{x} dx = \frac{\zeta_0(s)}{s}$$

holds for $\sigma \geq 1$. Hence

$$\int_1^\infty x^{-1} |E_2(x)| dx \leq \int_1^\infty x^{-2} N_0(x) \, dx = \zeta_0(1) < \infty. \qquad \square$$

12.5. Fundamental estimates

The study of $E_1(x)$ requires a more delicate argument. Here we develop estimates for use in the next section. We will see one motivation for the choice of $\mathcal{P}_B$: the considerable cancellation of $d(\Pi_B - d\Pi)$. We begin with bounds for

$$J_\ell(x) := \int_x^\infty t^{-\ell} d(\pi_B - \pi)(t), \quad \ell \in \mathbb{N}.$$

The reader is encouraged to review Figure 1 (§12.2) to see the contributions of the various intervals to $\mathcal{P}_B$; also recall the notation

$$A_n^* := \sqrt{k(n)}\, A_n.$$

LEMMA 12.9. *For sufficiently large n_0,*

$$|J_1(x)| \le \begin{cases} (\log^2 k(n_0))/(4\log^2 A_{n_0}) & \text{if } 1 \le x < A_{n_0}, \\ (\log k(n))/\log A_n & \text{if } A_n \le x < A_n^\star,\ n \ge n_0, \\ (\log^2 k(n+1))/(4\log^2 A_{n+1}) & \text{if } A_n^\star \le x < A_{n+1},\ n \ge n_0. \end{cases} \tag{12.14}$$

Also, for $\ell \ge 2$,

$$|J_\ell(x)| \le \begin{cases} A_{n_0}^{1-\ell}(\log k(n_0))/\log A_{n_0} & \text{if } 1 \le x < A_{n_0}, \\ 2A_n^{1-\ell}/\log A_n & \text{if } A_n \le x < A_n^\star,\ n \ge n_0, \\ A_{n+1}^{1-\ell}(\log k(n+1))/\log A_{n+1} & \text{if } A_n^\star \le x < A_{n+1},\ n \ge n_0. \end{cases} \tag{12.15}$$

PROOF. For $A_n^\star \le x < A_{n+1}$ with $n \ge n_0 - 1$ we have

$$J_1(x) = \sum_{m \ge n+1} \left(A_m^{-1} \left\lfloor \frac{A_m \log k(m)}{2\log A_m} \right\rfloor - \sum_{A_m < p \le A_m^\star} p^{-1} \right) := \sum_{m \ge n+1} S_1(m),$$

say. By (12.4),

$$S_1(m) = \frac{\log k(m)}{2\log A_m} + O(A_m^{-1}) - \left\{ \frac{\log k(m)}{2\log A_m} - \frac{\log^2 k(m)}{8\log^2 A_m} + O\left(\frac{\log k(m)}{\log^2 A_m}\right) \right\}.$$

It follows that

$$0 < J_1(x) = \sum_{m \ge n+1} \left\{ \frac{\log^2 k(m)}{8\log^2 A_m} + O\left(\frac{\log k(m)}{\log^2 A_m}\right) \right\} \le \frac{\log^2 k(n+1)}{4\log^2 A_{n+1}}.$$

This proves the third inequality of (12.14).

Next, say $A_n \le x < A_n^\star$ with $n \ge n_0$. For $x < t \le A_n^*$, $d\pi_B(t) = 0$ and hence

$$J_1(x) = -\sum_{x < p \le A_n^\star} p^{-1} + J_1(A_n^*). \tag{12.16}$$

By (12.4), the sum satisfies

$$0 \le \sum_{x < p \le A_n^\star} p^{-1} \le \sum_{A_n < p \le A_n^\star} p^{-1} \le \frac{\log k(n)}{2\log A_n} + O\left(\frac{\log^2 k(n)}{\log^2 A_n}\right),$$

and we have just seen that

$$0 \le J_1(A_n^*) \le \frac{\log^2 k(n+1)}{4\log^2 A_{n+1}}.$$

Together, these estimates imply that on this range

$$|J_1(x)| \le \frac{3\log k(n)}{4\log A_n}, \tag{12.17}$$

which proves the second inequality of (12.14).

For $1 \le x < A_{n_0}$, we have $d(\pi_B - \pi)(x) = 0$. Choose any number $z \in (A^*_{n_0-1}, A_{n_0})$, note that $J_1(x) = J_1(z)$, and apply the estimate for the range (A^*_n, A_{n+1}) with $n = n_0 - 1$. This completes the analysis of the case J_1.

Now we turn to (12.15), the case $\ell \ge 2$, and follow the steps used for J_1. For $A^\star_n \le x < A_{n+1}$ we have

$$J_\ell(x) = \sum_{m\ge n+1}\left(A_m^{-\ell}\left\lfloor\frac{A_m\log k(m)}{2\log A_m}\right\rfloor - \sum_{A_m<p\le A^\star_m} p^{-\ell}\right) := \sum_{m\ge n+1} S_\ell(m),$$

say. We claim that each $S_\ell(m)$ is positive, and hence we can obtain an upper bound for $J_\ell(x)$ simply by dropping the sum over primes. Indeed,

$$\sum_{A_m<p\le A^\star_m} p^{-\ell} \le \sum_{A_m<p<\infty} p^{-\ell} \sim \int_{A_m}^\infty \frac{dt}{t^\ell \log t} \sim \frac{A_m^{1-\ell}}{(\ell-1)\log A_m}, \tag{12.18}$$

and this is smaller than

$$\frac{A_m^{1-\ell}\log k(m)}{2\log A_m} - A_m^{-\ell} \le A_m^{-\ell}\left\lfloor\frac{A_m\log k(m)}{2\log A_m}\right\rfloor.$$

We find

$$0 < J_\ell(x) \le \sum_{m\ge n+1}\frac{A_m^{1-\ell}\log k(m)}{2\log A_m} \le \frac{A_{n+1}^{1-\ell}\log k(n+1)}{\log A_{n+1}}. \tag{12.19}$$

For $A_n \le x < A^*_n$ with $n \ge n_0$, we have

$$J_\ell(x) = -\sum_{x<p\le A^*_n} p^{-\ell} + J_\ell(A^*_n).$$

Now (12.18) and (12.19) give

$$|J_\ell(x)| \le \frac{\{1+o(1)\}A_n^{1-\ell}}{(\ell-1)\log A_n} + \frac{A_{n+1}^{1-\ell}\log k(n+1)}{\log A_{n+1}} < \frac{2A_n^{1-\ell}}{\log A_n}. \tag{12.20}$$

Finally, for $1 \le x < A_{n_0}$, we again use the estimate for $A^\star_n \le x < A_{n+1}$ with $n = n_0 - 1$. This completes the proof of (12.15). □

LEMMA 12.10. *For n_0 sufficiently large,*

$$c_2 := \int_1^\infty \left|\int_x^\infty t^{-1} d(\Pi_B - \Pi)(t)\right| \frac{dx}{x} \le 2\,\frac{\log^2 k(n_0)}{\log A_{n_0}}.$$

PROOF. Recall that

$$d(\Pi_B - \Pi)(t) = \sum_{\ell\ge 1}\frac{1}{\ell}\, d(\pi_B - d\pi)(t^{1/\ell}),$$

and we start by treating the case $\ell = 1$. By (12.14),

$$\int_1^{A_{n_0}} \left| \int_x^\infty t^{-1} d(\pi_B - \pi)(t) \right| \frac{dx}{x} \le \frac{\log^2 k(n_0)}{4 \log A_{n_0}},$$

$$\int_{A_n}^{\sqrt{k(n)}A_n} \left| \int_x^\infty t^{-1} d(\pi_B - \pi)(t) \right| \frac{dx}{x} \le \frac{\log^2 k(n)}{2 \log A_n},$$

$$\int_{\sqrt{k(n)}A_n}^{A_{n+1}} \left| \int_x^\infty t^{-1} d(\pi_B - \pi)(t) \right| \frac{dx}{x} \le \frac{\log^2 k(n+1)}{4 \log A_{n+1}}.$$

Hence

$$\int_1^\infty \left| \int_x^\infty t^{-1} d(\pi_B - \pi)(t) \right| \frac{dx}{x}$$
$$\le \frac{\log^2 k(n_0)}{4 \log A_{n_0}} + \sum_{n \ge n_0} \left(\frac{\log^2 k(n)}{2 \log A_n} + \frac{\log^2 k(n+1)}{4 \log A_{n+1}} \right) \le \frac{\log^2 k(n_0)}{\log A_{n_0}}.$$

Also, for $\ell \ge 2$,

$$\int_1^\infty \left| \frac{1}{\ell} \int_x^\infty t^{-1} d(\pi_B - \pi)(t^{1/\ell}) \right| \frac{dx}{x} = \int_1^\infty \left| \frac{1}{\ell} \int_{x^{1/\ell}}^\infty u^{-\ell} d(\pi_B - \pi)(u) \right| \frac{dx}{x}$$
$$= \int_1^\infty \left| \int_y^\infty u^{-\ell} d(\pi_B - \pi)(u) \right| \frac{dy}{y}.$$

Arguing as in the case $\ell = 1$, but now using (12.15), we see that the right hand side of the last equation is at most

$$A_{n_0}^{1-\ell} \log k(n_0) + \sum_{n \ge n_0} \left(\frac{A_n^{1-\ell} \log k(n)}{\log A_n} + A_{n+1}^{1-\ell} \log k(n+1) \right) < 2A_{n_0}^{1-\ell} \log k(n_0).$$

Hence

$$c_2 = \int_1^\infty \left| \int_x^\infty \sum_{\ell \ge 1} \frac{1}{\ell} t^{-1} d(\pi_B - \pi)(t^{1/\ell}) \right| \frac{dx}{x}$$
$$\le \frac{\log^2 k(n_0)}{\log A_{n_0}} + 2 \sum_{\ell \ge 2} A_{n_0}^{1-\ell} \log k(n_0) \le \frac{2 \log^2 k(n_0)}{\log A_{n_0}}.$$ □

12.6. Proof of the Optimality Theorem

It remains to study $E_1(x)$, defined in (12.10).

LEMMA 12.11. *For $x > 1$,*

$$E_1(x) \ll \frac{k(x)}{\log x}. \tag{12.21}$$

PROOF. We have, for $\ell \ge 1$,

$$\left| \int_x^\infty t^{-1}\, d(\Pi_B - \Pi)^{*\ell}(t) \right| \le \int_x^\infty t^{-1}\, d\Pi_0^{*\ell}(t) \le \frac{k(x)}{\log x} \int_x^\infty t^{-1} \frac{\log t}{k(t)}\, d\Pi_0^{*\ell}(t)$$

since $(\log t)/k(t)$ is increasing. Hence, for $x > 1$,

$$|E_1(x)| \le \frac{k(x)}{\log x} \int_x^\infty t^{-1} \frac{\log t}{k(t)} \sum_{\ell \ge 1} \frac{1}{\ell!}\, d\Pi_0^{*\ell}(t) \le c_3 \frac{k(x)}{\log x},$$

where

$$c_3 := \int_1^\infty t^{-1} \frac{\log t}{k(t)}\, dN_0(t),$$

a finite quantity by Lemma 12.7. □

LEMMA 12.12. *We have*

$$\int_1^\infty x^{-1} |E_1(x)|\, dx < \infty. \tag{12.22}$$

PROOF.

$$\int_1^\infty |E_1(x)| \frac{dx}{x} \le \int_1^\infty \sum_{\ell \ge 1} \frac{1}{\ell!} \left| \int_x^\infty t^{-1}\, d(\Pi_B - \Pi)^{*\ell}(t) \right| \frac{dx}{x} = I_1 + I_2,$$

say, where

$$I_1 := \int_1^\infty \sum_{1 \le \ell \le \log x / \log A_{n_0}} \frac{1}{\ell!} \left| \int_x^\infty t^{-1}\, d(\Pi_B - \Pi)^{*\ell}(t) \right| \frac{dx}{x}$$

and

$$I_2 := \int_1^\infty \sum_{\ell > \log x / \log A_{n_0}} \frac{1}{\ell!} \left| \int_x^\infty t^{-1}\, d(\Pi_B - \Pi)^{*\ell}(t) \right| \frac{dx}{x}.$$

Say $\ell > \log x / \log A_{n_0}$, i.e., $A_{n_0}^\ell > x$. Since $d(\Pi_B - \Pi)(t) = 0$ for $t < A_{n_0}$, $d(\Pi_B - \Pi)^{*\ell}(t) = 0$ for $1 \le t \le x$, and the inner integral of I_2 becomes

$$\int_1^\infty t^{-1}\, d(\Pi_B - \Pi)^{*\ell}(t) = \left\{ \int_1^\infty t^{-1}\, d(\Pi_B - \Pi)(t) \right\}^\ell = c_1^\ell.$$

Hence

$$I_2 \le \int_1^\infty \sum_{\ell > \log x / \log A_{n_0}} \frac{|c_1|^\ell}{\ell!} \frac{dx}{x} = \sum_{\ell \ge 1} \frac{|c_1|^\ell}{\ell!} \int_1^{A_{n_0}^\ell} \frac{dx}{x} = |c_1| e^{|c_1|} \log A_{n_0}. \tag{12.23}$$

Exchanging the sum and outer integral in I_1, we obtain

$$I_1 \le \sum_{\ell \ge 1} \frac{1}{\ell!} K(\ell),$$

where

$$K(\ell) := \int_{A_{n_0}^\ell}^\infty x^{-1} \left| \int_x^\infty t^{-1}\, d(\Pi_B - \Pi)^{*\ell}(t) \right| dx.$$

We have at once $K(1) \le c_2$ (from Lemma 12.10). For $\ell \ge 2$,

$$\begin{aligned} K(\ell) &\le \int_{A_{n_0}^\ell}^\infty \left(\int_{A_{n_0}^{\ell-1}}^\infty \left| \int_{x/v}^\infty u^{-1}\, d(\Pi_B - \Pi)(u) \right| v^{-1} d\Pi_0^{*\ell-1}(v) \right) \frac{dx}{x} \\ &= \int_{A_{n_0}^{\ell-1}}^\infty \left(\int_{A_{n_0}^\ell}^\infty \left| \int_{x/v}^\infty u^{-1}\, d(\Pi_B - \Pi)(u) \right| \frac{dx}{x} \right) v^{-1} d\Pi_0^{*\ell-1}(v). \end{aligned}$$

Letting $x/v = y$, the inner integrals on the right hand side become

$$\int_{A_{n_0}^\ell / v}^\infty \left| \int_y^\infty u^{-1}\, d(\Pi_B - \Pi)(u) \right| \frac{dy}{y} \le \int_1^\infty \left| \int_y^\infty u^{-1}\, d(\Pi_B - \Pi)(u) \right| \frac{dy}{y}.$$

By Lemma 12.10, the last expression is the finite quantity c_2. Therefore,

$$K(\ell) \le c_2 \int_{A_{n_0}^{\ell-1}}^{\infty} v^{-1}\, d\Pi_0^{*\ell-1}(v) = c_2 \Big(\int_{A_{n_0}}^{\infty} v^{-1}\, d\Pi_0(v) \Big)^{\ell-1} \le c_2\, c_4^{\ell-1},$$

where $c_4 := \int_1^\infty x^{-1}\, d\Pi_0(x)$; by Lemma 12.5 this also is finite. Hence

$$I_1 \le c_2 + \sum_{\ell \ge 2} \frac{1}{\ell!} c_2 c_4^{\ell-1} \le c_2(1 + e^{c_4})/2. \tag{12.24}$$

Now (12.22) follows from (12.23) and (12.24). □

Conclusion of the Proof. Property (1) of the theorem was established in Lemma 12.6, and Property (3) in Lemma 12.3. It remains only to show Property (2). By (12.8) and (12.9) we have

$$|N(x) - Ax|x^{-1} = |E(x)| \le x^{-1} + |E_1(x)| + |E_2(x)|.$$

Then, by (12.21), (12.12), and (1) of Lemma 12.2,

$$|E_1(x)| + |E_2(x)| \ll k(x)/\log x \ll f(x)/\log x.$$

Finally, by (12.13) and (12.22),

$$\int_1^\infty x^{-1}|E(x)|\, dx \le \int_1^\infty x^{-1}(x^{-1} + |E_1(x)| + |E_2(x)|)\, dx < \infty. \qquad \square$$

12.7. Notes

The main result of this section comes from the authors' article [**DZ13b**].

§12.6. For convenience, we wrote $|c_1|$ in (12.23). In fact, $c_1 > 0$, as the reader might show.

CHAPTER 13

Beurling's PNT

Our raison d'être

SUMMARY. We give a proof of Beurling's PNT for g-numbers and show this result is optimal in the class Beurling considered.

13.1. Introduction

The main aim of Beurling's g-number investigation was to give conditions on a g-integer system $\mathcal{N}$ implying that the generalized version of the PNT holds, i.e., $\pi(x) \sim x/\log x$ as $x \to \infty$. His result was the following.

THEOREM 13.1. *Suppose that $\mathcal{N}$ has a counting function satisfying*

$$N(x) = N_{\mathcal{N}}(x) = Ax + O(x(\log ex)^{-\gamma}) \tag{13.1}$$

for some $A > 0$. If $\gamma > 3/2$, then the PNT holds.

By Chebyshev's bounds $\psi(x) \ll x$; thus $\psi(x) \sim \Pi(x)\log x$ as $x \to \infty$ by integration by parts, and it suffices prove the PNT in the form $\psi(x) \sim x$. We first show that the g-zeta function is nonzero on $\{s : \Re s = 1\}$; our argument, starting with an extension of the classical 3-4-1 inequality via the Fejér kernel, follows that of Beurling. The second part is to apply the Wiener-Ikehara theorem, for which we use results established in Chapter 10. We remark that our tauberian method is somewhat simpler than Beurling's because we use ψ rather than $d\psi$.

Beurling's result is optimal in the class of g-number systems having a counting function of the form (13.1), as we shall see below: there are examples satisfying (13.1) with $\gamma = 3/2$ for which the PNT fails. The first of these examples is continuous; the second has discrete g-integers.

Note that if (13.1) holds for a g-number system $\mathcal{N}$ with some number γ, then the relation holds *a fortiori* for a number $\gamma' < \gamma$. *To simplify some calculations, we henceforth assume that* $3/2 < \gamma \le 7/4$.

13.2. A lower bound for $|\zeta(\sigma + it)|$

We prove the nonvanishing of $\zeta(1 + it)$ for $t \neq 0$ by combining two relations: the first, which we give here, is a lower bound for $|\zeta(\sigma + it)|$; in the next section we show that $\zeta(s)$ satisfies a Lipschitz condition of a positive order.

In Chapter 10 we introduced the Fejér kernel on $\mathbb{R}$; here we use a corresponding family of 2π periodic functions. For L a positive integer and x real, set

$$K_L(x) := \sum_{-L \le \ell \le L} \Big(1 - \frac{|\ell|}{L}\Big) e^{i\ell x} = 1 + 2\sum_{\ell=1}^{L-1} \Big(1 - \frac{\ell}{L}\Big) \cos \ell x\,.$$

LEMMA 13.2. *For each $L \in \mathbb{N}$ and $x \in \mathbb{R}$, $K_L(x) \ge 0$.*

PROOF. We have

$$0 \le \left|\sum_{\ell=1}^{L} e^{i\ell x}\right|^2 = \left(\sum_{\ell=1}^{L} e^{i\ell x}\right)\left(\sum_{k=1}^{L} e^{-ikx}\right) = \sum_{\ell=1}^{L}\sum_{k=1}^{L} e^{i(\ell-k)x}$$
$$= L + \sum_{\ell=1}^{L-1}(L-\ell)\left(e^{i\ell x}+e^{-i\ell x}\right) = LK_L(x). \qquad \square$$

We use the nonnegativity of K_L to derive a generalization of the classical bound $\zeta(\sigma)^3|\zeta(\sigma+it)|^4|\zeta(\sigma+2it)| \ge 1$ ("3-4-1 inequality"), which is used in many proofs of nonvanishing of the Riemann zeta function on $\{\Re s = 1\}$.

LEMMA 13.3. *For each integer $L \ge 2$, real t, and $\sigma > 1$,*

$$\zeta(\sigma)^L \prod_{\ell=1}^{L-1} |\zeta(\sigma + i\ell t)|^{2L-2\ell} \ge 1.$$

PROOF. We show that

$$L\log\zeta(\sigma) + 2\sum_{\ell=1}^{L-1}(L-\ell)\log|\zeta(\sigma+i\ell t)| \ge 0. \tag{13.2}$$

Recall from (1.5) that

$$\log\zeta(\sigma+it) = \int_1^\infty u^{-\sigma}e^{-it\log u}d\Pi(u).$$

Taking real parts and using the evenness of cos, we get

$$\log|\zeta(\sigma+it)| = \int_1^\infty u^{-\sigma}\cos(t\log u)\,d\Pi(u).$$

Thus the left side of (13.2) can be written as

$$L\int_1^\infty u^{-\sigma}\left\{1 + 2\sum_{\ell=1}^{L-1}\left(1-\frac{\ell}{L}\right)\cos(\ell t\log u)\right\}d\Pi(u).$$

Now the expression in curly brackets is $K_L(t\log u)$, which is nonnegative. Also, $d\Pi \ge 0$. This establishes (13.2) and hence the claim of the lemma. $\square$

Next, we show that $|\zeta(\sigma+it)|$ cannot approach 0 too swiftly as $\sigma \to 1+$.

LEMMA 13.4. *Let $\mathcal{N}$ be a g-number system which satisfies* (13.1) *with $\gamma > 3/2$. Let t be any nonzero real number and L any integer exceeding 1. There exists a positive number $C = C_{\mathcal{N}}(L,t)$ for which*

$$|\zeta(\sigma+it)| \ge C(\sigma-1)^{1/2+1/(2L)}, \quad 1 < \sigma \le 2. \tag{13.3}$$

PROOF. Once again, set

$$g(s) := \frac{1}{A}\int_1^\infty x^{-s-1}\{N(x)-Ax\}\,dx = \frac{\zeta(s)}{As} - \frac{1}{s-1}. \tag{11.7 bis}$$

The integral for g converges absolutely and uniformly for $\{\Re s \ge 1\}$. It defines there a continuous function with $|g(s)| \le H$ for some constant H, and we can write

$$\zeta(s) = A + \frac{A}{s-1} + Asg(s), \quad \Re s \ge 1. \tag{13.4}$$

In particular, for $1 < \sigma \le 2$ and fixed $t \neq 0$,

$$\zeta(\sigma) \le \frac{2A(H+1)}{\sigma-1} := \frac{A'}{\sigma-1},$$

$$|\zeta(\sigma+i\ell t)| \le \frac{A}{2|t|} + A + (2+L|t|)AH := A'', \quad 2 \le \ell \le L.$$

Now use Lemma 13.3 with $L+1$ in place of L and insert the preceding estimates. We find

$$\begin{aligned} 1 &\le \zeta(\sigma)^{L+1}|\zeta(\sigma+it)|^{2L}|\zeta(\sigma+2it)|^{2L-2}\cdots|\zeta(\sigma+Lit)|^2 \\ &\le \left(\frac{A'}{\sigma-1}\right)^{L+1}|\zeta(\sigma+it)|^{2L}(A'')^{2+4+\cdots+2L-2}, \end{aligned}$$

and, with $C = (A')^{-(L+1)/(2L)}(A'')^{-(L-1)/2}$,

$$|\zeta(\sigma+it)| \ge C(\sigma-1)^{1/2+1/(2L)}, \quad 1 < \sigma \le 2. \qquad \square$$

13.3. Nonvanishing of $\zeta(1+it)$

The next step is to show that zeta satisfies Lipschitz condition of order larger than 1/2. This result implies that if $\zeta(s_0) = 0$, then $|\zeta(s)|$ is quite small for s near s_0, which will lead to the desired contradiction. (In the classical case, one simply notes at this point that zeta is analytic and thus has a bounded derivative.)

LEMMA 13.5. *Suppose $\mathcal{N}$ is a g-number system that satisfies* (13.1) *for some $\gamma \in (3/2, 7/4]$. Let $0 < \delta < 1 < T$ be given and set*

$$\Omega = \Omega(\delta, T) := \{s = \sigma + it : 1 \le \sigma \le 2, \ \delta \le |t| \le T\}.$$

There is a positive number $D = D_{\mathcal{N}}(\delta, T)$ such that, if $s_1, s_2 \in \Omega$, then

$$|\zeta(s_1) - \zeta(s_2)| \le D|s_1 - s_2|^{\gamma-1}.$$

PROOF. We use the representation of zeta from (13.4) to write

$$\zeta(s_1) - \zeta(s_2) = \left\{\frac{A}{s_1-1} - \frac{A}{s_2-1}\right\} + A\{s_1 g(s_1) - s_2 g(s_2)\} := I + II.$$

If $|s_1 - s_2| \le 1$, then

$$|I| = \frac{A|s_1-s_2|}{|s_1-1||s_2-1|} \le \frac{A}{\delta^2}|s_1-s_2|^{\gamma-1},$$

since $\gamma - 1 < 1$ by assumption. If, on the other hand, $|s_1 - s_2| > 1$, then

$$|I| \le \frac{A|s_1-1| + A|s_2-1|}{|s_1-1||s_2-1|} \le \frac{2A}{\delta} \le \frac{2A}{\delta}|s_1-s_2|^{\gamma-1}.$$

In any case,

$$|I| \le \frac{2A}{\delta^2}|s_1-s_2|^{\gamma-1}.$$

Also, using the definition of $g(s)$,

$$\begin{aligned} |II| &= \left|\int_1^\infty \{s_1 x^{-s_1} - s_2 x^{-s_2}\} x^{-1}\{N(x) - Ax\}\, dx\right| \\ &\le \int_1^\infty |s_1 x^{-s_1} - s_2 x^{-s_2}|\, B(\log ex)^{-\gamma} dx, \end{aligned}$$

for some $B > 0$. We estimate $|s_1x^{-s_1} - s_2x^{-s_2}|$ differently depending on whether or not $|s_1 - s_2|\log x < 1$.

Let $X = \exp(|s_1 - s_2|^{-1})$, and suppose first that $x < X$. We have

$$\begin{aligned} |s_1x^{-s_1} - s_2x^{-s_2}| &= \left| \int_{s_1}^{s_2} \frac{d}{du}(u\,x^{-u})\,du \right| \\ &\le |s_1 - s_2|x^{-1}\{1 + (T+2)\log x\} \\ &\le |s_1 - s_2|x^{-1}3T\log ex. \end{aligned}$$

Now

$$II_A := \int_1^X |s_1x^{-s_1} - s_2x^{-s_2}|\,B(\log ex)^{-\gamma}dx \le 3BT|s_1 - s_2| \int_1^X (\log ex)^{1-\gamma}\frac{dx}{x},$$

and the last integral is less than

$$\int_1^X (\log x)^{1-\gamma}\frac{dx}{x} = \frac{(\log X)^{2-\gamma}}{2-\gamma} = \frac{|s_1 - s_2|^{\gamma-2}}{2-\gamma} \le 4|s_1 - s_2|^{\gamma-2}.$$

Thus

$$II_A \le 12BT|s_1 - s_2|^{\gamma-1}.$$

On the other hand, for $x \ge X$, we use the inequality

$$|s_1x^{-s_1} - s_2x^{-s_2}| \le 2(2+T)x^{-1} < 6Tx^{-1}.$$

It follows that

$$\begin{aligned} II_B &:= \int_X^\infty |s_1x^{-s_1} - s_2x^{-s_2}|\,B(\log ex)^{-\gamma}dx \\ &< 6BT\int_X^\infty (\log x)^{-\gamma}\frac{dx}{x} = \frac{6BT}{\gamma-1}|s_1 - s_2|^{\gamma-1} \le 12BT|s_1 - s_2|^{\gamma-1}, \end{aligned}$$

and hence $|II| \le 24BT|s_1 - s_2|^{\gamma-1}$.

Finally, I and II together, and $D = 2A\delta^{-2} + 24BT$, give the claimed bound. □

We apply the last two lemmas to show that $\zeta(s) \ne 0$ on $\Re s = 1$.

THEOREM 13.6. *Let $\mathcal{N}$ be a g-number system which satisfies* (13.1) *for some $\gamma > 3/2$. The associated zeta function $\zeta(s)$ has no zeros in the closed half plane $\{s : \Re s \ge 1\}$.*

PROOF. We have shown already in Lemma 3.3 that $\zeta(s)$ is nonvanishing in the open half plane $\{s : \Re s > 1\}$ – indeed, a much weaker condition was used there. Now we assume that $\zeta(1 + it_0) = 0$ for some $t_0 \ne 0$ (Why can we assume that $t_0 \ne 0$?), and show that this leads to a contradiction.

Taking $s_2 = 1 + it_0$ and $s_1 = \sigma + it_0$ with $1 < \sigma \le 2$ and $0 < \delta < |t_0|$ in Lemma 13.5, we have

$$|\zeta(\sigma + it_0)| \le D(\sigma - 1)^{\gamma-1}.$$

Next, take L to be an integer that is so large that $(2L)^{-1} < \gamma - 3/2$. We apply Lemma 13.4 with this L to obtain

$$|\zeta(\sigma + it_0)| \ge C(\sigma - 1)^{1/2+1/(2L)}, \quad 1 < \sigma \le 2.$$

Together, the last two relations imply that

$$(\sigma - 1)^{\gamma-3/2-1/(2L)} \ge C/D.$$

But this inequality is impossible for $\sigma - 1$ sufficiently small, since the exponent is positive. Thus $\zeta(1 + it_0) \neq 0$. □

13.4. An L^1 condition and conclusion of the proof

We complete the proof of Theorem 13.1 by applying Corollary 10.6 to conclude that $\psi(e^x)e^{-x} \to 1$ as $x \to \infty$, i.e., the PNT holds. For this, we show that

$$G(s) := -\frac{1}{s}\frac{\zeta'}{\zeta}(s) - \frac{1}{s-1} = \int_0^\infty \psi(e^u)\, e^{-su} du - \frac{1}{s-1}$$

satisfies an L^1 Cauchy condition. We shall use an L^2 hypothesis here that is slightly weaker than that of Theorem 13.1, in order to be able to use the same result for the punchline in the proof of the PNT given in Chapter 15.

LEMMA 13.7. *Suppose the zeta function of a g-number system has no zeros on the line* $\{\Re s = 1\}$ *and the integer counting function satisfies*

$$\int_1^\infty \left(\frac{N(x) - Ax}{x} \log x\right)^2 \frac{dx}{x} < \infty.$$

Then the function G*, defined above, satisfies*

$$\lim_{\sigma,\, \sigma' \to 1+} \int_{-2\lambda}^{2\lambda} |G(\sigma + it) - G(\sigma' + it)|\, dt = 0 \tag{10.8 bis}$$

for every $\lambda > 0$.

PROOF. We begin by expressing $G(s)$ in terms of

$$g(s) := \frac{1}{A}\int_1^\infty x^{-s-1}\{N(x) - Ax\}\, dx = \frac{\zeta(s)}{As} - \frac{1}{s-1}, \tag{11.7 ter}$$

a continuous function for $\sigma = \Re s \geq 1$ to which we can conveniently apply the hypothesis of the theorem. We claim that, for $\sigma > 1$,

$$G(s) = -\frac{1}{s} - \frac{1}{s^2} - \frac{Ag(s)}{(s-1)\zeta(s)} - \frac{Ag'(s)}{\zeta(s)}. \tag{13.5}$$

Indeed,

$$A(s-1)g(s) = (1 - s^{-1})\zeta(s) - A,$$

and if we differentiate each side of this formula and divide by $-(s-1)\zeta(s)$, we find

$$\begin{aligned} -\frac{A\{g(s) + (s-1)g'(s)\}}{(s-1)\zeta(s)} &= -\frac{(1 - s^{-1})\zeta'(s) + s^{-2}\zeta(s)}{(s-1)\zeta(s)} \\ &= -\frac{1}{s}\frac{\zeta'(s)}{\zeta(s)} - \frac{1}{s-1} + \frac{1}{s-1}(1 - s^{-2}) \\ &= G(s) + \frac{1}{s} + \frac{1}{s^2}, \end{aligned}$$

showing that (13.5) holds.

Writing $s = \sigma + it$ with $\sigma > 1$, we set

$$J_\sigma(t) := -\frac{1}{s} - \frac{1}{s^2} - \frac{Ag(s)}{(s-1)\zeta(s)}, \qquad K_\sigma(t) := \frac{1}{\zeta(s)}, \qquad L_\sigma(t) := Ag'(s).$$

By the triangle inequality,

$$\int_{-2\lambda}^{2\lambda} |G(\sigma + it) - G(\sigma' + it)|\, dt \leq I + II,$$

where

$$I = \int_{-2\lambda}^{2\lambda} |J_\sigma(t) - J_{\sigma'}(t)|\, dt, \quad II = \int_{-2\lambda}^{2\lambda} |L_\sigma(t)K_\sigma(t) - L_{\sigma'}(t)K_{\sigma'}(t)|\, dt.$$

We have $\lim_{\sigma,\,\sigma'\to 1+} I = 0$, since J_σ is continuous. (Recall that $(s-1)\zeta(s) \neq 0$ for $\Re s \geq 1$.) Also,

$$II \leq \int_{-2\lambda}^{2\lambda} |L_\sigma(t)\{K_\sigma(t) - K_{\sigma'}(t)\}|\, dt + \int_{-2\lambda}^{2\lambda} |\{L_\sigma(t) - L_{\sigma'}(t)\}K_{\sigma'}(t)|\, dt := II_a + II_b,$$

say. Now $K_\sigma(t)$ is uniformly continuous on

$$\{s = \sigma + it : 1 \leq \sigma \leq 2,\ -2\lambda \leq t \leq 2\lambda\},$$

and thus, as σ, $\sigma' \to 1+$,

$$II_a = o(1)\int_{-2\lambda}^{2\lambda} |L_\sigma(t)|\, dt, \qquad II_b \ll \int_{-2\lambda}^{2\lambda} |L_\sigma(t) - L_{\sigma'}(t)|\, dt.$$

It remains to show that $\lim_{\sigma,\sigma'\to 1+} II_b = 0$; this Cauchy condition implies that $\lim_{\sigma\to 1+} \int_{-2\lambda}^{2\lambda} |L_\sigma|$ exists, and so $II_a \to 0$ also.

By the Cauchy-Schwarz inequality,

$$\int_{-2\lambda}^{2\lambda} |L_\sigma(t) - L_{\sigma'}(t)|\, dt \leq \left\{\int_{-2\lambda}^{2\lambda} dt\right\}^{1/2} \cdot \left\{\int_{-2\lambda}^{2\lambda} |L_\sigma(t) - L_{\sigma'}(t)|^2\, dt\right\}^{1/2},$$

and thus, for fixed $\lambda > 0$, it suffices to show that

$$III := \int_{-2\lambda}^{2\lambda} |L_\sigma(t) - L_{\sigma'}(t)|^2\, dt$$

goes to 0 as σ, $\sigma' \to 1+$.

Making a change of variable in the integral (11.7) for g and differentiating, we obtain the Fourier integral representation

$$L_\sigma(t) = Ag'(\sigma + it) = \int_0^\infty e^{-itu}\, \phi_\sigma(u)\, du,$$

where

$$\phi_\sigma(u) := -e^{-\sigma u} u\{N(e^u) - Ae^u\}.$$

By hypothesis,

$$\int_0^\infty \phi_1(u)^2\, du < \infty, \tag{13.6}$$

and a small calculation, using the Cauchy-Schwarz inequality, shows that ϕ_σ is integrable on $(0, \infty)$ for $\sigma > 1$.

The function L_σ is the Fourier transform of ϕ_σ, and we have

$$L_\sigma(t) - L_{\sigma'}(t) = \int_0^\infty e^{-itu}\, \{\phi_\sigma(u) - \phi_{\sigma'}(u)\}\, du.$$

By (a weak form of) Plancherel's identity ([**Kn68**], Ch. 6),

$$III \leq 2\pi \int_0^\infty \{\phi_\sigma(u) - \phi_{\sigma'}(u)\}^2\, du.$$

The last integrand is dominated by the integrable function $\phi_1(u)^2$ and it tends to 0 pointwise as σ, $\sigma' \to 1+$, so by the dominated convergence theorem, $III \to 0$. Thus (10.8) holds. □

If we insert the result of the preceding lemma into Corollary 10.6, we can conclude that $\psi(e^x)\,e^{-x} \to 1$ as $x \to \infty$. As we noted at the beginning of the chapter, $\psi(x) \sim \Pi(x)\log x$. Thus $\pi(x) \sim \Pi(x) \sim x/\log x$, and the PNT holds for the g-number system $\mathcal{N}$. This completes the proof of Theorem 13.1.

13.5. Optimality – a continuous example

We now show that Theorem 13.1 is optimal in the class of g-number systems $\mathcal{N}$ with counting function

$$N(x) = N_{\mathcal{N}}(x) = Ax + O(x(\log ex)^{-\gamma}). \tag{13.1 bis}$$

Following Beurling, we give an example of a g-number system satisfying (13.1) with $\gamma = 3/2$ for which the PNT fails. Thus the exponent $3/2$ is optimal in this class of g-number systems. (We shall show in Chapter 15 that the PNT can be established under an L^2 condition that is weaker than the hypothesis in Theorem 13.1.)

EXAMPLE 13.8. (A wobbly prime counting function.) For $x \geq 1$, let a "wobbly g-prime counting function" be defined by

$$\pi_w(x) := \int_1^x \frac{1-\cos(\log t)}{\log t}\,dt. \tag{13.7}$$

The integrand is asymptotic to $\frac{1}{2}\log t$ as $t \to 1+$, so the integral is convergent. We shall show that $\pi_w(x)$ has the desired prime counting properties, and the "g-integer" system it generates satisfies (13.1).

First, $\pi_w(1) = 0$ and $\pi_w \uparrow$; these properties hold also for the weighted function

$$\Pi_w(x) := \pi_w(x) + \frac{1}{2}\pi_w(x^{1/2}) + \frac{1}{3}\pi_w(x^{1/3}) + \dots. \tag{13.8}$$

Thus π_w qualifies as a (continuous) g-prime counting function. Next,

$$\pi_w(x) := \pi_w(e) + \mathrm{li}(x) - \mathrm{li}(e) - \int_e^x \frac{\cos(\log t)}{\log t}\,dt,$$

where

$$\mathrm{li}(x) := \lim_{\epsilon\to 0+}\left\{\int_0^{1-\epsilon} + \int_{1+\epsilon}^x \frac{du}{\log u}\right\} = \mathrm{li}\,e + \int_e^x \frac{du}{\log u}.$$

By integrating twice by parts,

$$\pi_w(x) = \mathrm{li}(x) - \frac{x}{2\log x}\{\sin(\log x) + \cos(\log x)\} + O\Big(\frac{x}{\log^2 x}\Big),$$

or after application of a trigonometric identity,

$$\pi_w(x)/\,\mathrm{li}(x) = 1 - \sin(\pi/4 + \log x)/\sqrt{2} + o(1), \quad x \to \infty.$$

Thus the PNT does not hold for π_w.

Now we turn to the g-integers it generates.

THEOREM 13.9. *Let π_w be defined by (13.7) and let Π_w be the companion function defined by (13.8). The associated g-integer counting function*

$$N_w(x) := \int_{1-}^x \exp^* d\Pi_w$$

satisfies (13.1) with $\gamma = 3/2$.

We have

$$N_w(x) = \int_{1-}^{x} \exp^*\{d\pi_w + d\Pi_2\} \tag{13.9}$$

where

$$\Pi_2(x) := \frac{1}{2}\pi_w(x^{1/2}) + \frac{1}{3}\pi_w(x^{1/3}) + \dots .$$

We next analyze the two parts of the formula for N_w.

LEMMA 13.10. *For suitable constants c_0, c_1, we have*

$$N_1(x) := \int_{1-}^{x} \exp^* d\pi_w = x + \frac{c_0 x \cos(\log x)}{\log^{3/2} x} + \frac{c_1 x \cos(\log x)}{\log^{5/2} x} + O\Big(\frac{x}{\log^{7/2} x}\Big).$$

To establish Theorem 13.9, the weaker relation $N_1(x) = x + O(x/\log^{3/2} x)$ would suffice; the four-term form will be used in the next section.

LEMMA 13.11.

$$N_2(x) := \int_{1-}^{x} \exp^* d\Pi_2 \ll x^{1/2}.$$

By the homomorphic property of the exponential,

$$N_w(x) = \int_{1-}^{x} dN_1 * dN_2.$$

After proving the two lemmas, we shall establish the theorem by showing that $N_w(x)$ has the form of its bigger convolution factor, $N_1(x)$.

PROOF OF LEMMA 13.10. Recall

$$N_1(x) := \int_{1-}^{x} \exp^* d\pi_w \quad \text{and} \quad d\pi_w(x) := \frac{1 - \cos(\log x)}{\log x}\, dx.$$

Thus $N_1(x)$ is continuous on $(1, \infty)$. We first find its generating function. For $\Re s > 1$, write

$$\widehat{\pi_w}(s) := \int_1^{\infty} x^{-s} d\pi_w(x) = \int_1^{\infty} x^{-s}\Big\{\frac{1 - (x^i + x^{-i})/2}{\log x}\Big\}\, dx\,.$$

Then

$$\widehat{\pi_w}'(s) = -\int_1^{\infty} x^{-s}\Big\{1 - \frac{x^i + x^{-i}}{2}\Big\}\, dx = -\frac{1}{s-1} + \frac{1/2}{s-1-i} + \frac{1/2}{s-1+i},$$

and so

$$\begin{aligned}\widehat{\pi_w}(s) &= \frac{\log(s-1-i) + \log(s-1+i)}{2} - \log(s-1)\\ &= \log\Big\{\frac{\sqrt{(s-1-i)(s-1+i)}}{s-1}\Big\} = \log\Big\{\frac{\sqrt{(s-1)^2+1}}{s-1}\Big\}.\end{aligned}$$

We take the branch of the square root that is positive for s real and exceeding 1. It follows that

$$\widehat{N_1}(s) = \int_{1-}^{\infty} x^{-s}\{\exp^* d\pi_w\}(x) = \exp\int_1^{\infty} x^{-s} d\pi_w(x) = \frac{\sqrt{(s-1)^2+1}}{s-1}.$$

By the Mellin inversion formula, for any $x > 1$ and $a > 1$,

$$N_1(x) = \frac{1}{2\pi i}\int_{a-i\infty}^{a+i\infty} x^s\, \widehat{N_1}(s)\, \frac{ds}{s}\,.$$

To simplify convergence considerations, introduce

$$\phi(s) := \widehat{N_1}(s) - 1 = \frac{\sqrt{(s-1)^2+1} - (s-1)}{s-1}.$$

Rationalizing the numerator, we find

$$\phi(s) = \frac{1}{(s-1)\{\sqrt{(s-1)^2+1} + (s-1)\}} \ll \frac{1}{|s|^2}, \quad |s| \text{ large.} \tag{13.10}$$

We restate the Mellin formula as $N_1(x) = I + 1$, say, with

$$I := \frac{1}{2\pi i} \int_{a-i\infty}^{a+i\infty} x^s \, \phi(s) \, \frac{ds}{s}$$

and

$$\frac{1}{2\pi i} \int_{a-i\infty}^{a+i\infty} x^s \, \frac{ds}{s} = 1,$$

the last by a familiar calculation.

The integrand of I has singularities at 0, 1, and $1 \pm i$. We deform the contour to the vertical line $\{s : \Re s = 1/2\}$, with loops taken to avoid crossing the rays $(-\infty - i, 1 - i)$ and $(-\infty + i, 1 + i)$. By the residue theorem, the contribution to I from the pole of $\phi(s)$ at $s = 1$ is x. By the bound $\phi(s) \ll 1/|s|^2$ from (13.10) and the fact that $|x^s| = x^{1/2}$ on the vertical line, the contribution to I from the vertical portion of the new contour is $O(x^{1/2})$. (We could have extended the contour further to the left, but this would not be useful for the remainder of this analysis.)

We next calculate the loop integral about $1 + i$. On a circle of radius ϵ about this point, $\phi(s)$ is bounded, and thus the integral converges to 0 as $\epsilon \to 0$. It remains to calculate the integral along the segments

$$L_\pi = 1 + i + re^{i\pi}, \quad L_{-\pi} = 1 + i + re^{-i\pi},$$

with the former traversed from $r = 0$ to $r = 1/2$ and latter from 1/2 to 0. Writing

$$\widehat{N_1}(s)/s = (s - 1 - i)^{1/2} f(s)$$

where $f(s)$ is analytic for $|s - 1 - i| \le 1/2$, we have on this disc

$$\begin{aligned} \phi(s)/s &= (\widehat{N_1}(s) - 1)/s \\ &= -1/s + c_0'(s-1-i)^{1/2} + c_1'(s-1-i)^{3/2} + O(|s-1-i|^{5/2}) \end{aligned}$$

for c_0, c_1' suitable constants. The integral of $-1/s$ over the two segments cancels out, so it remains to study the contribution of the remaining terms.

We have

$$\begin{aligned}
J_{1/2} &:= \frac{c_0}{2\pi i}\Big\{\int_{L_{-\pi}} + \int_{L_\pi}\Big\} x^s (s-1-i)^{1/2}\, ds \\
&= \frac{c_0}{2\pi i}\,(e^{\pi i/2} - e^{-\pi i/2})\, x^{1+i} \int_0^{1/2} x^{-r} r^{1/2}\, dr \\
&= \frac{c_0\, x^{1+i}}{\pi}\Big\{\int_0^\infty - \int_{1/2}^\infty\Big\} e^{-r\log x} r^{1/2}\, dr \\
&= \frac{c_0\, x^{1+i}}{\pi \log^{3/2} x}\Big\{\int_0^\infty - \int_{(\log x)/2}^\infty\Big\} e^{-u} u^{1/2}\, du \\
&= \frac{c_0 \Gamma(3/2)\, x^{1+i}}{\pi \log^{3/2} x} + O(x^{1/2}),
\end{aligned}$$

where Γ denotes the Euler Gamma function.

Also,

$$J_{3/2} := \frac{c_1'}{2\pi i}\Big\{\int_{L_{-\pi}} + \int_{L_\pi}\Big\} x^s (s-1-i)^{3/2}\, ds,$$

and a similar calculation yields

$$J_{3/2} = -\frac{c_1' \Gamma(5/2)\, x^{1+i}}{\pi \log^{5/2} x} + O(x^{1/2}).$$

Then, with L representing either L_π or $L_{-\pi}$, we have the error term

$$\begin{aligned}
&\int_L |x^s|\,|s-1-i|^{5/2} \frac{ds}{|s|} \ll x \int_{r=0}^{1/2} x^{-r} r^{5/2} dr \\
&\quad \le x \int_{r=0}^\infty e^{-r\log x} r^{5/2} dr = \frac{x}{\log^{7/2} x}\int_0^\infty e^{-u} u^{5/2}\, du \ll \frac{x}{\log^{7/2} x},
\end{aligned}$$

since the last integral is $\Gamma(7/2)$, a finite quantity.

The same calculations apply for the loop integral about $1-i$, and noting that $x^i + x^{-i} = 2\cos(\log x)$, we obtain the claimed formula for $N_1(x)$. □

PROOF OF LEMMA 13.11. By Lemma 2.10 and a weak form of Lemma 13.10, we have first

$$N_3(x) := \int_{1-}^x \exp^*\Big\{\frac{1}{2} d\pi_w(t^{1/2})\Big\} = \int_{1-}^{\sqrt{x}} \exp^*\Big\{\frac{1}{2} d\pi_w\Big\} \le \int_{1-}^{\sqrt{x}} \exp^* d\pi_w \ll \sqrt{x}.$$

Now

$$\begin{aligned}
N_2(x) &:= \int_{1-}^x dN_3 * \exp^*\Big\{\frac{1}{3} d\pi_w(t^{1/3}) + \frac{1}{4} d\pi_w(t^{1/4}) + \dots\Big\} \\
&\ll \int_{1-}^x \sqrt{\frac{x}{t}}\, \exp^*\Big\{d\pi_w(t^{1/3}) + d\pi_w(t^{1/4}) + \dots\Big\}.
\end{aligned}$$

If we extend the integration to infinity and use Lemma 2.10 again, we find

$$\begin{aligned} N_2(x) &\ll \sqrt{x}\int_{1-}^{\infty} \exp^*\Big\{t^{-1/2}d\pi_w(t^{1/3}) + t^{-1/2}d\pi_w(t^{1/4}) + \dots\Big\} \\ &= \sqrt{x}\exp\int_{1}^{\infty} t^{-3/2}d\pi_w(t)\cdot\exp\int_{1-}^{\infty} t^{-2}d\pi_w(t)\cdots \\ &\le \sqrt{x}\exp\int_{1}^{\infty}\{t^{-3/2}+t^{-2}+t^{-5/2}+\dots\}\,d\pi_w(t) \\ &= \sqrt{x}\exp\int_{1}^{\infty}\frac{t^{-3/2}}{1-t^{-1/2}}\,\frac{1-\cos(\log t)}{\log t}\,dt\,. \end{aligned}$$

We note that

$$\frac{1-\cos(\log t)}{\log t} \le \begin{cases} (\log t)/2, & 1 < t \le e \\ 2, & t > e. \end{cases}$$

Hence the last integral is at most

$$\int_1^e \frac{t^{-3/2}}{1-t^{-1/2}}\,\frac{\log t}{2}\,dt + \int_e^{\infty}\frac{2t^{-3/2}\,dt}{1-e^{-1/2}} \ll 1,$$

since $\log t/(1-t^{-1/2})$ is bounded on $(1, e]$. It follows that $N_2(x) \ll \sqrt{x}$. □

PROOF OF THEOREM 13.9. First note that since $N_2(x) \ll \sqrt{x}$, integration by parts gives

$$\int_{1-}^{y}\frac{dN_2(t)}{t} = \int_{1-}^{\infty}\frac{dN_2(t)}{t} - \int_{y}^{\infty}\frac{dN_2(t)}{t} = A + O(y^{-1/2}), \quad y \to \infty,$$

for some constant A.

Apply the hyperbola method, writing

$$N_w(x) = \int_{1-}^{x} dN_1 * dN_2 = I + II$$

with

$$I := \int_{1-}^{\sqrt{x}} N_1(x/t)\,dN_2(t), \qquad II := \int_{1-}^{\sqrt{x}}\{N_2(x/t) - N_2(\sqrt{x})\}\,dN_1(t).$$

Now

$$\begin{aligned} I &= \int_{1-}^{\sqrt{x}}\left\{\frac{x}{t} + O\left(\frac{x/t}{\log^{3/2} x/t}\right)\right\}dN_2(t) \\ &= Ax + O(x^{3/4}) + O\left(\frac{x}{\log^{3/2} x}\int_{1-}^{\sqrt{x}} t^{-1}\,dN_2(t)\right) \\ &= Ax + O\left(\frac{x}{\log^{3/2} x}\right) \end{aligned}$$

with $A = \int_{1-}^{\infty} t^{-1}\,dN_2(t)$. Next, by another integration by parts,

$$II = \int_{1-}^{\sqrt{x}} O\big(\sqrt{x/t}\big)\,dN_1(t) = O\big(x^{3/4}\big)\,.$$

The two calculations together give the claimed formula for $N_w(x)$. □

13.6. Optimality – a discrete example

In the theory of g-numbers presented here, discreteness of primes has been of minor importance: most of our results and several examples apply also to continuous number distributions. Still, it is psychologically satisfying to have examples using discrete g-primes. Here, we convert the continuous example of the last section into a discrete one by a construction using the floor function. (In Chapter 17, below, a continuous example is made discrete by a stochastic process.)

We start with the wobbly g-prime counting function π_w of the last section and define a collection of g-primes by

$$\mathcal{P}_d := \{p_n := \pi_w^{-1}(n)\},$$

or $\pi_d(x) = \lfloor \pi_w(x) \rfloor$. Since $\pi_d(x) = \pi_w(x) + O(1)$, the result of the last section shows that the PNT does not hold for $\mathcal{P}_d$. Now we show that $\mathcal{N}_d$, the collection of g-integers generated by $\mathcal{P}_d$, satisfies the counting condition (13.1).

Our argument follows that of the last section. With

$$\Pi_d(x) := \pi_d(x) + \frac{1}{2}\pi_d(x^{1/2}) + \frac{1}{3}\pi_d(x^{1/3}) + \cdots =: \pi_d(x) + \Pi_r(x),$$

say, we have

$$N_d(x) := \int_{1-}^{x} \exp^* d\Pi_d = \int_{1-}^{x} \big(\exp^* d\pi_d\big) * \big(\exp^* d\Pi_r\big)\,. \tag{13.11}$$

As before, we analyze each of the last two convolution factors.

We have $\pi_d(x) \le \pi_w(x)$ for all x, and both functions are nondecreasing. It follows from Lemma 2.9 that $\int_1^x d\pi_d^{*n} \le \int_1^x d\pi_w^{*n}$ holds for all positive integers n and $x \ge 1$. Hence

$$N_r(x) := \int_{1-}^{x} \exp^* d\Pi_r \le \int_{1-}^{x} \exp^* d\Pi_2 =: N_2(x),$$

and, by Lemma 13.11, $N_r(x) \ll \sqrt{x}$.

Our main job is to show, analogously with Lemma 13.10,

LEMMA 13.12.

$$N_4(x) := \int_{1-}^{x} \exp^* d\pi_d = e^k x + O(x/\log^{3/2} x)\,,$$

where $k := \int_1^\infty t^{-1}(d\pi_d - d\pi_w)(t)$ *(a negative number).*

Here is an outline of our argument: by the definitions of N_1 and N_4,

$$N_4(x) = e^k \int_{1-}^{x} dN_1 * \exp^*(d\pi_d - d\pi_w - k\delta_1).$$

Applying the Dirichlet hyperbola method, $N_4(x) = I + II$, say, with

$$I := e^k \int_{1-}^{\sqrt{x}} N_1(x/t)\, \exp^*(d\pi_d - d\pi_w - k\delta_1)(t),$$

$$II := e^k \int_{1-}^{\sqrt{x}} \left\{ \int_{\sqrt{x}}^{x/s} \exp^*(d\pi_d - d\pi_w - k\delta_1) \right\} dN_1(s).$$

The main contribution to $N_4(x)$ comes from I, whose evaluation starts with Lemma 13.10. Then, we use the homomorphic property (2.16) of the multiplication operator T^α to obtain $T^{-1}\exp^* dF = \exp^*\{T^{-1}dF\}$ for any $dF \in d\mathcal{V}$. We find

$$I = e^k x \int_{1-}^{\sqrt{x}} \Big\{1 + \frac{c_0 \cos(\log x/t)}{\log^{3/2}(ex/t)} + \frac{c_1 \cos(\log x/t)}{\log^{5/2}(ex/t)} + O\Big(\log^{-7/2}\frac{ex}{t}\Big)\Big\} \exp^* d\nu(t), \tag{13.12}$$

with

$$d\nu := T^{-1}\{d\pi_d - d\pi_w - k\delta_1\}.$$

We now give growth estimates for ν, its convolution powers, and its total variation function. Let $\nu_1(x) := \nu(x)$ and for $n \geq 2$,

$$\nu_n(x) := \int_{1-}^{x} (d\nu)^{*n} = \int_{1-}^{x} \nu_{n-1}(x/t)\, d\nu(t).$$

LEMMA 13.13. *There exist positive constants A_0, A_1, and B_0 such that for all $x > 1$ and all positive integers n,*

$$|\nu_n(x)| \leq nA_0(2\log\log ex + A_1)^{n-1}x^{-1/n} \tag{13.13}$$

and

$$\int_{1-}^{x} |d\nu| \leq 2\log\log ex + B_0. \tag{13.14}$$

PROOF. We establish (13.13) by induction. First,

$$\nu_1(x) = \int_{1-}^{x} u^{-1}\{d\pi_d - d\pi_w - k\delta_1\}(u).$$

The definition of k implies that $\nu_1(\infty) = 0$; thus

$$\nu_1(x) = -\int_{x}^{\infty} u^{-1}\{d\pi_d - d\pi_w\}(u).$$

We have omitted the term $k\delta_1$ since $x > 1$. Integrating by parts and recalling that $-1 < \pi_d(u) - \pi_w(u) \leq 0$, we see that $-1/x \leq \nu_1(x) \leq 1/x$ for all $x > 1$. Thus the case $n = 1$ is established, provided that $A_0 \geq 1$.

Before we go further, we show that (13.14) holds. Indeed, since $k < 0$,

$$\begin{aligned}
\int_{1-}^{x} |d\nu| &= \int_{1-}^{x} u^{-1}\{d\pi_d + d\pi_w + |k|\delta_1\}(u) \\
&= 2\int_{1}^{x} u^{-1} d\pi_w(u) + \int_{1-}^{x} u^{-1}\{d\pi_d - d\pi_w - k\delta_1\}(u) \\
&= 2\int_{1}^{x} \frac{1 - \cos(\log u)}{u \log u}\, du + \nu(x) \\
&\leq 2\int_{0}^{\log x} \frac{1 - \cos v}{v}\, dv + \frac{1}{x} \leq 2\log\log ex + B_0.
\end{aligned}$$

(We have used the facts that the last integrand is bounded for v near 0 and, by integration by parts, $\int_1^w \cos v\, dv/v$ is bounded for all $w > 1$.)

Now suppose that case n of (13.13) holds, and apply the hyperbola method again, this time with $y = x^{1/(n+1)}$ and $z = x/y$. Assuming also that $A_1 > B_0$, we have

$$\nu_{n+1}(x) = \iint_{st\le x} (d\nu)^{*n}(s)\, d\nu(t) =: I_1 + I_2 - I_3,$$

say, with

$$\begin{aligned}|I_1| &= \Big| \int_{1-}^{y} \nu_n(x/t)\, d\nu(t) \Big| \le nA_0(2\log\log ex + A_1)^{n-1}(x/y)^{-1/n} \int_{1-}^{y} |d\nu| \\ &\le nA_0(2\log\log ex + A_1)^{n-1} x^{-1/(n+1)}(2\log\log ex + B_0)\,,\end{aligned}$$

$$\begin{aligned}|I_2| &= \Big| \int_{1-}^{z} \nu(x/t)(d\nu)^{*n}(t) \Big| \le \frac{A_0}{x/z} \Big(\int_{1-}^{z} |d\nu| \Big)^n \\ &\le A_0 x^{-1/(n+1)}(2\log\log ex + B_0)^n\ ,\end{aligned}$$

$$\begin{aligned}|I_3| = |\nu(y)\, \nu_n(z)| &\le \frac{A_0}{y} nA_0(2\log\log ex + A_1)^{n-1} z^{-1/n} \\ &\le nA_0^2 x^{2/(n+1)}(2\log\log ex + A_1)^{n-1}\end{aligned}$$

Now

$$\begin{aligned}|I_1| + |I_2| + |I_3| &\le A_0 x^{-1/(n+1)}(2\log\log ex + A_1)^{n-1} \\ &\qquad \times \{(n+1)(2\log\log ex + B_0) + nA_0 x^{-1/(n+1)}\} \\ &\le (n+1)A_0 x^{-1/(n+1)}(2\log\log ex + A_1)^n,\end{aligned}$$

if we choose $A_1 = B_0 + A_0$. □

Recall

$$d\nu(t) := t^{-1}\{d\pi_d(t) - d\pi_w(t) - k\delta_1\} \ \text{ and } \ \nu_n(x) := \int_{1-}^{x} d\nu^{*n}.$$

We now apply the bounds of the last lemma.

LEMMA 13.14. *For $y > 1$ and each $k > 0$ we have the estimates*

$$\begin{aligned} f(y) &:= -1 + \int_{1-}^{y} \exp^* d\nu \ll \log^{-k} y\,, \\ F(y) &:= \int_{1-}^{y} \exp^*\{d\pi_d - d\pi_w - k\delta_1\} \ll y\log^{-3} y \\ h(y) &:= \int_{1-}^{y} \exp^* |d\nu| \ll \log^2 y. \end{aligned}$$

PROOF. For $y > 1,\ f(y) = \sum_{n=1}^{\infty} \nu_n(y)/n!$. Break the sum at

$$N := \lfloor (8+k)(\log\log ey + A_1) \rfloor,$$

with constants coming from estimates of the preceding lemma. We find

$$\sum_{1}^{N} \le A_0 y^{-1/N} \exp(2\log\log ey + A_1) \ll \log^{-k} y$$

(with much to spare!) and

$$\sum_{N+1}^{\infty} \le A_0 \sum_{N}^{\infty} (2\log\log ey + A_1)^n/n! \le 2A_0(2\log\log ey + A_1)^N/N!$$
$$\le 2A_0\Big(\frac{e}{8+k}\Big)^N \le 2A_0 e^{-N} \ll \log^{-k} y.$$

The two estimates give $f(y) \ll \log^{-k} y$.

For the second estimate, write

$$F(y) = \int_{1-}^{y} T\exp^* d\nu = \int_{1-}^{y} \{\delta_1 + t\, df(t)\},$$

integrate by parts, and use the preceding estimate for f with $k = 3$. We find $F(y) \ll y\log^{-3} y$.

For the estimate of h, recall that

$$|d\nu| = T^{-1} d\pi_d + T^{-1} d\pi_w + |k|\delta_1\,,$$

and thus

$$\exp^*|d\nu| = T^{-1}dN_d * T^{-1}dN_w * \exp^*(|k|\delta_1)\,.$$

Since $\pi_d(y) \le \pi_w(y)$, we have $N_d(y) \le N_w(y)$, and then, by Lemma 2.1,

$$\int_{1-}^{y} T^{-1}dN_d \le \int_{1-}^{y} T^{-1}dN_w$$

for all $y \ge 1$. Also, the integral of $\exp^*(|k|\delta_1)$ is $e^{|k|}$, a constant. Thus we have

$$\int_{1-}^{y} \exp^*|d\nu| \ll \int_{1-}^{y} T^{-1}N_d * T^{-1}dN_w$$
$$\le \int_{1-}^{y} \big(T^{-1}dN_w\big)^{*2} \le \Big(\int_{1-}^{y} T^{-1}dN_w\Big)^2 \ll \log^2 y,$$

the last since $N_w(x) \ll x$ by Theorem 13.9. □

PROOF OF LEMMA 13.12. We now approximate $N_4(x)$ by studying I and II. Recall that $N_1 \uparrow$, $N_1(x) \ll x$, and $x/s > \sqrt{x}$ in our s-range. Also, we estimate $\int_{1-}^{y} \exp^*(d\pi_d - d\pi_w - k\delta_1)$ using the preceding lemma. We find

$$II := e^k \int_{1-}^{\sqrt{x}} \Big\{ \int_{\sqrt{x}}^{x/s} \exp^*(d\pi_d - d\pi_w - k\delta_1) \Big\}\, dN_1(s)$$
$$= \int_{1-}^{\sqrt{x}} O\Big(\frac{x}{s}\log^{-3}\frac{x}{s}\Big)\, dN_1(s) = O(x\log^{-2} x).$$

Now we turn to

(13.12 bis)
$$I = e^k x \int_{1-}^{\sqrt{x}} \Big\{1 + \frac{c_0\cos(\log x/t)}{\log^{3/2}(ex/t)} + \frac{c_1\cos(\log x/t)}{\log^{5/2}(ex/t)} + O\Big(\log^{-7/2}\frac{ex}{t}\Big)\Big\} \exp^* d\nu(t),$$

The main contribution to I comes from the term 1: by Lemma 13.14,

$$I_1 := e^k x \int_{1-}^{\sqrt{x}} \exp^* d\nu(t) = e^k x\{1 + f(\sqrt{x})\} = e^k x + O(x\log^{-3/2} x).$$

For the O- term, use the bound on h from Lemma 13.14:

$$I_4 := x\int_{1-}^{\sqrt{x}} O\Big(\log^{-7/2}\frac{ex}{t}\Big)\exp^* d\nu(t)$$
$$\ll x\log^{-7/2} x\int_{1-}^{\sqrt{x}} \exp^*|d\nu| \ll x\log^{-3/2} x\,.$$

The remaining parts of (13.12 bis) can be expressed as a linear combination of four terms of the form

$$I_{a,b} := x\int_{1-}^{\sqrt{x}} (x/t)^{ai}\log^{-b}(ex/t)\exp^* d\nu(t),$$

with $a = \pm 1$, and $b = 3/2,\ 5/2$. The representation $\exp^* d\nu = \delta_1 + df$ gives

$$I_{a,b} = x^{1+ai}\log^{-b}(ex) + x^{1+ai}\int_{1-}^{\sqrt{x}} t^{-ai}\log^{-b}(ex/t)\,df(t).$$

Then, integrating by parts, and using the bound on $f(y)$ from the last lemma, we find $I_{a,b} \ll x\log^{-b}(ex)$. Together, these estimates imply that $N_4(x) = e^k x + O(x\log^{-3/2} x)$. □

Conclusion of the example. It remains to establish the formula for $N_d(x)$. We can simply repeat *mutatis mutandis* the argument that concluded the proof of Theorem 13.9. For that result, we had

$$N_w(x) = \int_{1-}^{x} \exp^*(d\pi_w + d\Pi_2),$$

with (Lemmas 13.10 and 13.11)

$$\int_{1-}^{y} \exp^* d\pi_w = y + O(y\log^{-3/2} y)\,, \qquad \int_{1-}^{y} \exp^* d\Pi_2 \ll y^{1/2},$$

while here we have

$$N_d(x) = \int_{1-}^{x} \exp^*(d\pi_d + d\Pi_r)$$

with

$$\int_{1-}^{y} \exp^* d\pi_d = e^k y + O(y\log^{-3/2} y)\,, \qquad \int_{1-}^{y} \exp^* d\Pi_r \ll y^{1/2}.$$

Thus we have $N_d(x) = cx + O(x\log^{-3/2} x)$ for a suitable constant c. □

13.7. Notes

§13.1. Theorem13.1 was proved by Beurling in [**Be37**]. It is given also in [**BD69**] and [**MV07**]. A small extension appears in [**Di69**].

Another proof of the PNT under a weaker assumption than (13.1) has been given by J. C. Schlage-Puchta and J. Vindas [**SPV12**]. Their hypothesis is a Cesaro version of Beurling's, namely, for some positive integer m and real $\gamma > 3/2$,

$$\int_1^x \frac{N(t) - at}{t}\Big(1 - \frac{t}{x}\Big)^m dt = O\Big(\frac{x}{\log^\gamma x}\Big), \quad x \to \infty.$$

The authors give an example where their condition holds and Beurling's fails.

§13.4. Our proof of Theorem 13.1 cites the tauberian theorem of Chapter 10. An alternative approach would be to extend results from Chapter 11, for here

$$E(x) := x^{-1}\{N(x) - Ax\} \ll (\log x)^{-\gamma}$$

holds with $\gamma > 3/2$, whence (11.1), the basic condition of that chapter, is satisfied. However, that path would retrace arguments of Chapter 10.

The results of Lemma 13.7 and Corollary 10.6 substantiate the remark in [**BD69**], p. 199, that if the zeta function has no zeros with real part 1, then the PNT can be derived from an L^2 hypothesis.

§13.5. We emphasize that Beurling's result is optimal in the class of counting functions having the form (13.1); as we shall show in Chapter 15, the PNT holds under a weaker L^2 condition.

§13.6. The reader is invited to consider why in proving Lemma 13.12 we needed more terms from Lemma 13.10 than we did in proving Theorem 13.9.

CHAPTER 14

Equivalences to the PNT

Vive la différence

SUMMARY. In classical prime number theory there are several asymptotic formulas that are called "equivalent" to the PNT. Here we investigate conditions under which Beurling analogues of these relations do or do not hold.

14.1. Introduction

In classical prime number theory, there are several asymptotic formulas said to be "equivalent" to the PNT. These include

$$\psi(x) \sim x, \tag{14.1}$$

$$M(x) := \sum_{n \le x} \mu(n) = o(x), \tag{14.2}$$

$$m(x) := \sum_{n \le x} \mu(n)/n = o(1), \tag{14.3}$$

$$\int_1^x \frac{d\psi(t)}{t} = \sum_{n \le x} \frac{\Lambda(n)}{n} = \log x - \gamma + o(1). \tag{14.4}$$

Here γ is Euler's constant. The sense of the equivalence is that these statements can be derived from one another without complex variable or Fourier theory. Further remarks on this notion are given in the Notes at the end of the chapter.

Each of the preceding formulas has an interpretation for g-numbers, which we discuss in the next section. In this chapter we investigate conditions under which the Beurling versions of these relations do or do not imply one other. Some of the implications hold unconditionally, while others are not valid without further hypotheses. Because the relations all imply each other in the classical case, we expect them to do the same under conditions that approximate the classical ones, e.g. if the g-integer system has a positive density A or, perhaps, that $N(x) - Ax$ is suitably small.

14.2. Implications

First, we set out the Beurling version of the relations (14.1) – (14.4). The psi relation (14.1) is equivalent to $\Pi(x) \sim x/\log x$ by integration by parts. The formula

$$\psi_1(x) := \int_1^x \frac{d\psi(t)}{t} = \log x + c_1 + o(1), \quad x \to \infty, \tag{14.5}$$

is an analogue of (14.4) for g-numbers. We call this a "sharp Mertens relation." Note that $c_1 \ne -\gamma$ in general. We can see this by making a g-number system containing all primes in $\mathbb{N}$ along with one additional g-prime λ. In such a system, (14.5) holds with $c_1 = -\gamma + (\log\lambda)/(\lambda - 1)$.

In classical number theory, $M(x)$ denotes the sum function of the Moebius μ function. The characteristic property of μ, as we noted in §3.3, is that it is the convolution inverse of the 1 function. For a g-number system with counting function N, we define $dM =: dN^{*-1}$, and, by Corollary 3.11, have $dM = \exp^*\{-d\Pi\}$. Also set $m(x) := \int_{1-}^{x} t^{-1} dM(t)$. Note that these definitions make sense even when dM is not discrete or when factorization into primes is not unique.

The following diagram shows the implications we shall establish between our several asymptotic formulas. Double arrows indicate unconditional implications. Single arrows indicate implications subject to an additional condition. Absence of vertical arrows indicates that we cannot prove implications without hypotheses which *themselves* imply the assertions $\psi(x) \sim x$ and $\psi_1(x) = \log x + c_1 + o(1)$.

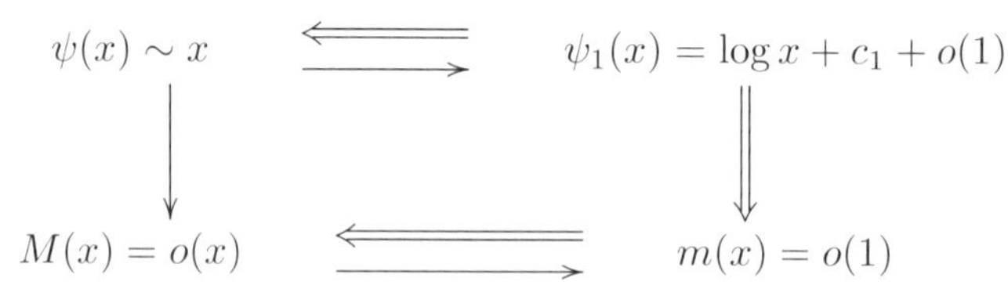

Figure 2. Implications between four formulas

14.3. Sharp Mertens relation and the PNT

One relation between the sharp Mertens formula and the PNT is true unconditionally:

PROPOSITION 14.1. *If a g-number system satisfies* (14.5), *then the PNT holds.*

PROOF. We verify (14.1) by applying integration by parts:

$$\begin{aligned}\psi(x) &= \int_1^x t\, d\psi_1(t) = x\psi_1(x) - \int_1^x \psi_1(t)\, dt \\ &= x\log x + c_1 x + o(x) - \int_1^x \{\log t + c_1 + o(1)\}\, dt \\ &= x + o(x).\end{aligned}$$

□

Next we show that (14.5) actually implies somewhat more than the PNT.

PROPOSITION 14.2. *Suppose that a sharp Mertens-type relation holds for a g-number system. Then*

$$\int_1^x \{\psi(t) - t\}t^{-2}\, dt$$

converges to a finite limit as $x \to \infty$.

PROOF. Again apply integration by parts, this time to $\int d\psi(t)/t$, to get

$$\psi_1(x) = \frac{\psi(x)}{x} + \int_1^x \frac{\psi(t)}{t^2} dt = 1 + o(1) + \log x + \int_1^x \{\psi(t) - t\}t^{-2}\, dt.$$

Since $\psi_1(x) - \log x$ has a limit as $x \to \infty$, so does the last integral on the right hand side. □

The last proposition shows that proof of (14.5) requires more than the PNT alone. Note for later use that divergence of $\int_1^\infty \{\psi(t) - t\}t^{-2}\,dt$ could occur when $\psi(t) - t$ is too large, for example if it exceeds $t/\log t$, or for certain oscillatory cases. Now we give a condition under which the PNT does imply (14.5).

THEOREM 14.3. *Suppose that the PNT holds and that, for some constant $A > 0$,*

$$|N(x) - Ax| \leq xD(x), \quad x \geq 1, \tag{14.6}$$

where D is right continuous, monotone decreasing, and satisfies

$$\int_1^\infty x^{-1} D(x)\,dx < \infty. \tag{14.7}$$

Then the sharp Mertens relation (14.5) *holds as well.*

COROLLARY 14.4. *Suppose that the PNT holds and that* (14.7) *is satisfied with*

$$D(x) := \sup_{y \geq x} |N(y) - Ay|/y.$$

Then (14.5) *holds.*

PROOF OF THEOREM 14.3. The key for our argument is the formula

$$\int_1^x \log t\,dN(t) = \int_1^x dN * d\psi,$$

the integral form of Chebyshev's identity (3.3) for g-numbers. The left hand side, integrated by parts, becomes

$$N(x)\log x - \int_1^x \frac{N(y)}{y}dy = Ax\log x - Ax + o(x), \tag{14.8}$$

since $D(y) = o(1/\log ey)$ by Lemma 9.6, and

$$\int_1^x y^{-1}N(y)\,dy = A(x-1) + \int_1^x O(D(y))\,dy = Ax + o(x/\log x).$$

For a measure $d\alpha$ on $[1, \infty)$ and Dirac point mass δ_1 at 1, we have

$$\begin{aligned} \int_{1-}^x (\delta_1 + du) * d\alpha &= \iint_{st \leq x} (\delta_1 + ds)\,d\alpha(t) \\ &= \int_{1-}^x \int_{1-}^{x/t} (\delta_1 + ds)\,d\alpha(t) = \int_{-1}^x \frac{x}{t}\,d\alpha(t). \end{aligned}$$

Thus, upon adding and subtracting $(A\delta_1 + Adt) * d\psi$ on the right hand side of the Chebyshev identity, it becomes

$$\int_1^x \frac{Ax}{t}\,d\psi(t) + \int_1^x (dN - A\delta_1 - A\,dt) * d\psi. \tag{14.9}$$

Now we show that

$$I := \int_1^x (dN - A\delta_1 - A\,dt) * d\psi = cx + o(x)$$

for some constant c. We rewrite I as $I_1 + I_2$, with

$$I_1 := \int_{1-}^x (dN - A\delta_1 - A\,dt) * (\delta_1 + dt)$$

and

$$I_2 := \int_{1-}^{x} (dN - A\delta_1 - A\,dt) * (d\psi - \delta_1 - dt).$$

We have

$$\begin{aligned} I_1 &= \int_{1-}^{x} \frac{x}{t}(dN - A\delta_1 - A\,dt) = N(x) - Ax + x\int_1^x \frac{N(t) - At}{t^2}dt \\ &= O(xD(x)) + x(c + o(1)) = cx + o(x), \end{aligned}$$

with

$$c = \int_1^\infty \frac{N(t) - ct}{t^2}dt.$$

(The integral is convergent, since it is dominated by $\int_1^\infty t^{-1}D(t)\,dt$.)

Then apply Axer's Theorem 7.2 to estimate I_2. For $x \geq 1$, take

$$A(x) := N(x) - Ax, \quad B(x) := \psi(x) - x.$$

We have

$$|A(x)| \leq xD(x),$$

with $D(x)$ satisfying the conditions of Case (B). Also, $B(x) = o(x)$ by the PNT and $B_v(x) = \psi(x) + x \ll x$. By the theorem, $I_2 = o(x)$. Combining (14.8) and (14.9) with the approximation of I, we get

$$Ax\log x - Ax + o(x) = Ax\psi_1(x) + cx + o(x),$$

which is equivalent to (14.5). □

We conclude this section by mentioning a special case of Theorem 14.3.

COROLLARY 14.5. *Suppose that the PNT holds and that* (14.6) *is satisfied with*

$$D(x) := C\log^{-\gamma} x, \quad x \geq x_0 \tag{14.10}$$

with $C > 0$ *and* $\gamma > 1$. *Then* (14.5) *holds.*

14.4. Optimality of the sharp Mertens theorem

We observed in Proposition 14.1 that the PNT is needed in order to establish the sharp Mertens formula (14.5), but, as we saw in Proposition 14.2, the PNT by itself is not sufficient for this purpose. Theorem 14.3 and its corollaries provide conditions that yield (14.5). Are those conditions excessive? In particular, we ask whether perhaps we could prove the theorem under the weaker hypothesis

$$N(x) = cx + O(x/(\log x)). \tag{14.11}$$

The following example shows that the answer to this question is No.

PROPOSITION 14.6. *PNT and* $N(x) = Ax + O(x/\log x) \not\Rightarrow$ (14.5).

PROOF. We give an example in which the first two conditions hold but the sharp Mertens relation fails. As the prime density, take

$$d\Pi(u) := \frac{1 - u^{-1}}{\log u}du + \Big(\frac{1 - u^{-1}}{\log u}\Big)^2 du. \tag{14.12}$$

First note that the PNT holds for $d\Pi$. Indeed, for $x > e$,

$$\Pi(x) = c + \int_e^x \Big\{\frac{1}{\log u} + O\Big(\frac{1}{\log^2 u}\Big)\Big\}\,du = \frac{x}{\log x} + O\Big(\frac{x}{\log^2 x}\Big).$$

We next observe that the sharp Mertens relation (14.5) does not hold for this prime density. Consider

$$\psi(x) = \int_1^x \log u \, d\Pi(u) = \int_1^x (1 - u^{-1}) \, du + \int_1^x \frac{(1 - u^{-1})^2}{\log u} \, du = x + f(x)$$

with

$$f(x) := \int_1^x \frac{(1 - u^{-1})^2}{\log u} \, du - \log x \; - 1 \sim x / \log x.$$

By the contrapositive form of Proposition 14.2, (14.5) does not hold for this g-number system.

It remains to show that (14.11) holds for this example. Let

$$d\alpha(u) = \Big(\frac{1 - u^{-1}}{\log \; u}\Big)^2 \frac{du}{u}.$$

Combining the convolution identities (3.1), (2.5), Lemma 3.10, (3.7), and Lemma 2.15, we get

$$\begin{aligned} N(x) &= \int_{1-}^x \exp^* \Big\{ \frac{1 - u^{-1}}{\log u} du + \Big(\frac{1 - u^{-1}}{\log u}\Big)^2 du \Big\} \\ &= \int_{1-}^x (\delta_1 + du) * \exp^*\{u \, d\alpha(u)\} = \int_{1-}^x \frac{x}{u} \exp^*\{u \, d\alpha(u)\} \\ &= x \int_{1-}^x \exp^* d\alpha = x \Big(1 + \sum_{n=1}^\infty \frac{1}{n!} \int_1^x d\alpha^{*n} \Big) = e^c \, x - R(x), \end{aligned}$$

where

$$c := \int_1^\infty d\alpha \quad \text{and} \quad R(x) := x \sum_{n=1}^\infty \frac{1}{n!} \int_x^\infty d\alpha^{*n}.$$

By the definition of convolution,

$$\int_x^\infty d\alpha^{*n} = \int \cdots \int_{s_1 \cdots s_n > x} d\alpha(s_1) \cdots d\alpha(s_n).$$

At least one of the s_i in the last integral exceeds $x^{1/n}$, and the remaining s_j lie in $[1, \infty)$. Letting in succession $s_1, s_2, \ldots s_n$ assume the maximal value, we have

$$\int_x^\infty d\alpha^{*n} \le n \Big(\int_1^\infty d\alpha \Big)^{n-1} \int_{x^{1/n}}^\infty d\alpha \le n \, c^{n-1} \int_{x^{1/n}}^\infty \frac{du}{u \log^2 u} = \frac{n^2 \, c^{n-1}}{\log x},$$

and hence

$$0 \le R(x) < \sum_{n=1}^\infty \frac{n^2 \, c^{n-1}}{n!} \frac{x}{\log x} = O\Big(\frac{x}{\log x}\Big).$$

This establishes (14.11). □

14.5. Implications between $M(x) = o(x)$ and $m(x) = o(1)$

PROPOSITION 14.7. *Suppose $\mathcal{N}$ is a g-number system for which $m(x) = o(1)$. Then $M(x) = o(x)$.*

PROOF. Integrate by parts:

$$M(x) = \int_{1-}^x t \, dm(t) = x \, m(x) - \int_1^x m(t) \, dt = o(x).$$ □

The converse implication need not be true without some condition on the density of integers.

PROPOSITION 14.8. $M(x) = o(x) \not\Rightarrow m(x) = o(1)$.

PROOF. We give the following example. Let $N_r(x) = \lfloor x \rfloor$, the counting function of rational integers, and let $dN := T^{-1}dN_r$. This is a g-number system which has integer density 0 and convolution inverse function

$$M(x) = \int_{1-}^{x} T^{-1}dM_r = \sum_{n\le x} \frac{1}{n}\,\mu(n).$$

By classical prime number theory, this $M(x) \to 0$ as $x \to \infty$, and so $M(x) = o(x)$ as $x \to \infty$. On the other hand,

$$m(x) := \int_{1-}^{x} T^{-1}dM = \sum_{n\le x} \frac{1}{n^2}\,\mu(n) = \frac{6}{\pi^2} + o(1) \neq o(1). \qquad \square$$

However, if $N(x) - Ax$ is small enough, then $M(x) = o(x) \implies m(x) = o(1)$.

PROPOSITION 14.9. *Suppose $\mathcal{N}$ is a g-number system for which $M(x) = o(x)$ and for some $A > 0$,*

$$|At - N(t)| \le tD(t) \text{ with } D\downarrow \text{ and } \int_1^\infty D(t)t^{-1}dt < \infty.$$

Then $m(x) = o(1)$.

PROOF. We use Axer's Theorem. For $x \ge 1$, write

$$x\,m(x) = \int_{1-}^{x} \frac{x}{t}\,dM(t) = A^{-1} + \int_{1-}^{x} \{\delta_1 + dt - A^{-1}dN\} * dM.$$

Evaluate the last integral using Form (B) of Theorem 7.2. With $A(t) := t - A^{-1}N(t)$ and $B(t) := M(t)$, the conditions are satisfied, and we conclude that the integral is $o(x)$ and thus $m(x) = o(1)$. $\square$

14.6. Connections of the PNT with $M(x) = o(x)$

PROPOSITION 14.10. *Let $\mathcal{N}$ be a g-number system for which the PNT holds and, further, suppose $\mathcal{N}$ has log density. Then $M(x) = o(x)$.*

PROOF. By Corollary 7.5, $\mathcal{N}$ has density; let A denote the density. Applying Proposition 3.15 to the relation $dM = \exp^*\{-d\Pi\}$, we find

$$LdM = \exp^*\{-d\Pi\} * L\{-d\Pi\} = -dM * d\psi.$$

Now add and subtract terms $(\delta_1 + dt) * dM$ and $A^{-1}dN * dM$ $(= A^{-1}\delta_1)$ in the last formula and integrate to obtain

$$\int_1^x LdM = \int_{1-}^{x} \Big\{\frac{x}{t} - \psi\Big(\frac{x}{t}\Big)\Big\}\,dM(t) + \int_{1-}^{x} \Big\{\frac{1}{A}N\Big(\frac{x}{t}\Big) - \frac{x}{t}\Big\}\,dM(t) - \frac{1}{A} \tag{14.13}$$
$$=: I + II - 1/A\,.$$

The left side of (14.13) equals

$$M(x)\log x - \int_1^x \frac{M(t)}{t}\,dt \;=\; M(x)\log x + O(x),$$

since $|\exp^*\{-d\Pi\}| \le \exp^* d\Pi$ and thus $|M(t)| \le N(t) \ll t$.

On the right side of (14.13), the PNT and the $|dM|$ bound give

$$|I| \le \int_{1-}^{x} o(x/t)\,dt = o(x \log x).$$

Similarly, the density condition implies that $A^{-1}N(x/t) - x/t = o(x/t)$, whence $II = o(x \log x)$ also. Thus $M(x)\log x = o(x \log x)$. □

Does the last proposition have a converse? The answer is No: We give an example of a g-number system with a density and satisfying $M(x) = o(x)$ for which the PNT does not hold. Our method is based on the Wiener-Ikehara Theorem and the observation that

$$dN + dM = \exp^*\{d\Pi\} + \exp^*\{-d\Pi\} = 2\sum_{n=0}^{\infty} d\Pi^{*2n}/(2n)! \ge 0. \tag{14.14}$$

LEMMA 14.11. *Assume $\mathcal{N}$ is a g-number system with density A and zeta function $\zeta(s)$ and that both $\zeta(\sigma+it) - A/(\sigma+it-1)$ and $1/\zeta(\sigma+it)$ converge in L_1 norm on every fixed interval $-T \le t \le T$ as $\sigma \to 1+$. Then $M(x) = o(x)$.*

PROOF. By the density condition, $N(e^x) \sim Ae^x$. We shall show that also

$$F(x) := N(e^x) + M(e^x) \sim Ae^x.$$

Set

$$G(s) := \int_0^\infty e^{-sx} F(x)\,dx - \frac{A}{s-1} = \frac{1}{s}\Big\{\zeta(s) - \frac{A}{s-1} + \zeta(s)^{-1} - A\Big\}.$$

We apply Corollary 10.6, the W-I Theorem. By (14.14), F is increasing; also condition (10.7) holds by the hypothesis on ζ and $1/\zeta$. Thus $F(x)e^{-x} \to A$ as $x \to \infty$, and so $M(e^x) = o(e^x)$. □

PROPOSITION 14.12. *$M(x) = o(x)$ and the condition*

$$N(x) = N_w(x) = Ax + O(x(\log ex)^{-3/2}) \tag{14.15}$$

do not imply the PNT.

PROOF. We shall show that $M(x) = o(x)$ but not the PNT holds for the g-number system $\mathcal{N}_w$ from §13.5 that satisfies (14.15). (This condition is clearly more demanding than density alone.) Our argument is based on the last lemma. First note, by (14.15), that $\zeta(s) - A/(s-1)$ can be extended to $\{s : \sigma \ge 1\}$ as a continuous function.

We spell out the form of the function $1/\zeta$ by using results from §13.5. We have

$$dN := \exp^* d\Pi_w = \exp^*\{d\pi_w + d\Pi_2\} = dN_1 * dN_2,$$

with $N_1(x)$ satisfying (14.15), with a different value of A, and $N_2(x) \ll x^{1/2}$. Thus $\zeta(s) = \widehat{N_1}(s)\widehat{N_2}(s)$ for $\Re s > 1$. Here

$$\widehat{N_1}(s) = \int_{1-}^{\infty} x^{-s}\{\exp^* d\pi_w\}(x) = \exp\int_1^\infty x^{-s} d\pi_w(x) = \frac{\sqrt{(s-1)^2+1}}{s-1}$$

(using the branch of the square root that is positive for s real and $s > 1$) and

$$\widehat{N_2}(s) = \int_{1-}^{\infty} x^{-s}\{\exp^* d\Pi_2\}(x) = \exp\int_{1-}^{\infty} x^{-s} d\Pi_2(x),$$

the last a function that is analytic and non zero on $\{s : \sigma > 1/2\}$. Thus

$$\frac{1}{\zeta(s)} = \frac{s-1}{\sqrt{(s-1-i)(s-1+i)}}\widehat{N_2}(s)^{-1},$$

which has an L_1 limit as $\sigma \to 1+$ on any bounded t interval. Applying the last lemma, we conclude that $M(x) = o(x)$. □

PROPOSITION 14.13. *$M(x) = o(x)$ and density* $\not\Rightarrow \psi_1(x) = \log x + c_1 + o(1)$.

PROOF. If the implication *were* true, we would have

$$M(x) = o(x) \text{ and density} \Rightarrow \psi_1(x) = \log x + c_1 + o(1) \Rightarrow \psi(x) \sim x$$

(the last implication by Proposition 14.1), which we have just shown false. □

14.7. Sharp Mertens relation and $m(x) = o(1)$

THEOREM 14.14. *Suppose $\mathcal{N}$ is a g-number system for which*

$$\psi_1(x) := \int_1^x \frac{d\psi(t)}{t} = \log x + c_1 + o(1) \tag{14.5 bis}$$

holds. Then $m(x) = o(1)$.

PROOF. We begin as in the proof of Proposition 14.10, but multiply through the Chebyshev-type formula by the homomorphism T^{-1}. We find

$$L\,dm = LT^{-1}dM = -T^{-1}dM * T^{-1}d\psi = -dm * d\psi_1.$$

Integration and use of the ψ_1 formula yield

$$\int_1^x \log t\, dm(t) = -\int_1^x \psi_1(x/t)\, dm(t) = -\int_1^x \{\log(x/t) + c_1 + o(1)\}\, dm(t)$$

or (cancelling $\int_1^x \log t\, dm(t)$ and using the bound $|dm(t)| \le dN(t)/t$)

$$m(x)\log x = -c_1 m(x) + o\Big(\int_1^x T^{-1}dN\Big) = -c_1 m(x) + o(\log x).$$

Thus

$$m(x) = \frac{o(\log x)}{\log x + c_1} = o(1).$$ □

The last theorem does not have a converse.

PROPOSITION 14.15. *$m(x) = o(1)$ and density* $\not\Rightarrow \psi_1(x) = \log x + c_1 + o(1)$.

PROOF. Again, it suffices to show that $m(x) = o(1)$ and density $\not\Rightarrow \psi(x) \sim x$. Our construction is a modification of Example 13.8 from §13.5: to reduce calculation, we this time define a "wobbly" g-number system $\mathcal{N}$ by setting

$$\Pi(x) := \int_1^x \frac{1-\cos(\log t)}{\log t}\, dt.$$

As in §13.5, the PNT does not hold, since

$$\Pi(x)/\operatorname{li}(x) = 1 - \sin(\pi/4 + \log x)/\sqrt{2} + o(1), \quad x \to \infty.$$

Also, by Lemma 13.10, the integer-counting function satisfies

$$N(x) := \int_{1-}^x \exp^* d\Pi = x + \frac{c_0 x \cos(\log x)}{\log^{3/2} x} + O\Big(\frac{x}{\log^{5/2} x}\Big),$$

and thus this g-integer system has density.

It remains to show that $m(x) = o(1)$ as $x \to \infty$. From the proof of Lemma 13.10,

$$\zeta_{\mathcal{N}}(s) = \zeta(s) = \frac{\sqrt{(s-1)^2+1}}{s-1},$$

with the branch of the square root that is positive for s real and exceeding 1. Now

$$1/\zeta(s+1) = \int_{1-}^{\infty} u^{-s-1} dM(u) = \int_{1-}^{\infty} u^{-s} dm(u).$$

and by the Mellin inversion formula, for any real number $b > 0$,

$$\frac{1}{2}\{m(x+) + m(x-)\} = \lim_{T\to\infty} \frac{1}{2\pi i} \int_{b-iT}^{b+iT} \frac{x^s\, ds}{s\,\zeta(s+1)}$$

or

$$m(x) = \lim_{T\to\infty} \frac{1}{2\pi i} \int_{b-iT}^{b+iT} \frac{x^s\, ds}{\sqrt{s^2+1}} + O(1/x)\,.$$

The integrand of the last integral has singularities at $\pm i$. Arguing as in the proof of Lemma 13.10, we deform the contour to the vertical line $\{s\colon \Re s = -1/2\}$, with loops taken about $\pm i$ to avoid crossing the rays $(-\infty - i,\, -i)$ and $(-\infty + i,\, i)$.

We claim first that the contribution of the integral along the vertical line is of order $x^{-1/2}$. We see this by writing the integral as

$$\int_{-1/2-i\infty}^{-1/2+i\infty} \frac{x^s}{s}\, ds + \int_{-1/2-i\infty}^{-1/2+i\infty} x^s \left\{ \frac{1}{\sqrt{s^2+1}} - \frac{1}{s} \right\} ds\,,$$

with the understanding that the last integral is broken at the points $-1/2 \pm i$. The first integral is 0 by the residue argument used in proving the Perron formula, and the second integral has order

$$\int_0^{\infty} x^{-1/2}(t+1)^{-3}\, dt \ll x^{-1/2}.$$

We next calculate the loop integral I_i about i. The details are close to those of Lemma 13.10, so we will be brief. In the disc of radius 1/2 about i, we have

$$\frac{1}{\sqrt{s^2+1}} = \frac{1}{\sqrt{s-i}}\,\frac{1}{\sqrt{s+i}} = \frac{1}{\sqrt{s-i}}\,\frac{1}{\sqrt{2i}}\Big\{1 + O(|s-i|)\Big\}.$$

The bottom of the loop is the line segment $s = i + re^{-\pi i}$ with r going from 1/2 to (a small number) ϵ, then a circle about i of radius ϵ, and finally a line segment $s = i + re^{\pi i}$ along the top segment with r going from ϵ to 1/2. The integrand has order $\epsilon^{-1/2}$ on the small circle and the arc length is $2\pi\epsilon$, so this portion of the integral goes to 0 with ϵ. Combining the contributions from the two horizontal line segments, we find

$$I_i = \frac{x^i}{\pi\sqrt{2i}} \int_0^{1/2} e^{-r\log x}\{r^{1/2} + O(r^{3/2})\}\,\frac{dr}{r} \ll \int_0^{\infty} e^{-r\log x} r^{1/2}\,\frac{dr}{r} = \frac{\Gamma(1/2)}{\sqrt{\log x}}$$

with Γ the Euler gamma function. A similar bound holds for the integral taken about $-i$. Thus $m(x) = o(1)$. □

14.8. Notes

§14.1. The word "equivalent" deserves a word of explanation because it is not used in a mathematical sense. Shortly after the classical PNT was proved, several other asymptotic formulas were established that were deducible from one another by simple real variable arguments ([**BD04**], [**La09**], [**Na00**]). These propositions were considered equivalent with respect to this property. While this notion was reasonable then, its logical basis completely disappeared with the discovery of elementary proofs of the PNT. Still, we continue to use this word to describe a convenient grouping of propositions.

§14.2. We have shown that no implications exist between some formulas in Figure 2 that are not connected with arrrows, unless we make additional hypotheses that are at least as strong as those used in proving the PNT.

§14.3. Theorem 14.3 and the optimality example of §14.4 were established by the authors in [**DZ12**].

Assuming that the zeta function of $\mathcal{N}$ has right hand residue A at $s = 1$ and the sharp Mertens relation (14.5) holds, then by the second part of Theorem 7.3, we have $N(x) \sim Ax$. Also, assuming (14.5), Proposition 14.2 implies that

$$\int_1^\infty \{\psi(x) - x\}\, x^{-2}\, dx$$

converges. Assuming (14.5), can we conclude that

$$\int_1^\infty \{N(x) - Ax\}\, x^{-2}\, dx$$

converges? We do not know the answer to this question, but if one assumes a rather stronger hypothesis than (14.5), the answer is Yes by Theorem 7.13.

§14.5. Another hint that the condition $m(x) = o(1)$ is stronger than $M(x) = o(x)$ is provided by the following evaluations of the Riemann zeta function for $\sigma \to 1+$, each made using one of these conditions:

$$\frac{1}{\zeta(\sigma)} = \int_{1-}^\infty x^{-\sigma} dM(x) = \sigma \int_1^\infty x^{-\sigma} \frac{M(x)}{x}\, dx = o\Big(\frac{1}{\sigma - 1}\Big),$$

$$\frac{1}{\zeta(\sigma)} = \int_{1-}^\infty x^{-(\sigma-1)} dm(x) = (\sigma - 1)\int_1^\infty x^{-\sigma} m(x)\, dx = o(1).$$

§14.7. We could have deduced Proposition 14.12 from Proposition 14.15.

CHAPTER 15

Kahane's PNT

What lies beyond optimal?

SUMMARY. Following Kahane, the PNT is established under an L^2 condition on the error term of $N(x)$ that is weaker than Beurling's condition. The argument uses the Poisson summation formula and Schwartz functions.

15.1. Introduction

Beurling's g-number problem is to find weak conditions on the integer counting function $N = N_{\mathcal{N}}$ of a g-integer system $\mathcal{N}$ under which the generalized version of the PNT holds, i.e. $\pi(x) \sim x/\log x$ as $x \to \infty$. For $\mathcal{N}$ with a positive density A, we write throughout this chapter

$$N(x) = N_{\mathcal{N}}(x) = Ax + xE(x) \tag{15.1}$$

with $E(x) \to 0$ as $x \to \infty$ (see Notes for a comment on this condition). As we showed in Chapter 13, the hypothesis

$$E(x) = O(\log^{-\gamma} x) \quad \text{with} \quad \gamma > 3/2 \tag{15.2}$$

yields the PNT, and this result is optimal in the class of error terms of this type.

P. T. Bateman and first author conjectured that the weaker L^2-condition

$$\int_1^\infty (E(x)\log x)^2 x^{-1}\,dx < \infty \tag{15.3}$$

sufficed to establish the PNT [**BD69**]. Late in the twentieth century, this conjecture was established by J.-P. Kahane by an ingenious argument in a series of papers [**Ka96**] – [**Ka98**]. Here we give a proof of this theorem, based on Kahane's work, but presented in a classical framework with full details.

Recall that the key step in Beurling's proof of the PNT was to establish the nonvanishing of his zeta function on the line $\sigma = 1$; this is the main task here as well. We show

THEOREM 15.1. *If the counting function of a g-integer system $\mathcal{N}$ satisfies* (15.3), *then the zeta function of $\mathcal{N}$ has no zeros on the vertical line $\Re s = 1$.*

Then we apply the Wiener-Ikehara theorem, in essentially the same way as in Theorem 13.1, to establish

THEOREM 15.2. *If $N_{\mathcal{N}}(x)$ satisfies* (15.3), *then the PNT holds for $\mathcal{N}$.*

As a warm-up and for use in deriving the Poisson summation formula, we show that (15.3) insures that a g-number system satisfies Chebyshev bounds:

LEMMA 15.3. *Assume that* (15.3) *holds. Then*

$$x/\log x \ll \pi(x) \ll x/\log x.$$

PROOF. The Cauchy-Schwarz inequality implies

$$\int_2^\infty x^{-1}|E(x)|\,dx \le \left(\int_2^\infty (E(x)\log x)^2 x^{-1}\,dx\right)^{1/2}\left(\int_2^\infty \frac{1}{x\log^2 x}\,dx\right)^{1/2} < \infty$$

and

$$\left|\int_1^x E(u)\log u\,du\right| \le \left(\int_1^x (E(u)\log u)^2 u^{-1}\,du\right)^{1/2}\left(\int_1^x u\,du\right)^{1/2} \ll x.$$

The bounds $x/\log x \ll \pi(x) \ll x/\log x$ then follow from Theorem 11.1. □

15.2. Zeros of the zeta function

Here is a sketch of how we prove Theorem 15.1. We first show, under conditions (15.1) and (15.3), that $\zeta(s)$ has at most a single pair of zeros, say $1\pm ia$, with $a>0$. If such a pair of zeros exists, then, as $\sigma \to 1+$,

$$\log|\zeta(\sigma+ia)| + \log|\zeta(\sigma-ia)| + \log\zeta(\sigma) = \log o(1) \to -\infty.$$

In §15.4, this estimate is captured by introducing a Schwartz function and applying the Poisson summation formula; the result is given in (15.22). (This is the device that replaces the Fejér trigonometric polynomial in Beurling's argument.) Finally, with some delicate Fourier analysis, the left hand side of (15.22) is shown to have a finite lower bound uniformly for all $\sigma > 1$, in contradiction to (15.22). We conclude that it is impossible for $\zeta(s)$ to have any zeros on $\Re s = 1$.

Condition (15.3) can be restated as

$$c := \int_0^\infty \{E(e^u)u\}^2\,du < \infty. \tag{15.4}$$

We shall henceforth assume that

$$\int_U^\infty \{E(e^u)u\}^2\,du > 0 \tag{15.5}$$

for every $U > 0$. Otherwise, $E(x) = 0$ for all $x \ge e^U$ and hence Beurling's condition (15.2) would be satisfied. In this case Theorems 13.6 and 13.1 would hold, establishing Theorems 15.1 and 15.2 at once.

With $s := \sigma + it$, as usual, write the function g of (11.7) as a Fourier transform

$$g(s) := \frac{1}{A}\int_1^\infty x^{-s}E(x)\,dx = \frac{1}{A}\int_0^\infty e^{-itu}e^{-(\sigma-1)u}E(e^u)\,du.$$

Then (with $g'(\sigma+it)$ as a shorthand for $\{(1/i)d/dt\}g(\sigma+it)$) we have

$$g'(\sigma+it) = \frac{-1}{A}\int_0^\infty e^{-itu}e^{-(\sigma-1)u}E(e^u)u\,du,$$

and, by Plancherel's identity for Fourier transforms, we have uniformly for $\sigma > 1$,

$$\int_{-\infty}^\infty |g'(\sigma+it)|^2\,dt \le \frac{1}{2\pi A^2}\int_0^\infty \{E(e^u)u\}^2\,du = \frac{c}{2\pi A^2}. \tag{15.6}$$

Now we examine the Lipschitz behavior of $g(s)$ for $\Re s = 1+$. By (11.7),

$$\frac{\zeta(s)}{As} = \frac{1}{s-1} + g(s), \tag{15.7}$$

and near a hypothetical zeta zero, $g(s)$ is close to $-1/(s-1)$.

LEMMA 15.4. *Assume that* (15.3) *holds. Then there exists a positive constant* $\eta \le 1$ *such that, whenever* $1 \le \sigma' \le \sigma$ *and* $|\sigma - \sigma' + ih| \le \eta$, *we have*

$$|g(\sigma' + it) - g(\sigma + i\{t + h\})| \le 4c^{1/4}|\sigma - \sigma' + ih|^{1/2}\Big(\int_K^\infty \{E(e^u)u\}^2 du\Big)^{1/4}, \tag{15.8}$$

where K *satisfies*

$$\frac{K}{\left(\int_K^\infty \{E(e^u)u\}^2\, du\right)^{1/2}} = \frac{1}{c^{1/2}|\sigma - \sigma' + ih|}. \tag{15.9}$$

Moreover, $K \to \infty$ *as* $|\sigma - \sigma' + ih| \to 0$.

In any case, the left hand side of (15.8) *is at most*

$$c_1 := \frac{2}{A}\int_0^\infty |E(e^u)|\, du.$$

REMARK 15.5. The right hand side of (15.8) depends only on the distance between the two points, not on the point $\sigma' + it$ itself. This is essential for the proof of the lower bound (15.14) below.

PROOF. We first note that, by (15.5), the left side of (15.9) is a continuous and increasing function of K; thus K satisfying (15.9) is well-defined.

Now we write

$$\begin{aligned} &g(\sigma' + it) - g(\sigma + i\{t + h\}) \\ &\quad = A^{-1}\int_0^\infty e^{-itu}e^{-(\sigma'-1)u}(1 - e^{-(\sigma-\sigma'+ih)u})\, E(e^u)\, du \\ &\quad =: A^{-1}(I_1 + I_2), \end{aligned} \tag{15.10}$$

say, where I_1 and I_2 are the partial integrals over the intervals $[0, K]$ and $[K, \infty)$ respectively and K satisfies condition (15.9). Let $\Delta := \sigma - \sigma' + ih$, and note that, as $\Delta \to 0$, we have $K \to \infty$ and $K|\Delta| = o(1)$. Thus there exists a positive number $\eta \le 1$ such that $K|\Delta| \le 1$ for $|\Delta| \le \eta$. For $0 < u < K$ we then have

$$|1 - e^{-u\Delta}| \le |u\Delta|\Big(1 + \frac{1}{2!} + \frac{1}{3!} + \dots\Big) < 2|u\Delta|.$$

Insert this estimate into I_1 and apply the Cauchy-Schwarz inequality to get

$$|I_1| \le 2\int_0^K |E(e^u)||\Delta|u\, du \le 2|\Delta|c^{1/2}K^{1/2}.$$

Also,

$$|I_2| \le 2\int_K^\infty |E(e^u)|\, du \le 2\Big(\int_K^\infty \{E(e^u)u\}^2 du\Big)^{1/2} K^{-1/2}.$$

Note that

$$|\sigma - \sigma' + ih|\, c^{1/2}K^{1/2} = \Big(\int_K^\infty \{E(e^u)u\}^2 du\Big)^{1/2} K^{-1/2}$$

for K satisfying (15.9). Therefore we arrive at (15.8).

The final claim of the lemma is immediate. □

We use the last lemma to show the consequences of $\zeta(s)$ having a zero on $\sigma = 1$.

LEMMA 15.6. *Assume* (15.3) *and let* $a > 0$.
(i) If $\zeta(1+ia) = 0$, *then* $\zeta(\sigma+ia) = o((\sigma-1)^{1/2})$ *as* $\sigma \to 1+$,
(ii) $\zeta(s)$ *has at most one zero on the half line* $1+it$, $t > 0$,
(iii) If $\zeta(s)$ *has zeros* $1 \pm ia$, *then, as* $\epsilon \to 0+$,

$$\int_0^\infty e^{-(1+\epsilon)v}\, d\pi(e^v) + \int_0^\infty e^{-(1+\epsilon)v} e^{iav}\, d\pi(e^v) + \int_0^\infty e^{-(1+\epsilon)v} e^{-iav}\, d\pi(e^v) \to -\infty. \tag{15.11}$$

PROOF. To show (i), apply the last lemma with $\sigma' = 1, t = a$, and $h = 0$. We have $K \to \infty$ as $\sigma \to 1+$ and

$$\int_K^\infty \{E(e^u)u\}^2\, du \to 0.$$

Then, by (15.7) and (15.8),

$$\begin{aligned} |\zeta(\sigma+ia)| &= |\zeta(\sigma+ia) - \zeta(1+ia)| \\ &\le \frac{(\sigma-1)A}{|ia(\sigma-1+ia)|} + A(\sigma-1)|g(1+ia)| \\ &\qquad + A|\sigma+ia||g(1+ia) - g(\sigma+ia)| \\ &= o(\{\sigma-1\}^{1/2}). \end{aligned}$$

To show (ii), suppose on the contrary that $\zeta(s)$ has two zeros $1+it_1$ and $1+it_2$ with $0 < t_1 < t_2$. Let

$$k(x) := 1 + \cos xt_1 + \cos xt_2 + (1/2)\{\cos x(t_2+t_1) + \cos x(t_2-t_1)\}.$$

Note that

$$k(x) = (1+\cos xt_1)(1+\cos xt_2) \ge 0.$$

Hence

$$\int_1^\infty x^{-\sigma} k(\log x)\, d\Pi(x) \ge 0, \quad \text{for} \quad \sigma > 1. \tag{15.12}$$

On the other hand, the left hand side of (15.12) equals

$$\begin{aligned} &\log\zeta(\sigma) + \log|\zeta(\sigma+it_1)| + \log|\zeta(\sigma+it_2)| \\ &\quad + (1/2)\log|\zeta(\sigma+i\{t_2+t_1\})| + (1/2)\log|\zeta(\sigma+i\{t_2-t_1\})|. \end{aligned}$$

By (15.7),

$$\log\zeta(\sigma) = \log\frac{1}{\sigma-1} + O(1)$$

and, by (i),

$$\log|\zeta(\sigma+it_j)| = (1/2)\log(\sigma-1) + \log o(1), \quad j = 1, 2,$$

as $\sigma \to 1+$. Also, $\zeta(\sigma+i(t_2\pm t_1)) \ll 1$ because $g(s)$ is uniformly bounded on the closed half-plane $\sigma \ge 1$. Therefore the left hand side of (15.12) is

$$\le \log\frac{1}{\sigma-1} + \log(\sigma-1) + \log o(1) + O(1) \to -\infty$$

as $\sigma \to 1+$. This contradicts (15.12).

To show (iii), assume that $\zeta(s)$ has zeros $1 \pm ia$. (If $\zeta(1+ia) = 0$, the reflection principle implies that $\zeta(1-ia) = 0$.) We write

$$\zeta(s) = \exp\left\{ \int_1^\infty x^{-s}\, d\pi(x) \right\} \exp\{Q(s)\},$$

where

$$Q(s) := \int_1^\infty \{\log(1 - x^{-s})^{-1} - x^{-s}\}\, d\pi(x)$$

is a holomorphic function on $\{s : \sigma > 1/2\}$. Note that $Q(s)$ and $Q'(s)$ are bounded for $\sigma \geq \frac{1}{2} + \delta$ for each fixed $\delta > 0$. Thus we have for $\epsilon > 0$,

$$\int_0^\infty e^{-itv} e^{-(1+\epsilon)v}\, d\pi(e^v) = \log\zeta(1 + \epsilon + it) - Q(1 + \epsilon + it). \tag{15.13}$$

Now, by (i), the left hand side of (15.11) equals

$$\begin{aligned}\{\log\zeta(1+\epsilon) - Q(1+\epsilon)\} &+ \{\log\zeta(1+\epsilon - ia) - Q(1+\epsilon - ia)\} \\ &+ \{\log\zeta(1+\epsilon+ia) - Q(1+\epsilon+ia)\} \\ &= \log o(1) + O(1) \to -\infty\end{aligned}$$

as $\epsilon \to 0+$. □

15.3. A lower bound for $|\zeta(\sigma + it)|$

The assumption that $\zeta(s)$ has zeros $1 \pm ia$ implies, besides (15.11), that $|\zeta(s)|$ satisfies a lower bound on the half plane $\sigma > 1$.

LEMMA 15.7. *Assume that* (15.3) *holds and that* $\zeta(s)$ *has zeros* $s = 1 \pm ia$ *with* $a > 0$. *Then there exists a positive constant* B *such that*

$$|\zeta(1 + \epsilon + it)| > B|t|^{-8} \tag{15.14}$$

uniformly for $|t| \geq 3a/2$ *and* $0 < \epsilon \leq 1$.

PROOF. We apply the same idea used in proving (ii) of the last lemma. Let

$$k(x) := \Big(1 + \frac{4}{3}\cos ax + \frac{1}{3}\cos 2ax\Big)(1 + \cos tx).$$

Since $k(x) \geq 0$, we have

$$\int_1^\infty x^{-\sigma} k(\log x)\, d\Pi(x) \geq 0, \quad \sigma > 1. \tag{15.15}$$

On the other hand,

$$\begin{aligned}\int_1^\infty & x^{-\sigma} k(\log x)\, d\Pi(x) \\ &= \log\zeta(\sigma) + \frac{4}{3}\log|\zeta(\sigma + ia)| + \frac{1}{3}\log|\zeta(\sigma + i2a)| + \log|\zeta(\sigma + it)| \\ &+ \frac{2}{3}(\log|\zeta(\sigma + i\{a+t\})| + \log|\zeta(\sigma + i\{a-t\})|) \\ &+ \frac{1}{6}(\log|\zeta(\sigma + i\{2a+t\})| + \log|\zeta(\sigma + i\{2a-t\})|).\end{aligned} \tag{15.16}$$

By (15.7), we have

$$\begin{aligned}\log \zeta(\sigma) &= -\log(\sigma-1)+O(1),\\ \log|\zeta(\sigma+i2a)| &= O(1),\\ \log|\zeta(\sigma+i(ma\pm t))| &\le \log|t|+O(1), \quad m=1,2,\end{aligned}$$

for sufficiently large $|t|$ and $1<\sigma\le 2$. Also, by a weak form of (15.11),

$$\log|\zeta(\sigma+ia)| \le \frac{1}{2}\log(\sigma-1)+O(1).$$

Therefore, by (15.15) and (15.16), we arrive at

$$-\frac{1}{3}\log(\sigma-1)+\log|\zeta(\sigma+it)|+\frac{5}{3}\log|t|+O(1)\ge 0$$

and hence

$$\log|\zeta(\sigma+it)| \ge \frac{1}{3}\log(\sigma-1)-\frac{5}{3}\log|t|+O(1) \ge \frac{1}{3}\log(\sigma-1)-\frac{11}{6}\log|t|$$

for sufficiently large $|t|$ and $1<\sigma\le 2$. It follows, for $\sigma\ge 1+|t|^{-\alpha}$, that

$$\log|\zeta(\sigma+it)| \ge -\left(\frac{\alpha}{3}+\frac{11}{6}\right)\log|t|,$$

i.e.

$$|\zeta(\sigma+it)| \ge |t|^{-(\alpha/3+11/6)}.$$

On the other hand, for $1<\sigma<1+|t|^{-\alpha}$ with $\alpha>2$, by (15.7) and (15.8),

$$\begin{aligned}&|\zeta(\sigma+it)-\zeta(1+|t|^{-\alpha}+it)|\\ &\quad\le A|t|^{-\alpha-2}+O(|t|^{-\alpha})+2A|t|\left|g(\sigma+it)-g(1+|t|^{-\alpha}+it)\right|\\ &\quad\le O(|t|^{-\alpha})+8c^{1/4}|t|(1+|t|^{\ \alpha}-\sigma)^{1/2}o(1)=o(|t|^{-\alpha/2+1})\end{aligned}$$

as $|t|\to\infty$. Hence, by taking $\alpha=18$, we find

$$|\zeta(\sigma+it)| \ge |t|^{-(\alpha/3+11/6)}+o\left(\frac{1}{|t|^{\alpha/2-1}}\right) \ge |t|^{-8}$$

uniformly for $|t|\ge t_0>3a/2$, $1<\sigma\le 2$.

Finally, for $3a/2\le|t|\le t_0$, $1<\sigma\le 2$, we have $\zeta(\sigma+it)\ne 0$. It follows that $|\zeta(\sigma+it)|>B|t|^{-8}$ with some $B>0$, uniformly for $|t|\ge 3a/2$, $1<\sigma\le 2$. □

15.4. A Schwartz function and Poisson summation

Next, by introducing a Schwartz function, we reformulate the hypothetical inequality (15.11) in terms of the Poisson summation formula (see (15.21) and (15.22)). As we noted above, this device replaces the Fejér trigonometric polynomials used in Beurling's PNT proof.

For the remainder of this chapter, let $\phi(u)$ denote a real-valued even *Schwartz function* on $\mathbb{R}$, i.e. one with the properties that it has derivatives of all orders and for all nonnegative integers k and ℓ, the k-th derivative satisfies

$$\phi^{(k)}(u)=O_{k,\ell}(|u|^{-\ell}),\ k=0,1,2,\ldots,\quad \text{as}\quad |u|\to\infty. \tag{15.17}$$

Informally, a Schwartz function is one that, along with all derivatives, vanishes rapidly at $\pm\infty$. We first show that the Fourier transform preserves the Schwartz property. For the rest of the chapter, '$\widehat{X}$' denotes the *Fourier transform* of X.

LEMMA 15.8. *Let $\phi(u)$ denote a real-valued even Schwartz function on $\mathbb{R}$. Its Fourier transform*

$$\widehat{\phi}(t) := \int_{-\infty}^{\infty} e^{-itu}\phi(u)\,du$$

is real-valued, even, and has the following properties:

(i) $\widehat{\phi} \subset C^{\infty}(\mathbb{R})$,

(ii) (15.17) *holds for* $(\widehat{\phi})^{(k)}$ *for all nonnegative integers k and ℓ,*

and

(iii) it satisfies the inversion formula

$$\phi(u) = \frac{1}{2\pi}\int_{-\infty}^{\infty} e^{-itu}\widehat{\phi}(t)\,dt.$$

SKETCH OF PROOF. The complex conjugate of $\widehat{\phi}(t)$ is

$$(\widehat{\phi}(t))^{-} = \int_{-\infty}^{\infty} e^{itu}\phi(u)du = \int_{-\infty}^{\infty} e^{-itv}\phi(-v)dv = \int_{-\infty}^{\infty} e^{-itv}\phi(v)dv = \widehat{\phi}(t);$$

thus $\widehat{\phi}$ is real-valued. A similar argument shows $\widehat{\phi}$ is even. Differentiation of the integral for $\widehat{\phi}$ is valid, since the differentiated integral is in L^1. The inversion formula we quote from [**Kn68**], Ch. 6. (Note that the inversion formula is valid with the negative sign in the exponential because ϕ is even.) The remaining relations are easy to establish using calculus. □

We next introduce a function whose Fourier transform is essentially the product of $\log\zeta(1+\epsilon+iat)$ and a Schwartz function.

LEMMA 15.9. *For $\epsilon > 0$, the function*

$$H_\epsilon(u) := \int_0^{\infty} \widehat{\phi}(u-v)e^{-(1+\epsilon)v}\,d\pi(e^v) \tag{15.18}$$

is real-valued and satisfies

$$\phi(at)\int_{-\infty}^{\infty} e^{-iatu}e^{-(1+\epsilon)u}\,d\pi(e^u) = \frac{1}{2\pi}\int_{-\infty}^{\infty} e^{-iatu}H_\epsilon(u)\,du. \tag{15.19}$$

PROOF. We note that, by (ii) of Lemma 15.8,

$$\begin{aligned}\int_{-\infty}^{\infty}\int_0^{\infty} &|\widehat{\phi}(u-v)|e^{-(1+\epsilon)v}\,d\pi(e^v)\,du\\ &= \int_0^{\infty} e^{-(1+\epsilon)v}\left(\int_{-\infty}^{\infty}|\widehat{\phi}(u-v)|\,du\right)d\pi(e^v)\\ &= \int_0^{\infty} e^{-(1+\epsilon)v}\left(\int_{-\infty}^{\infty}|\widehat{\phi}(w)|dw\right)d\pi(e^v) < \infty.\end{aligned}$$

Hence we can exchange the order of the iterated integration on the right hand side of (15.19). Then the equality (15.19) follows by a simple manipulation. □

The function H_ϵ has the following elementary properties.

LEMMA 15.10. *Assume $0 < \epsilon \le 1$ and $\pi(x) \ll x/\log x$. Then we have*

(i) $H_\epsilon \in C^{\infty}(\mathbb{R})$;

(ii) For nonnegative integers k and ℓ,

$$H_\epsilon^{(\ell)}(u) = \begin{cases} O_{k,\ell}\big(u^{-k} + e^{-\epsilon u/2}u^{-1}\big) & \text{for } u \geq 2, \\ O_\ell(1) & \text{for } |u| < 2, \\ O_{k,\ell}(|u|^{-k}) & \text{for } u \leq -2, \end{cases}$$

where the $O-$constants are independent of ϵ;

(iii) For nonnegative integers ℓ and m, $|u|^m H_\epsilon^{(\ell)}(u) \in L^1(\mathbb{R})$.

PROOF. The lemma is proved by calculus. Given u, take $K > |u|$ and suppose first $v \geq K+1$. Then, by (ii) of Lemma 15.8 we have for fixed $k \geq 2$

$$(\widehat{\phi})'(u-v)e^{-(1+\epsilon)v} \ll \frac{e^{-(1+\epsilon)v}}{(v-K)^k},$$

and the last function is integrable on $[K+1, \infty)$ with respect $d\pi(e^v)$. Also, for $|u| < K$ and $0 \leq v \leq K+1$, the left hand side of the last display is bounded. It follows that

$$H_\epsilon'(u) = \int_0^\infty (\widehat{\phi})'(u-v)e^{-(1+\epsilon)v} d\pi(e^v)$$

and $H_\epsilon'(u) \in C(\mathbb{R})$, since the differentiated integrand is absolutely integrable. Similarly, we can show that

$$H_\epsilon^{(\ell)}(u) = \int_0^\infty (\widehat{\phi})^{(\ell)}(u-v)e^{-(1+\epsilon v} d\pi(e^v)$$

and $H_\epsilon^{(\ell)}(u) \in C(\mathbb{R})$. This proves (i).

To prove (ii), it suffices to show that the bounds asserted there are in fact satisfied by

$$\overline{H}_\epsilon^{(\ell)}(u) := \int_0^\infty |(\widehat{\phi})^{(\ell)}(u-v)|e^{-(1+\epsilon)v}\, d\pi(e^v). \tag{15.20}$$

For $|u| < 2$, the bound on $(\widehat{\phi})^{(\ell)}(u-v)$ given by (ii) of Lemma 15.8 insures that $\overline{H}_\epsilon^{(\ell)}(u) = O_\ell(1)$.

Next, suppose that $u > 2$, and split the integral for $\overline{H}_\epsilon^{(\ell)}(u)$ into three partial integrals I_1, I_2 and I_3 extending over intervals $(0, u-1]$, $(u-1, u+1]$, and $(u+1, \infty)$ respectively. We have, by (ii) of Lemma 15.8, integration by parts, and application of the Chebyshev upper bound,

$$\begin{aligned} |I_1| &\ll \int_0^{u-1} \frac{e^{-(1+\epsilon)v}\, d\pi(e^v)}{(u-v)^{k+1}} \\ &= e^{-(1+\epsilon)(u-1)}\pi(e^{u-1}) + \int_0^{u-1} \pi(e^v)\Big(\frac{1}{(u-v)^{k+1}} - \frac{k+1}{(u-v)^{k+2}}\Big)e^{-(1+\epsilon)v}\, dv \\ &\ll \frac{e^{-\epsilon u}}{u} + \int_1^{u-1} \frac{e^{-\epsilon v} dv}{v(u-v)^{k+1}} \ll_k \frac{e^{-\epsilon u/2}}{u} + \frac{\log u}{u^{k+1}} \end{aligned}$$

with $k \geq 2$, since the last integral is bounded by

$$\begin{aligned} &\int_1^{u/2} \frac{dv}{v(u-v)^{k+1}} + \int_{u/2}^{u-1} \frac{e^{-\epsilon v}\, dv}{v(u-v)^{k+1}} \\ &\qquad \ll_k \frac{1}{u^{k+1}}\int_1^{u/2} \frac{dv}{v} + \frac{e^{-\epsilon u/2}}{u}\int_{u/2}^{u-1} \frac{dv}{(u-v)^{k+1}}. \end{aligned}$$

An analogous but simpler calculation gives

$$|I_3| \ll \int_{u+1}^{\infty} \frac{e^{-(1+\epsilon)v}\, d\pi(e^v)}{(v-u)^{k+1}} \ll e^{-\epsilon u} \int_{u+1}^{\infty} \frac{dv}{v(v-u)^{k+1}} \ll e^{-\epsilon u} u^{-1}.$$

Also, since $(\widehat{\phi})^{(\ell)}$ is bounded for each ℓ,

$$|I_2| \ll_\ell \int_{u-1}^{u+1} e^{-(1+\epsilon)v}\, d\pi(e^v) \le e^{-(u-1)(1+\epsilon)} \pi(e^{u+1}) \ll e^{-\epsilon u} u^{-1}.$$

This establishes the bound for $\overline{H}_\epsilon^{(\ell)}(u)$ for $u \ge 2$.

We handle $u < -2$ similarly: by (ii) of Lemma 15.8, the integral for $\overline{H}_\epsilon^{(\ell)}(u)$ is

$$\begin{aligned} &\ll_{k,\ell} \int_0^\infty \frac{e^{-(1+\epsilon)v} d\pi(e^v)}{(v+|u|)^{k+1}} \\ &= \int_0^\infty \pi(e^v) \left(\frac{1+\epsilon}{(v+|u|)^{k+1}} + \frac{k+1}{(v+|u|)^{k+2}} \right) e^{-(1+\epsilon)v}\, dv \\ &\ll \frac{1}{|u|^{k+1}} + \int_1^\infty \frac{dv}{(v+|u|)^{k+1}} \ll \frac{1}{|u|^k}. \end{aligned}$$

This finishes the proof of (ii). Note that all the O-constants are independent of ϵ.

Finally, we turn to (iii). To show that $|u|^m H_\epsilon^{(\ell)}(u) \in L^1(\mathbb{R})$ for any nonnegative integers ℓ, m and $\epsilon > 0$, we use the estimates of $|H_\epsilon^{(\ell)}(u)|$ from (ii). We have

$$\begin{aligned} \int_{-\infty}^{-\infty} |u|^m |H_\epsilon^{(\ell)}(u)|\, du &\ll \int_{-\infty}^{-2} |u|^{m-k}\, du + \int_{-2}^{2} |u|^m\, du \\ &\quad + \int_2^\infty u^m \left(u^{-k} + e^{-\epsilon u/2} u^{-1}\right) du < \infty, \end{aligned}$$

provided that k is taken suitably large in (ii). Therefore $|u|^m H_\epsilon^{(\ell)}(u) \in L^1(\mathbb{R})$. $\square$

To investigate the connection between (15.11) and (15.14), we fix a Schwartz function $\phi(t)$ satisfying the additional conditions (i) $\phi(t) = 1$ on $[-3a/2, 3a/2]$ and (ii) ϕ has support contained in $(-2a, 2a)$. For $0 < \epsilon \le 1$, let

$$M_\epsilon(t) := \phi(at) \int_0^\infty e^{-iatv} e^{-(1+\epsilon)v}\, d\pi(e^v).$$

By Lemma 15.9, $M_\epsilon(t)$ is the Fourier transform of $H_\epsilon(u/a)/(2\pi a)$. Then, by (iii) of Lemma 15.10, $|H_\epsilon(u/a)| + |(H_\epsilon(u/a))'| \in L^1(\mathbb{R})$. Hence we have by the Poisson summation formula ([**BD04**], Appendix A3; [**MV07**], Appendix D2)

$$\sum_{m \in \mathbb{Z}} M_\epsilon(m) = \sum_{m \in \mathbb{Z}} H_\epsilon(2\pi m/a)/a. \tag{15.21}$$

We note that $M_\epsilon(m) = 0$ for $|m| \ge 2$. Hence the left hand sides of (15.21) and (15.11) are equal, and we have another description of the contribution of $\log \zeta(s)$ near the pole and purported zeros of zeta.

LEMMA 15.11. *Assume that* (15.3) *is satisfied and that* $\zeta(s)$ *has zeros* $1 \pm ia$. *Let* H_ϵ^- *denote the negative part of* H_ϵ. *Then*

$$\sum_{n \ge 1} H_\epsilon^-(2\pi n/a) \to -\infty \quad as \quad \epsilon \to 0+. \tag{15.22}$$

PROOF. By (15.11), the left hand side of (15.21) tends to $-\infty$ as $\epsilon \to 0+$. Then, by (ii) of Lemma 15.10, the right hand side of (15.21) is at least

$$\sum_{n\geq 1} H_\epsilon^-(2\pi n/a) + O(1).$$

Hence (15.22) follows. □

15.5. Estimating the sum of a series by an improper integral

For further analysis of (15.22), let

$$h_\epsilon(u) := uH_\epsilon(u/a)/(2\pi a).$$

By (15.22),

$$\sum_{n\geq 1} \frac{h_\epsilon^-(2\pi n)}{n} \to -\infty \quad \text{as} \quad \epsilon \to 0+, \tag{15.23}$$

provided that (15.3) is satisfied and $\zeta(s)$ has zeros at $1 \pm ia$. We now establish some analytic properties of h_ϵ.

LEMMA 15.12. *Assume that* $0 < \epsilon \leq 1$ *and* $\pi(x) \ll x/\log x$. *We have*
(i) $h_\epsilon(u) \in L^1(\mathbb{R}) \cap C^\infty(\mathbb{R})$
(ii) $h_\epsilon(u) \to 0$ *as* $|u| \to \infty$;
(iii) There is a constant C *such that* $|h_\epsilon(u)|$, $|h_\epsilon'(u)| \leq C$ *hold uniformly for* $-\infty < u < \infty$ *and* $0 < \epsilon \leq 1$.

PROOF. (i) follows directly from (i) and (iii) of Lemma 15.10.
To show (ii), we use the bound (ii) of the same lemma:

$$h_\epsilon(u) = O(e^{-\epsilon|u|/2} + |u|^{-k}).$$

For (iii), note that the preceding bound applies as well for $h_\epsilon'(u)$. It remains to deal with small u, say $|u| \leq 2$. In this case, for each ℓ, we have the bounds

$$|H_\epsilon(u)|, \, |H_\epsilon'(u)| \ll_\ell \int_0^\infty (1+v)^{-\ell} e^{-v} d\pi(e^v),$$

independently of $\epsilon \in (0, 1]$. For $\ell = 2$, say, the last integral converges to a finite number. Thus $h_\epsilon(u)$ and $h_\epsilon'(u)$ are bounded in this range too, so the claimed absolute bounds hold. □

We next estimate the sum of a series involving $|h_\epsilon^-|^3$ by a definite integral, using a delicate calculation. For this, introduce

$$S_j = S_{j,\epsilon} := \{n \in \mathbb{N} : h_\epsilon^-(2\pi n) \in [-2^{-j}, -2^{-j-1})\}, \quad j = 1, 2, \ldots$$

and

$$S_0 = S_{0,\epsilon} := \{n \in \mathbb{N} : h_\epsilon^-(2\pi n) \in [-C, -2^{-1})\}.$$

with C given in (iii) of Lemma 15.12. By (ii) of the same lemma, S_j and S_0 are all finite sets. Also, we set

$$K = C + 3 \tag{15.24}$$

and define intervals

$$I_n(\alpha) = I_{j,n}(\alpha) := \left[2n\pi - \alpha\, 2^{-j-1}/K, \, 2n\pi + \alpha\, 2^{-j-1}/K\right]$$

for $n \in S_j$, $j = 0, 1, 2, \ldots$, and $\alpha \in (0, 1]$.

Let $\tau(u)$ be a real-valued Schwartz function with support $[-1, 1]$ such that $\tau(u)$ is even and nonnegative with $\int \tau(u)\,du = 1$, and let $\tau_\delta(u) = \frac{1}{\delta}\tau(u/\delta)$ with $\delta > 0$.

LEMMA 15.13. *Assume that* $\pi(x) \ll x/\log x$. *Let* $\delta = \delta_j := 2^{-j-3}/K$, *where* K *is defined in* (15.24), *and let* $0 < \epsilon \le 1$. *Then*

$$\sum_{n\in S_j}\int_{I_n(3/4)} |-h_\epsilon^- * \tau_\delta(u)|^2\,du \ge \frac{2^{-6}}{K^4}\sum_{n\in S_j}|h_\epsilon^-(2n\pi)|^3 \tag{15.25}$$

for $j = 0, 1, 2, \ldots$.

PROOF. Recalling that $|h_\epsilon'(u)| < K$, by Lemma 15.12 (iii), we see first that

$$h_\epsilon^-(u) = h_\epsilon(u) < -2^{-j-1} + K|u - 2n\pi| \le 0 \tag{15.26}$$

for $u \in I_n(1)$. Then, for $u \in I_n(1/2)$,

$$-h_\epsilon^-(u) > 2^{-j-1} - K\frac{2^{-j-1}}{2K} = 2^{-j-2}.$$

Hence, for $u \in I_n(1/4)$,

$$\begin{aligned} -h_\epsilon^- * \tau_\delta(u) &= \int_{|u-v|\le\delta} \tau_\delta(u-v)(-h_\epsilon^-(v))\,dv \\ &\ge \int_{|u-v|\le\delta} \tau_\delta(u-v)2^{-j-2}\,dv = 2^{-j-2}\int_{|w|\le\delta}\tau_\delta(w)\,dw = 2^{-j-2}. \end{aligned}$$

It follows that

$$\int_{I_n(3/4)} (-h_\epsilon^- * \tau_\delta(u))^2\,du \ge 2^{-2j-4}\frac{2^{-j-2}}{K}, \tag{15.27}$$

where $2^{-j-2}/K$ is the length of $I_n(1/4) \subset I_n(3/4)$. For $n \in S_j$, $j = 1, 2, \ldots$, $|h_\epsilon^-(2n\pi)| \le 2^{-j}$, and for $n \in S_0$, $|h_\epsilon^-(2n\pi)| \le C < K$. Hence (15.25) follows from (15.27). We remark that K^4 is extravagant in (15.25) for $j = 1, 2, \ldots$, for there the factor K suffices. ∎

Recall from Lemma 15.9 that

$$M_\epsilon(t) = \phi(at)\int_0^\infty e^{-iatv}e^{-(1+\epsilon)v}\,d\pi(e^v) = \int_{-\infty}^\infty e^{-itu}h_\epsilon(u)u^{-1}du.$$

To remove the effects of the pole of $\zeta(s)$ at $s = 1$ and its supposed zeros at $s = 1\pm ia$, introduce

$$\begin{aligned} m_\epsilon(t) &:= (1-\phi(at))\int_0^\infty e^{-iatv}e^{-(1+\epsilon)v}\,d\pi(e^v) \\ &= \int_{-\infty}^\infty e^{-itu}\{e^{-(1+\epsilon)u/a}\,d\pi(e^{u/a}) - h_\epsilon(u)u^{-1}du\}. \end{aligned} \tag{15.28}$$

Then $m_\epsilon(t) = 0$ for $t \in [-3/2, 3/2]$. Also, since $h_\epsilon(u) \in L^1(\mathbb{R})$, we have

$$m_\epsilon'(t) = -i\int_{-\infty}^\infty e^{-itu}\{ue^{-(1+\epsilon)u/a}d\pi(e^{u/a}) - h_\epsilon(u)du\}.$$

Therefore

$$i\,m_\epsilon'(t)\widehat{\tau}_\delta(t) = \int_{-\infty}^\infty e^{-itu}\{f_\epsilon(u) - h_\epsilon * \tau_\delta(u)\}\,du, \tag{15.29}$$

where

$$f_\epsilon(u) := \int_0^\infty \tau_\delta(u-v) v e^{-(1+\epsilon)v/a} d\pi(e^{v/a}).$$

Note that $f_\epsilon(u) \geq 0$.

We now come to a key inequality for showing the uniform boundedness of $\sum_{n\geq 1} h_\epsilon^-(2n\pi)/n$.

LEMMA 15.14. *Under the assumptions of the last lemma,*

$$\frac{1}{2\pi}\int_{-\infty}^{\infty} |m_\epsilon'(t)\widehat{\tau}_\delta(t)|^2\, dt \geq \frac{2^{-6}}{K^4}\sum_{n\in S_j} |h_\epsilon^-(2n\pi)|^3, \quad j = 0,1,2,\ldots. \tag{15.30}$$

PROOF. It suffices to show that

$$\int_{-\infty}^{\infty} |m_\epsilon'(t)\widehat{\tau}_\delta(t)|^2\, dt \geq 2\pi \sum_{n\in S_j}\int_{I_n(3/4)} |-h_\epsilon^- * \tau_\delta(u)|^2\, du \tag{15.31}$$

for $j = 0,1,2,\ldots$; then (15.30) follows from (15.25).

We have

$$\int_{-\infty}^{\infty} f_\epsilon(u) du = \int_{-\infty}^{\infty} \tau_\delta(u) du \cdot \int_0^\infty v e^{-(1+\epsilon)v/a} d\pi(e^{v/a}) < \infty.$$

Since $f_\epsilon \in L^1(\mathbb{R})$ and f_ϵ is bounded, $f_\epsilon \in L^2(\mathbb{R})$. Also, $h_\epsilon(u) \in L^2(\mathbb{R})$ for the same reason. Then $h_\epsilon * \tau_\delta(u) \in L^2(\mathbb{R})$ too. Hence, from (15.29) and the Plancherel identity for Fourier transforms, the left hand side of (15.31) equals

$$2\pi \int_{-\infty}^{\infty} |f_\epsilon(u) - h_\epsilon * \tau_\delta(u)|^2\, du \geq 2\pi \sum_{n\in S_j}\int_{I_n(3/4)} |f_\epsilon(u) - h_\epsilon * \tau_\delta(u)|^2\, du, \tag{15.32}$$

since the sets $\{I_n(3/4) : n \in S_j\}$ are pairwise disjoint.

For $u \in I_n(3/4)$ and $|v| \leq \delta$, we have $u - v \in I_n(1)$ and so by (15.26),

$$-h_\epsilon^- * \tau_\delta(u) = -\int_{|v|\leq\delta} \tau_\delta(v)\, h_\epsilon^-(u-v)\, dv \geq 0.$$

Also, $f_\epsilon \geq 0$. Thus

$$\int_{I_n(3/4)} |f_\epsilon(u) - h_\epsilon * \tau_\delta(u)|^2\, du \geq \int_{I_n(3/4)} |-h_\epsilon^- * \tau_\delta(u)|^2\, du, \tag{15.33}$$

and (15.31) follows from (15.32) and (15.33). □

15.6. Conclusion of the proof

In the remainder of the argument, we use the zeta lower bound (15.14) to establish uniform boundedness of the left side of (15.23) for $0 < \epsilon \leq 1$; this is the sought-after contradiction.

LEMMA 15.15. *Assume* (15.3) *and let* $0 < \delta \leq 1$. *Then, we have uniformly for* $0 < \epsilon \leq 1$,

$$\int_{-\infty}^{\infty} |m_\epsilon'(t)\widehat{\tau}_\delta(t)|^2\, dt \ll \delta^{-18}. \tag{15.34}$$

PROOF. The Fourier transform $\widehat{\tau}_\delta(t)$ of the Schwartz function τ_δ is itself a Schwartz function, and so for any positive n it satisfies

$$|\widehat{\tau}_\delta(t)| = |\widehat{\tau}(\delta t)| = O_n(\{\delta|t|\}^{-n}), \quad \text{as} \quad |t| \to \infty.$$

Also, recall that $m'_\epsilon(t) = 0$ for $|t| < 3/2$, and $|\widehat{\tau}_\delta|$ and $|m'_\epsilon|$ are even, because each of $\widehat{\tau}_\delta$ and m_ϵ is the Fourier transform of a real function. Therefore

$$\int_{-\infty}^{\infty} |m'_\epsilon(t)\widehat{\tau}_\delta(t)|^2\,dt = 2\int_{3/2}^{\infty} |m'_\epsilon(t)\widehat{\tau}_\delta(t)|^2\,dt \le \frac{T_n}{\delta^{2n}} \int_{3/2}^{\infty} t^{-2n}|m'_\epsilon(t)|^2\,dt \tag{15.35}$$

for some constant T_n. Differentiate the first relation in (15.28), obtaining

$$m'_\epsilon(t) = \int_0^\infty e^{-iatu}e^{-(1+\epsilon)u}\{(-iau)(1-\phi(at)) - a\phi'(at)\}\,d\pi(e^u).$$

Also, recall that $1-\phi(at)$ and $\phi'(at)$ are each bounded. It follows that

$$|m'_\epsilon(t)|^2 \ll \left|\int_0^\infty e^{-iatu}e^{-(1+\epsilon)u}\,d\pi(e^u)\right|^2 + \left|\int_0^\infty e^{-iatu}e^{-(1+\epsilon)u}u\,d\pi(e^u)\right|^2. \tag{15.36}$$

For $|t| \ge 3/2$, by (15.13) and (15.7), the first term on the right hand side of the last inequality is bounded by

$$2(|\log\zeta(1+\epsilon+iat)|^2 + |Q(1+\epsilon+iat)|^2) = O(\log^2|t|).$$

Here we use the lower bound (15.14) to show the second term is at most

$$2\left(\left|\frac{\zeta'(1+\epsilon+iat)}{\zeta(1+\epsilon+iat)}\right|^2 + |Q'(1+\epsilon+iat)|^2\right) \ll 1 + |t|^{16}|\zeta'(1+\epsilon+iat)|^2.$$

Also, by (15.7), for $|t| \ge 3/2$,

$$|\zeta'(1+\epsilon+iat)|^2 \ll 1 + |t|^2|g'(1+\epsilon+iat)|^2.$$

The last inequalities, together with (15.35) and (15.36) give

$$\int_{-\infty}^{\infty} |m'_\epsilon(t)\widehat{\tau}_\delta(t)|^2\,dt \le \frac{T'_n}{2\pi}\,\delta^{-2n}\int_{3/2}^{\infty} t^{16-2n}\{1+t^2|g'(1+\epsilon+iat)|^2\}\,dt.$$

Choosing $n=9$ in the last expression, we obtain, by (15.6),

$$\int_{-\infty}^{\infty} |m'_\epsilon(t)\widehat{\tau}_\delta(t)|^2\,dt \le T'\delta^{-18}\{1 + c/(2\pi a A^2)\}.$$

□

The next step approaches the boundedness of $\sum_{n\ge1} h_\epsilon^-(2n\pi)/n$.

LEMMA 15.16. *Under the assumptions of the last lemma there are a positive integer k and a constant Z, both independent of ϵ, such that*

$$\sum_{n\ge1} |h_\epsilon^-(2n\pi)|^k \le Z. \tag{15.37}$$

PROOF. Combining (15.30) and (15.34), we obtain, uniformly for $0 < \epsilon \le 1$ and $j = 0, 1, 2, \ldots$,

$$\sum_{n\in S_j} |h_\epsilon^-(2n\pi)|^3 < Z_1\int_{-\infty}^{\infty} |m'_\epsilon(t)\widehat{\tau}_\delta(t)|^2\,dt < Z_2\,\delta^{-18} \tag{15.38}$$

For $j = 0, 1, 2, \ldots$, choose $\delta = \delta_j := 2^{-j-3}/K$, as in Lemma 15.13. We obtain

$$\sum_{n\in S_j} |h_\epsilon^-(2n\pi)|^3 < Z_2\,(2^{j+3}K)^{18} = Z_3\,2^{18j}.$$

Also, $|h_\epsilon(2n\pi)| \le 2^{-j}$ for $n \in S_j$, $j = 1, 2, \ldots$, so we find for each index $j \ge 1$

$$\sum_{n \in S_j} |h_\epsilon^-(2n\pi)|^{22} < Z_3\, 2^{-j}. \tag{15.39}$$

The preceding bound holds also for $n \in S_0$, and we have $|h_\epsilon(2n\pi)| \le K$. Thus

$$\sum_{n \in S_0} |h_\epsilon^-(2n\pi)|^{22} < Z_3\, K^{19} = Z_4.$$

Finally,

$$\sum_{n \ge 1} |h_\epsilon^-(2n\pi)|^{22} = \sum_{j \ge 0} \sum_{n \in S_j} |h_\epsilon^-(2n\pi)|^{22} < Z_4 + Z_3 \sum_{j \ge 1} 2^{-j} =: Z. \qquad \square$$

Proof of Theorem 15.1. By Hölder's inequality and (15.37), we have

$$\left| \sum_{n \ge 1} \frac{h_\epsilon^-(2n\pi)}{n} \right| \le \left(\sum_{n \ge 1} |h_\epsilon^-(2n\pi)|^k \right)^{1/k} \left(\sum_{n \ge 1} \frac{1}{n^{k/(k-1)}} \right)^{(k-1)/k} \le P, \tag{15.40}$$

where the constant P is independent of ϵ. This contradicts (15.23). Therefore, $\zeta(s)$ has no zeros on the line $\sigma = 1$. $\square$

Proof of Theorem 15.2. The argument here is the same as that used for the conclusion of Theorem 13.1. The hypothesis (15.3) enables us to apply Lemma 13.7. We have

$$\lim_{\sigma, \sigma' \to 1+} \int_{-2\lambda}^{2\lambda} |G(\sigma + it) - G(\sigma' + it)|\, dt = 0 \tag{10.8 ter}$$

for every $\lambda > 0$, where

$$G(s) := -\frac{1}{s} \frac{\zeta'}{\zeta}(s) - \frac{1}{s-1}.$$

Then, an application of Corollary 10.6 implies that the PNT holds for the g-number system $\mathcal{N}$. This completes the proof of Theorem 15.2. $\square$

15.7. Notes

§15.1. If $E(x)$ satisfies (15.3), then $\int_2^\infty x^{-1} |E(x)|\, dx$ is finite by the first estimate of Lemma 15.3. It follows that $\mathcal{N}$ has density and $E(x) = o(1)$ by Proposition 7.7.

Bateman and the first author showed in [**BD69**] that if condition (15.3) is satisfied and if $\zeta(s)$ has no zeros on the line $\Re s = 1$ then the PNT holds. This raised the question of whether (15.3) might in fact imply the nonvanishing of $\zeta(s)$ on the line, and so suffice for the PNT.

Further improvements of Kahane's theorem were given by the second author in [**Zh15**]. Let

$$E_0(x) := (N(x) - Ax)/x \quad \text{and} \quad E_k(x) := x^{-1} \int_1^x E_{k-1}(t)\, dt, \;\; k \ge 1.$$

One of the results asserts that if

$$\int_1^\infty \{E_k(x) \log x\}^2 x^{-1} dx < \infty$$

for some $k \ge 1$, then the PNT holds. (Kahane's theorem is the $k = 0$ case.)

CHAPTER 16

PNT with Remainder

What's left to say?

SUMMARY. The PNT remainder is the difference $\Delta(x) := |\pi(x) - \mathrm{li}(x)|$. Here we establish "Nyman type" and "de la Vallée Poussin type" estimates for Δ, starting from different assumptions on the g-number counting function $N(x)$.

16.1. Introduction

In this chapter, we shall establish the PNT with remainder terms of two forms. The first is $\Delta(x) \ll x \log^{-k} x$, which we call a "Nyman type remainder term," after its discoverer, and the second is $\Delta(x) \ll x \exp\{-c \log^{\beta} x\}$, which we call a "dlVP type remainder term," after Ch. J. de la Vallée Poussin, who first gave such a bound.

The rational integers differ from g-integers in that the former are equally spaced. This property allows the Riemann zeta function $\zeta_R(s)$ to be extended to an analytic function on $\mathbb{C}$ with a simple pole at $s = 1$. These facts and known zero-free regions for zeta yield a dlVP type error term with β any number less than $3/5$. This result was achieved in 1958 and is still essentially the best one known.

The situation for a g-number system $\mathcal{N}$ can be quite different, depending on the distribution of $\mathcal{N}$: if we assume that $N(x) = Ax + O(x^{\theta})$ with some number $\theta \in [0, 1)$, then the zeta function $\zeta = \zeta_{\mathcal{N}}$ can be analytically continued to the half plane $\{\sigma > \beta\}$. In this case, estimates of Δ with β equal to $1/2$ can be obtained by mimicking classical contour shifting methods. We shall not pursue this case further here. In the next chapter, we shall show with an example that such a bound is optimal.

The hypotheses made on $\mathcal{N}$ in this chapter do not allow extension of zeta to the left of the line $\{\sigma = 1\}$. In the first case we assume that $N(x) = Ax + O(x \log^{-\gamma} x)$, and in the second one, $N(x) = Ax + O(x \exp\{-c \log^{\alpha} x\})$. In both situations we apply an inversion method to a high order derivative of (essentially) a Mellin transform of Δ. In the first case we use an analogue of the Wiener-Ikehara method, but with a Gaussian function in place of the triangular function; and in the second case, we use the Plancherel identity. For one side of the resulting integral identities we estimate the derivative of the zeta function expression, and on the other side, obtain a pointwise bound for $\Delta(x)$ by simple tauberian arguments.

The zeta function estimates in the two cases are similar; we treat these together in the next section. The parts of the arguments that differ are presented separately in the last two sections.

16.2. Two general lemmas

Suppose that a g-number system $\mathcal{N}$ has integer counting function

$$N(x) = Ax + O(x\log^{-\gamma} x) \tag{16.1}$$

with constants $A > 0$ and $\gamma > 3$. As this is the weaker of our two conditions, results established for this case are valid in the dlVP case as well.

We introduce an auxiliary function

$$f(s) := (s-1)\zeta(s)/s, \quad \sigma > 1,$$

which will be very useful in both cases. By integration by parts

$$f(s) = A + (s-1)\int_1^\infty x^{-s-1}(N(x) - Ax)\,dx. \tag{16.2}$$

This formula shows that $f(s)$ is well-defined and differentiable in the closed half-plane $\{\sigma \geq 1\}$ and is bounded near 1. We begin by establishing a lower bound for $|f|$ using upper bounds upon f and its derivatives.

LEMMA 16.1. *Suppose that*

$$|f^{(\ell)}(\sigma + it)| \leq Bg(t)^{\ell+1}, \quad \ell = 0, 1, \tag{16.3}$$

holds with some constant $B > 0$ and some positive valued function $g(t)$ uniformly for $1 < \sigma \leq 2$, $t \in \mathbb{R}$, and that $g(t)$ satisfies

$$1 \leq g(nt) \leq Dg(t), \quad n = 1, 2, \ldots, 7, \tag{16.4}$$

with some constant $D \geq 1$ for all $t \in \mathbb{R}$. Then

$$|f(\sigma + it)|^{-1} \leq Cg(t)^\lambda \tag{16.5}$$

holds with constants $C > 0$ and $\lambda = 5$ for $1 < \sigma \leq 2$ and all $t \in \mathbb{R}$.

PROOF. It is convenient to treat separately the case of small values of $|t|$. We see from (16.2) that $f(s)$ is continuous on $\{\sigma \geq 1\}$ and nonvanishing at 1, so $|f(s)|$ has a positive lower bound on a sufficiently small closed semidisc about 1. Also, $\zeta(s) \neq 0$ on the open half plane $\{\sigma > 1\}$, so $f(s)$ has the same property. It follows that $|f(s)| \geq \epsilon$ for $1 \leq \sigma \leq 2$ and $|t| \leq t_0$ for sufficiently small $\epsilon > 0$ and $t_0 > 0$. In this case, (16.5) holds with any $\lambda \geq 0$.

For $|t| \geq t_0$, we are away from the pole of zeta, and, since $\zeta(s) = sf(s)/(s-1)$, an estimate for either $f(s)$ or $\zeta(s)$ translates to an estimate for the other. In this case, it suffices to show that

$$|\zeta(\sigma + it)| \geq C_1 g(t)^{-\lambda}. \tag{16.6}$$

We use the classical method of first giving a lower bound for $|\zeta(\sigma + it)|$ for all $\sigma \in [\sigma_1, 2]$ with a suitable $\sigma_1 > 1$ and then extending the result to all $\sigma \in (1, 2]$ with the aid of an upper bound on $|\zeta'(\sigma + it)|$. By (16.3),

$$|\zeta^{(\ell)}(\sigma + it)| \leq C_2 g(t)^{\ell+1}, \quad \ell = 0, 1. \tag{16.7}$$

We remark that the classical 3-4-1 inequality,

$$\zeta^3(\sigma)|\zeta(\sigma + it)|^4|\zeta(\sigma + i2t)| \geq 1,$$

will yield (16.6) with $\lambda = 7$. However, we shall obtain a better bound with the aid of a more complicated trigonometric polynomial.

There exists a cosine polynomial (see Notes)

$$p(\theta) = a_0 + a_1 \cos\theta + a_2 \cos 2\theta + \cdots + a_7 \cos 7\theta$$

with the properties

(i) $p(\theta) \geq 0$ for all $\theta \in \mathbb{R}$,

(ii) $a_1 > a_0 > 0$ and $a_k \geq 0$ for $2 \leq k \leq 7$,

(iii) $$S := (a_0 + a_1 + \cdots + a_7)/(a_1 - a_0) < 5.91.$$

Property (i) implies that

$$\zeta^{a_0}(\sigma)|\zeta(\sigma + it)|^{a_1} \cdots |\zeta(\sigma + 7it)|^{a_7} \geq 1 \tag{16.8}$$

for all $\sigma > 1$. It follows, from (16.7) with $\ell = 0$, property (ii), and (16.4), that

$$|\zeta(\sigma + it)| \geq C_3(\sigma - 1)^{a_0/a_1} g(t)^{-(a_2+\cdots+a_7)/a_1}$$

with some positive constant C_3; we may assume that $C_2 > C_3$. Therefore, for $1 + \delta g(t)^{-\eta} \leq \sigma \leq 2$ with $\delta > 0$ and $\eta > 0$, both to be specified,

$$|\zeta(\sigma + it)| \geq C_3 \delta^{a_0/a_1} g(t)^{-(a_0\eta+a_2+\cdots+a_7)/a_1}. \tag{16.9}$$

Now suppose that $1 < \sigma \leq 1 + \delta g(t)^{-\eta}$. By (16.7) with $\ell = 1$, we have

$$\left|\zeta(\sigma + it) - \zeta(1 + \delta g(t)^{-\eta} + it)\right| \leq C_2 \delta g(t)^{-\eta+2}. \tag{16.10}$$

Choose η to make the exponents of $g(t)$ in (16.9) and (16.10) the same and choose δ to make $C_3\delta^{a_0/a_1} = 2C_2\delta$, namely,

$$\eta = \frac{a_0 + a_1 + \cdots + a_7}{a_1 - a_0} + 1 = S + 1 \quad \text{and} \quad \delta = \left(\frac{C_3}{2C_2}\right)^{a_1/(a_1-a_0)}.$$

In this case, we obtain

$$\begin{aligned} |\zeta(\sigma + it)| &\geq |\zeta(1 + \delta g(t)^{-\eta} + it)| - \left|\zeta(\sigma + it) - \zeta(1 + \delta g(t)^{-\eta} + it)\right| \\ &\geq C_4 g(t)^{-S+1} \end{aligned} \tag{16.11}$$

with some constant $C_4 > 0$. The lemma with $\lambda = S + 1$ follows from the last inequality, (16.9), and property (iii) of $p(\theta)$. □

The following identity will enable us to combine bounds for $1/f(s)$ and for derivatives of $f(s)$ to obtain bounds for derivatives of f'/f.

LEMMA 16.2. *Assume that a function $h(s)$ is nonvanishing in a region and is n-times differentiable there for some positive integer n. Then*

$$\left(\frac{h'(s)}{h(s)}\right)^{(n-1)} = \sum_{k=1}^{n} (-1)^{k-1} P_{k,n}\left(\frac{h'(s)}{h(s)}, \frac{h''(s)}{h(s)}, \ldots, \frac{h^{(n)}(s)}{h(s)}\right), \tag{16.12}$$

where

$$P_{k,n}(x_1, \ldots, x_n) = \sum_{\mathbf{k}} C_{\mathbf{k},k} x_1^{k_1} \cdots x_n^{k_n}$$

is a homogeneous polynomial of degree k with positive integral coefficients $C_{\mathbf{k},k}$, where $\mathbf{k} = \{k_1, \ldots, k_n\}$ is an n-vector with components satisfying

$$k_1 + \cdots + k_n = k \quad \text{and} \quad k_1 + 2k_2 + \cdots + nk_n = n. \tag{16.13}$$

Moreover, we have

$$\sum_{k=1}^{n}\sum_{\mathbf{k}} C_{\mathbf{k},k} \le 2^{n-1}(n-1)!. \tag{16.14}$$

REMARK. Also, the coefficients $C_{\mathbf{k},k}$ satisfy

$$\sum_{k=1}^{n}(-1)^{k-1}\sum_{\mathbf{k}} C_{\mathbf{k},k} = 0,$$

but this property is not used in the sequel.

To simplify the notation, we will not write the argument s in the proof.

PROOF. The lemma is proved by induction on n. Plainly it is true for $n = 1$. For $n = 2$ we have

$$\left(\frac{h'}{h}\right)' = \frac{h''}{h} - \left(\frac{h'}{h}\right)^2,$$

which corresponds to the monomials

$$P_{1,2}(x_1, x_2) = C_{\{0,1\},1}x_2 = x_2, \quad P_{2,2}(x_1, x_2) = C_{\{2,0\},2}x_1^2 = x_1^2\,;$$

thus the formula holds here as well. Now suppose that (16.12) and (16.13) hold and that h is $n+1$-times differentiable. By the first condition in (16.13), we have

$$P_{k,n}\left(\frac{h'}{h}, \frac{h''}{h}, \dots, \frac{h^{(n)}}{h}\right) = \frac{P_{k,n}(h', h'', \dots, h^{(n)})}{h^k}.$$

Differentiating, we obtain

$$\begin{aligned}
\left(\frac{h'}{h}\right)^{(n)} &= \left\{\left(\frac{h'}{h}\right)^{(n-1)}\right\}' = \left\{\sum_{k=1}^{n}\frac{(-1)^{k-1}P_{k,n}(h', h'', \dots, h^{(n)})}{h^k}\right\}' \\
&= \sum_{k=1}^{n}\left\{(-1)^k k\frac{h'}{h}\frac{P_{k,n}(h', h'', \dots, h^{(n)})}{h^k}\right. \\
&\qquad \left. + \frac{(-1)^{k-1}DP_{k,n}(h', h'', \dots, h^{(n)})}{h^k}\right\} := I + II.
\end{aligned}$$

Here, $DP_{k,n}(h', \dots)$ represents $(d/ds)\{P_{k,n}(h'(s), \dots)\}$. In terms of the coefficients we have

$$\begin{aligned}
I &= \sum_{k=1}^{n}(-1)^k k\sum_{\mathbf{k}} C_{\mathbf{k},k}\left(\frac{h'}{h}\right)^{k_1+1}\left(\frac{h''}{h}\right)^{k_2}\cdots\left(\frac{h^{(n)}}{h}\right)^{k_n} \\
&= \sum_{k=2}^{n+1}(-1)^{k-1}(k-1)\sum_{\mathbf{k}} C_{\mathbf{k},k-1}\left(\frac{h'}{h}\right)^{k_1+1}\left(\frac{h''}{h}\right)^{k_2}\cdots\left(\frac{h^{(n)}}{h}\right)^{k_n}
\end{aligned}$$

and

$$\begin{aligned}
II = \sum_{k=1}^{n}\Bigg((-1)^{k-1}\sum_{\mathbf{k}} C_{\mathbf{k},k}\Bigg\{&k_1\left(\frac{h'}{h}\right)^{k_1-1}\left(\frac{h''}{h}\right)^{k_2+1}\cdots\left(\frac{h^{(n)}}{h}\right)^{k_n} \\
&+ k_2\left(\frac{h'}{h}\right)^{k_1}\left(\frac{h''}{h}\right)^{k_2-1}\left(\frac{h^{(3)}}{h}\right)^{k_3+1}\cdots\left(\frac{h^{(n)}}{h}\right)^{k_n} \\
&+ \cdots + k_n\left(\frac{h'}{h}\right)^{k_1}\cdots\left(\frac{h^{(n)}}{h}\right)^{k_n-1}\frac{h^{(n+1)}}{h}\Bigg\}\Bigg).
\end{aligned}$$

Therefore, if we set polynomials

$$\begin{aligned} &P_{k,n+1}(x_1,\dots,x_n,x_{n+1}) \\ &\quad := (k-1)\sum_{\mathbf{k}} C_{\mathbf{k},k-1}x_1^{k_1+1}x_2^{k_2}\cdots x_n^{k_n} \\ &\qquad + \sum_{\mathbf{k}} C_{\mathbf{k},k}\Big(k_1x_1^{k_1-1}x_2^{k_2+1}x_3^{k_3}\cdots x_n^{k_n} \\ &\qquad\qquad + k_2x_1^{k_1}x_2^{k_2-1}x_3^{k_3+1}\cdots x_n^{k_n} + \cdots \\ &\qquad\qquad + k_nx_1^{k_1}\cdots x_{n-1}^{k_{n-1}}x_n^{k_n-1}x_{n+1}\Big) \end{aligned} \tag{16.15}$$

for $k=1,\dots,n,n+1$, then

$$\left(\frac{h'}{h}\right)^{(n)} = \sum_{k=1}^{n+1}(-1)^{k-1}P_{k,n+1}\left(\frac{h'}{h},\dots,\frac{h^{(n)}}{h},\frac{h^{(n+1)}}{h}\right).$$

By (16.15) and the induction hypothesis, the polynomial $P_{k,n+1}$ has coefficients $C_{\mathbf{k}',k}$ (with $\mathbf{k}'=\{k_1,\dots,k_n,k_{n+1}\}$) that are all positive integers and also terms $x_1^{k_1}\cdots x_n^{k_n}x_{n+1}^{k_{n+1}}$ that satisfy

$$k_1+\cdots+k_n+k_{n+1}=k \quad\text{and}\quad k_1+2k_2+\cdots+nk_n+(n+1)k_{n+1}=n+1.$$

This proves (16.13).

Finally, let S_n denote the left hand side of (16.14). Note that (16.13) implies that $k=k_1+\cdots+k_n\le n$. Then, by (16.15) (with the first form of I),

$$S_{n+1}=\sum_{k=1}^{n+1}\sum_{\mathbf{k}'}C_{\mathbf{k}',k}=\sum_{k=1}^{n}\left(k\sum_{\mathbf{k}}C_{\mathbf{k},k}+\sum_{\mathbf{k}}C_{\mathbf{k},k}(k_1+\cdots+k_n)\right)<2nS_n.$$

Therefore, by the induction hypothesis, $S_{n+1}\le 2^n n!$. This proves (16.14). □

16.3. A Nyman type remainder term

We showed earlier that if $\gamma>1$ in (16.1) then Chebyshev bounds hold for $\mathcal{N}$ (Corollary 9.10), and if $\gamma>3/2$ then we have the PNT (Theorem 13.1). Now, for larger γ values, we establish the PNT with a remainder term.

THEOREM 16.3. *Assume that* (16.1) *holds with* $\gamma>3$. *Then*

$$\psi(x)=x+O\big(x(\log x)^{-(\gamma-3)/8}\big). \tag{16.16}$$

REMARKS. This remainder term is not optimal. (See Notes.) By integration by parts, the conclusion of the theorem can be restated as

$$\pi(x)=\mathrm{li}(x)+O\big(x(\log x)^{-(\gamma+5)/8}\big).$$

Combining Theorem 16.3 with the g-integer approximation Theorem 7.13 (A) we obtain

COROLLARY 16.4. *A Beurling generalized number system satisfies*

$$N(x)=Ax+O_k(x\log^{-k}x)$$

with some constant $A>0$ *and all large* $k\in\mathbb{N}$ *if and only if*

$$\pi(x)=\mathrm{li}(x)+O_\ell(x\log^{-\ell}x)$$

holds for all large $\ell\in\mathbb{N}$.

The method we apply in the proof of Theorem 16.3 has some similarities with that used for the Wiener-Ikehara Theorem. We exploit the "weight" produced by derivatives of $-(\zeta'/\zeta)(1+it)$ to obtain the remainder term; the required bounds for derivatives are established using condition (16.1). A second difference is that here we use a Gaussian convergence factor, which vanishes at infinity much faster than the Fejér kernel.

This is an outline of our argument. For a positive integer n of our choice, we bound the n-th derivative of $(\zeta'/\zeta)(s)$; because n will be fixed for given γ, estimates need not be uniform in this parameter. We start calculations with the identity

$$\begin{aligned}\frac{d^n}{ds^n}\Big(\frac{-\zeta'(s)}{s\zeta(s)}-\frac{1}{s-1}\Big) &= \int_1^\infty x^{-s}(-\log x)^n\frac{\psi(x)-x}{x}\,dx\\ &= \int_0^\infty e^{-(s-1)u}(-u)^n\frac{\psi(e^u)-e^u}{e^u}\,du\end{aligned}$$

for $\sigma=\Re s>1$. With numbers $y>0$ (large) and $\epsilon>0$ (small), multiply through the last formula by

$$e^{ity-\epsilon^2t^2/2}/(2\pi)$$

and integrate on t. The double integral below converges absolutely, which justifies switching the order of integration. We define functions $L=L_n(\sigma,y)$ and $R=R_n(\sigma,y)$ with

$$\begin{aligned}L:=&\frac{(-1)^n}{2\pi}\int_{-\infty}^\infty\Big\{\frac{-\zeta'(\sigma+it)}{(\sigma+it)\zeta(\sigma+it)}-\frac{1}{\sigma+it-1}\Big\}^{(n)}e^{ity-\epsilon^2t^2/2}dt\\ =&\int_{u=0}^\infty e^{-(\sigma-1)u}u^n\frac{\psi(e^u)-e^u}{e^u}\frac{1}{2\pi}\int_{t=-\infty}^\infty e^{it(y-u)}e^{-\epsilon^2t^2/2}dt\,du\\ =&\frac{1}{\sqrt{2\pi}\,\epsilon}\int_0^\infty e^{-(\sigma-1)u}u^n e^{-(y-u)^2/(2\epsilon^2)}\frac{\psi(e^u)-e^u}{e^u}du=:R.\end{aligned}$$

(In the last integral, the Fourier transform of the Gaussian function was used.)

Now let $\sigma\to 1+$. By the dominated convergence theorem,

$$R\to\frac{1}{\sqrt{2\pi}\,\epsilon}\int_0^\infty u^n e^{-(y-u)^2/(2\epsilon^2)}\frac{\psi(e^u)-e^u}{e^u}du=:R^*.$$

We have the normalization

$$\frac{1}{\sqrt{2\pi}\,\epsilon}\int_{-\infty}^\infty e^{-v^2/(2\epsilon^2)}dv=1. \tag{16.17}$$

Thus an average of $u^n(\psi(e^u)-e^u)/e^u$ is expressed in terms of the limit of L_n as $\sigma\to 1+$. It remains to choose a suitable n, estimate $\lim_{\sigma\to1+}L_n(\sigma,y)$, and extract pointwise upper and lower bounds from this average. We begin with the last task.

Bounds for $\psi(x)$

Lemma 16.5. *Suppose that the integer counting function of $\mathcal{N}$ satisfies* (16.1) *and that*

$$L_n\ll\epsilon^{-7(n+1)/\gamma}$$

holds for some natural number $n<\gamma-2$, uniformly for all $\sigma>1$, all $y\geq 2$, and $0<\epsilon<1$. Then, for $\eta=\eta(\gamma,n)>(7+\gamma)/n$,

$$\psi(x)=x+O\big(x\{\log x\}^{-\gamma/(7+\eta)}\big).$$

PROOF. Take n to be a positive integer satisfying $n < \gamma - 2$. (The reason for this restriction will appear in the proof of the next lemma.) The first step is to derive an upper bound for ψ. With a number δ to be chosen that is large compared with ϵ and satisfying $\delta = o(y)$, the ψ-part of the integral for R^* is expressible as

$$
\begin{aligned}
R_1^* &:= \frac{1}{\sqrt{2\pi}\,\epsilon}\int_0^\infty u^n e^{-(y-u)^2/(2\epsilon^2)}\,\psi(e^u)e^{-u}du \qquad (16.18)\\
&\ge \frac{1}{\sqrt{2\pi}\,\epsilon}\int_{y-\delta}^{y+\delta} u^n e^{-(y-u)^2/(2\epsilon^2)}\,\psi(e^u)e^{-u}du\\
&\ge \frac{\psi(e^{y-\delta})(y-\delta)^n}{\sqrt{2\pi}\,\epsilon\, e^{y+\delta}}\int_{y-\delta}^{y+\delta} e^{-(y-u)^2/(2\epsilon^2)}du.
\end{aligned}
$$

Using (16.17) and the Chebyshev O-bounds, we find

$$R_1^* \ge \psi(e^{y-\delta})(y-\delta)^n\, e^{-y-\delta} - \frac{O(y^n)}{\epsilon}\int_\delta^\infty e^{-v^2/(2\epsilon^2)}dv.$$

The last integral is smaller than

$$\frac{1}{\delta}\int_\delta^\infty ve^{-v^2/(2\epsilon^2)}dv = \frac{\epsilon^2}{\delta}\,e^{-\delta^2/(2\epsilon^2)} < \epsilon e^{-\delta^2/(2\epsilon^2)}; \qquad (16.19)$$

thus

$$R_1^* \ge \psi(e^{y-\delta})(y-\delta)^n\, e^{-y-\delta} + O(y^n e^{-\delta^2/(2\epsilon^2)}).$$

The rest of R^* is $-R_2^*$ with

$$R_2^* := \frac{1}{\sqrt{2\pi}\,\epsilon}\int_0^\infty u^n e^{-(y-u)^2/(2\epsilon^2)}du = \frac{1}{\sqrt{2\pi}\,\epsilon}\int_{-y}^\infty (y+v)^n e^{-v^2/(2\epsilon^2)}dv.$$

Expanding the binomial and using (16.17) and (16.19) again (but with y in place of δ), we see that

$$
\begin{aligned}
R_2^* &= y^n + O(\epsilon y^{n-1}e^{-y^2/(2\epsilon^2)}) + \frac{ny^{n-1}}{\sqrt{2\pi}\,\epsilon}\int_{-y}^\infty ve^{-v^2/(2\epsilon^2)}dv \qquad (16.20)\\
&\quad + \frac{n(n-1)}{2\sqrt{2\pi}\,\epsilon}y^{n-2}\int_{-y}^\infty v^2e^{-v^2/(2\epsilon^2)}dv + \dots .
\end{aligned}
$$

Since the integrand $ve^{-v^2/(2\epsilon^2)}$ is odd,

$$\int_{-y}^\infty ve^{-v^2/(2\epsilon^2)}dv = \int_y^\infty ve^{-v^2/(2\epsilon^2)}dv = \epsilon^2e^{-y^2/(2\epsilon^2)};$$

also, the last integral in (16.20) is less than

$$\int_{-\infty}^\infty v^2e^{-v^2/(2\epsilon^2)}dv = O(\epsilon^3).$$

Thus the first three terms of R_2^* equal

$$y^n + O\big(\epsilon^2y^{n-2}\big).$$

The remaining terms of R_2^* have lower powers of y and higher powers of ϵ, so the last estimate is valid for all of R_2^*.

The preceding bounds for R_1^* and R_2^* together with the condition for L_n give

$$\frac{\psi(e^{y-\delta})(y-\delta)^n}{e^{y+\delta}} \le y^n\{1+O(\epsilon^2y^{-2})+O(e^{-\delta^2/(2\epsilon^2)})\} + O(\epsilon^{-7(n+1)/\gamma}).$$

or

$$\frac{\psi(e^{y-\delta})}{e^{y-\delta}} \leq \Big(\frac{y}{y-\delta}\Big)^n e^{2\delta}\{1 + O(\epsilon^2 y^{-2}) + O(e^{-\delta^2/(2\epsilon^2)})\} + O\big(y^{-n}\epsilon^{-7(n+1)/\gamma}\big)$$

or

$$\frac{\psi(e^{y-\delta})}{e^{y-\delta}} \leq 1 + O\big(\delta + e^{-\delta^2/(2\epsilon^2)} + y^{-n}\epsilon^{-7(n+1)/\gamma}\big).$$

Taking $\delta := \epsilon\sqrt{2\log 1/\epsilon}$, we find

$$e^{-\delta^2/(2\epsilon^2)} = \epsilon.$$

Next, set

$$\epsilon := y^{-\gamma/(7+\{7+\gamma\}/n)};$$

with this choice, $y^{-n}\epsilon^{-7(n+1)/\gamma} = \epsilon$. Since $\delta > \epsilon$, these bounds give

$$\frac{\psi(e^{y-\delta})}{e^{y-\delta}} \leq 1 + O(\delta).$$

Finally, δ is connected with y by

$$\delta = \epsilon\sqrt{2\log 1/\epsilon} \ll y^{-\gamma/\{7+(7+\gamma)/n\}+o(1)}$$

for large y. Thus, we have

$$\psi(x) \leq x + O\big(x\{\log x\}^{-\gamma/(7+\eta)}\big)$$

provided that we take $\eta = \eta(\gamma, n) > (7+\gamma)/n$.

Now we turn to the lower bound for ψ, starting again with $R^* = R_1^* - R_2^*$ and using the same values for δ, ϵ and n. Note that we have shown already that

$$R_2^* = y^n + O(\epsilon^2 y^{n-2}).$$

To study R_1^*, recall (16.18) and the Chebyshev bound $\psi(x)/x < B$ for all $x \geq 1$. This time, write

$$\begin{aligned} R_1^* &:= \frac{1}{\sqrt{2\pi}\,\epsilon}\int_0^\infty u^n e^{-(y-u)^2/(2\epsilon^2)}\,\psi(e^u)e^{-u}du \\ &\leq \frac{1}{\sqrt{2\pi}\,\epsilon}\int_{y-\delta}^{y+\delta} u^n e^{-(y-u)^2/(2\epsilon^2)}\,\frac{\psi(e^u)}{e^u}du \\ &\quad + \frac{B}{\sqrt{2\pi}\epsilon}\int_{[0,\infty)\setminus(y-\delta,\,y+\delta)} u^n e^{-(y-u)^2/(2\epsilon^2)}\,du =: I + II. \end{aligned}$$

By (16.17), our main term is

$$I \leq \frac{\psi(e^{y+\delta})(y+\delta)^n}{\sqrt{2\pi}\,\epsilon\,e^{y-\delta}}\int_{y-\delta}^{y+\delta} e^{-(y-u)^2/(2\epsilon^2)}du \leq \frac{\psi(e^{y+\delta})(y+\delta)^n}{e^{y-\delta}}.$$

Using the binomial expansion

$$\begin{aligned} II &< \frac{2B}{\sqrt{2\pi}\epsilon}\int_{y+\delta}^\infty u^n e^{-(y-u)^2/(2\epsilon^2)}\,du = \frac{2B}{\sqrt{2\pi}\epsilon}\int_\delta^\infty (y+v)^n e^{-v^2/(2\epsilon^2)}\,dv \\ &= \frac{2B}{\sqrt{2\pi}\epsilon}\int_\delta^\infty \Big(y^n + ny^{n-1}v + \frac{n(n-1)}{2}y^{n-2}v^2 + \dots\Big)e^{-v^2/(2\epsilon^2)}\,dv, \end{aligned}$$

and from the upper bound calculations, we find

$$II \leq \epsilon y^n + O\big(\epsilon^2 y^{n-1}\big).$$

Thus

$$R_1^* \le \frac{\psi(e^{y+\delta})(y+\delta)^n}{e^{y-\delta}} + \epsilon y^n + O(y^{n-1}\epsilon^2),$$

and combining the bounds for R_1^* and R_2^* and that for L_n, we obtain

$$\frac{\psi(e^{y+\delta})(y+\delta)^n}{e^{y-\delta}} + \epsilon y^n - y^n + O(\epsilon^2 y^{n-1}) \ge -c\epsilon^{-7(n+1)/\gamma}.$$

Proceeding as before, we find

$$\frac{\psi(e^{y+\delta})}{e^{y+\delta}} \ge 1 - c_1\delta - c_2\epsilon^2 y^{-1} - c_3 y^{-n}\epsilon^{-7(n+1)/\gamma}$$

or

$$\psi(x) \ge x + O\big(x\{\log x\}^{-\gamma/(7+\eta)}\big).$$

The two inequalities for $\psi(x)$ give the claimed formula. □

BOUNDS FOR L_n

We need a bound for L_n for use in Lemma 16.5. Recall that the integrand of the integral defining L_n includes as a factor the nth derivative of

$$Z(s) := \frac{-\zeta'(s)}{s\zeta(s)} - \frac{1}{s-1}; \tag{16.21}$$

we have to estimate $|Z^{(n)}|$. For this, instead of $\zeta(s)$, we use $f(s) := (s-1)\zeta(s)/s$, and we give bounds for $1/f$ and derivatives of f along with the recipe for combining them.

To start, we write a convenient representation for Z in terms of the auxiliary function $f(s) = (s-1)\zeta(s)/s$. Since

$$\frac{1}{s}\Big\{\log\Big(\frac{s-1}{s}\zeta(s)\Big)\Big\}' = \frac{1}{s}\Big(\frac{\zeta'}{\zeta}(s) + \frac{1}{s-1} - \frac{1}{s}\Big) = \frac{\zeta'(s)}{s\zeta(s)} + \frac{1}{s-1} - \frac{1}{s} - \frac{1}{s^2},$$

we have

$$Z(s) = -\frac{f'(s)}{sf(s)} - \frac{1}{s} - \frac{1}{s^2}. \tag{16.22}$$

Also, we introduce here

$$g(t) := (|t|+e)^{1/\gamma}. \tag{16.23}$$

(In the next section, we choose g differently.)

We begin with upper bounds for the modulus of derivatives of f.

LEMMA 16.6. *If* (16.1) *is satisfied then*

$$|f^{(\ell)}(\sigma+it)| \le Bg(t)^{\ell+1} \tag{16.24}$$

with some constant $B > 0$, *uniformly for* $1 < \sigma \le 2$, $t \in \mathbb{R}$, *and all integers satisfying* $0 \le \ell < \gamma - 1$.

PROOF. We have

$$f(s) = \Big(1 - \frac{1}{s}\Big)\zeta(s) = \int_{1-}^{\infty} x^{-s}\Big(dN(x) - \frac{N(x)\,dx}{x}\Big) = \int_{1-}^{\infty} x^{-s} dN_1(x),$$

where

$$\begin{aligned} N_1(x) &:= \int_{1-}^{x} \Big(dN(u) - \frac{N(u)\,du}{u}\Big) \\ &= N(x) - Ax - \int_{1-}^{x} \frac{(N(u)-Au)\,du}{u} + A \ll x\log^{-\gamma} x. \end{aligned}$$

For any $X > 1$, write

$$f(s) = \int_{1-}^{X} x^{-s}\,dN_1(x) - N_1(X)X^{-s} + s\int_X^\infty x^{-s-1}N_1(x)\,dx.$$

Differentiating ℓ times ($\ell = 0$ is OK), we have by Leibniz's rule

$$\begin{aligned} (-1)^\ell f^{(\ell)}(s) &= \int_{-1}^{X} x^{-s}\log^\ell x\,dN_1(x) - N_1(X)X^{-s}\log^\ell X \\ &\quad + s\int_X^\infty x^{-s-1}\log^\ell x\,N_1(x)\,dx \\ &\quad - \ell\int_X^\infty x^{-s-1}\log^{\ell-1} x\,N_1(x)\,dx \\ &:= I + II + III + IV, \text{ say.} \end{aligned}$$

Integration by parts shows

$$|I| < \int_{1-}^{X} x^{-1}\log^\ell x\Big(dN(x) + \frac{N(x)\,dx}{x}\Big) \ll (\log X)^{\ell+1}$$

uniformly for $\sigma > 1$ and all real t. Also, we have $II \ll (\log X)^{\ell-\gamma}$. And

$$III \ll (|t| + e)\int_X^\infty x^{-1}\log^{\ell-\gamma} x\,dx \ll (|t|+e)\,(\log X)^{\ell+1-\gamma},$$

with the integral converging, since $\ell < \gamma - 1$. *This condition limits the order of zeta derivatives, whence the condition* $n < \gamma - 2$ *in Lemma* 16.5. Also, for $1 \le \ell < \gamma - 1$,

$$IV \ll \ell\int_X^\infty x^{-1}(\log x)^{\ell-1-\gamma}\,dx \ll (\log X)^{\ell-\gamma}.$$

Now II and IV are negligible, so we obtain

$$f^{(\ell)}(s) \ll (\log X)^{\ell+1} + (|t|+e)\,(\log X)^{\ell+1-\gamma}.$$

Setting $\log X = g(t)$, we find $f^{(\ell)}(\sigma + it) \ll g(t)^{\ell+1}$, the claimed bound. □

The function $g(t)$ of (16.23) satisfies the condition (16.4). It follows from (16.24) and Lemma 16.1 that

$$|f(\sigma+it)|^{-1} \le Cg(t)^\lambda$$

holds with constants $C > 0$ and $\lambda = 5$ for $1 < \sigma \le 2$ and all $t \in \mathbb{R}$.

LEMMA 16.7. *Assume that f satisfies*

(16.24 bis) $$|f^{(\ell)}(\sigma + it)| \le B_1 g(t)^{\ell+1}$$

for all nonnegative integers $\ell < \gamma - 1$ *and*

(16.5 bis) $$|f(\sigma+it)|^{-1} \le B_2 g(t)^5$$

with positive constants B_1 and B_2, uniformly for $1 < \sigma \le 2$ and all $t \in \mathbb{R}$. Then

$$\left| \left(\frac{f'(s)}{sf(s)} \right)^{(n-1)} \right| \ll \frac{\{Bng(t)^7\}^n}{|s|}.$$

holds with some constant $B > 1$ for $1 < \sigma \le 2$ and $1 \le n < \gamma - 1$.

PROOF. We apply Lemma 16.2 to estimate

$$\left| \left(\frac{f'(s)}{f(s)} \right)^{(n-1)} \right|.$$

By (16.24 bis) and (16.5 bis), we have for each natural number $j < \gamma - 1$,

$$\left| \frac{f^{(j)}(\sigma + it)}{f(\sigma + it)} \right| \le B_1 B_2\, g(t)^{j+6}. \tag{16.25}$$

Let $B_3 := B_1 B_2$; we can assume $B_3 > 1$. Thus, for $n \ge 1$ and nonnegative integers $k_1, \ldots, k_n$ satisfying $k_1 + \cdots + k_n = k$ and $k_1 + 2k_2 + \cdots + nk_n = n$, we have

$$\prod_{j=1}^{n} \left| \frac{f^{(j)}(\sigma + it)}{f(\sigma + it)} \right|^{k_j} \le B_3^k \prod_{j=1}^{n} \{g(t)^{j+6}\}^{k_j} \le B_3^n g(t)^{7n}.$$

(We have used the observation that $k \le n$.)

Hence, by Lemma 16.2, for $1 \le n < \gamma - 1$, all real t, and $1 < \sigma \le 2$,

$$\left| \left(\frac{f'(s)}{f(s)} \right)^{(n-1)} \right| \le \sum_{k=1}^{n} \sum_{\mathbf{k}} C_{\mathbf{k},k} \left(B_3 g(t)^7\right)^n \le (n-1)!\, \left(2B_3 g(t)^7\right)^n. \tag{16.26}$$

Now we find derivatives of $f'(s)/\{sf(s)\}$ using Leibniz's rule. For each $n \ge 1$,

$$\left| \left(\frac{f'(s)}{sf(s)} \right)^{(n-1)} \right| = \sum_{j=0}^{n-1} \binom{n-1}{j} \frac{(-1)^{n-1-j}(n-1-j)!}{s^{n-j}} \left(\frac{f'(s)}{f(s)} \right)^{(j)}.$$

Using (16.26) with j in place of $n-1$, we have

$$\begin{aligned} \left| \left(\frac{f'(s)}{sf(s)} \right)^{(n-1)} \right| &\le \sum_{j=0}^{n-1} \binom{n-1}{j} \frac{(n-1-j)!\, j!}{|s|^{n-j}} \{2B_3 g(t)^7\}^{j+1} \\ &\le \frac{(n-1)!}{|s|} \sum_{j=0}^{n-1} \{2B_3 g(t)^7\}^{j+1} \le \frac{n!}{|s|} \{2B_3 g(t)^7\}^n. \end{aligned}$$

Noting that $n! \le n^n$, we conclude that

$$\left| \left(\frac{f'(s)}{sf(s)} \right)^{(n-1)} \right| \le \frac{1}{|s|} \{2B_3 n g(t)^7\}^n. \qquad \square$$

PROOF OF THEOREM 16.3. Our next step is to show that the L_n condition of Lemma 16.5 is satisfied. Recall that

$$L_n := \frac{(-1)^n}{2\pi} \int_{-\infty}^{\infty} Z^{(n)}(\sigma + it) e^{ity - \epsilon^2 t^2/2} dt,$$

with

$$Z(s) := \frac{-\zeta'(s)}{s\zeta(s)} - \frac{1}{s-1} = \frac{-f'(s)}{sf(s)} - \frac{1}{s} - \frac{1}{s^2},$$

and $f(s) := (s-1)\zeta(s)/s$. If we adjusting the O-constant to account also for derivatives of $1/s$ and $1/s^2$, the last lemma implies that

$$|Z^{(j-1)}(s)| \ll |s|^{-1}\{Bjg(t)^7\}^j, \tag{16.27}$$

provided that $j < \gamma - 1$.

By the last estimate, with n in place of $j-1$ (and $n < \gamma - 2$),

$$\begin{aligned} L_n &\ll_\gamma \int_0^\infty \frac{g(t)^{7(n+1)}}{t+e} e^{-\epsilon^2 t^2/2} dt = \int_0^\infty (t+e)^{7(n+1)/\gamma - 1} e^{-\epsilon^2 t^2/2} dt \\ &\ll 1 + \int_0^\infty t^{7(n+1)/\gamma} e^{-\epsilon^2 t^2/2} dt/t \ll \epsilon^{-7(n+1)/\gamma}, \end{aligned}$$

which is the condition of Lemma 16.5.

Finally, we determine the exponent for the remainder term in Theorem 16.3. In Lemma 16.5 we showed that

$$\psi(x) = x + O\big(x(\log x)^{-E(\gamma)}\big),$$

provided that

$$E(\gamma) < \frac{\gamma}{7 + (7+\gamma)/n}$$

with $n < \gamma - 2$. Here we obtain an explicit bound for $E(\gamma)$. Take

$$n = \lceil \gamma - 3 \rceil,$$

an integer satisfying the conditions $\gamma - 3 \le n < \gamma - 2$. By a small computation,

$$\begin{aligned} \frac{\gamma}{7 + (7+\gamma)/n} &\ge \frac{\gamma}{7 + (7+\gamma)/(\gamma - 3)} \\ &= \frac{\gamma}{8} - \frac{5}{32} - \frac{35}{128\gamma - 224} > \frac{\gamma}{8} - \frac{3}{8} =: E(\gamma), \end{aligned}$$

provided that $\gamma > 3$. This completes the proof of Theorem 16.3. □

REMARK. The exponent $E(\gamma)$ can be increased slightly if γ is restricted to larger values; for example, for $\gamma \ge 11$ we have

$$\frac{\gamma}{8} - \frac{5}{32} - \frac{35}{128\gamma - 224} > \frac{\gamma}{8} - \frac{3}{16}.$$

In the other direction, there is a limit for this approach, as shown by the inequality

$$\frac{\gamma}{7 + (7+\gamma)/n} < \frac{\gamma}{7 + (7+\gamma)/(\gamma - 2)} = \frac{\gamma}{8} - \frac{9}{64} - \frac{63/64}{8\gamma - 7} < \frac{\gamma}{8} - \frac{9}{64}$$

for all $\gamma > 3$.

16.4. A dlVP-type remainder term

In this section, we study a g-number system $\mathcal{N}$ whose g-integer counting function satisfies

$$N(x) = Ax + O(x\exp\{-c(\log x)^\alpha\}) \tag{16.28}$$

with some constants $A > 0$, $c > 0$, and $\alpha \in (0, 1]$. Such a remainder term is smaller than those of the previous section, and we show that the g-prime counting function $\pi(x)$ of $\mathcal{N}$ has a smaller remainder term as well.

THEOREM 16.8. *If (16.28) is satisfied then*

$$\pi(x) = \mathrm{li}(x) + O(x\exp\{-c'(\log x)^{\beta}\}) \tag{16.29}$$

with $\beta = \alpha/(7+\alpha)$ and some constant $c' > 0$.

REMARK. In fact, (16.29) holds with a somewhat larger value of β, e.g. $\beta > \alpha/(6.91+\alpha)$, as will appear below.

Recall the function

$$\Pi_c(x) := \int_1^x \frac{1-u^{-1}}{\log u}\,du, \quad x \geq 1, \tag{1.9 bis}$$

a close relative of $\mathrm{li}(x)$. The proof of the theorem starts with the familiar equation

$$\frac{1}{s}\log\Big(\frac{s-1}{s}\,\zeta(s)\Big) = \int_1^\infty x^{-s-1}\{\Pi(x) - \Pi_c(x)\}\,dx, \quad \sigma > 1. \tag{16.30}$$

Differentiating k times gives

$$\frac{d^k}{ds^k}\Big\{\frac{1}{s}\log\Big(\frac{s-1}{s}\,\zeta(s)\Big)\Big\} = \int_1^\infty x^{-s-1}(-\log x)^k(\Pi(x) - \Pi_c(x))\,dx,$$

and, by the Plancherel identity with $s = \sigma + it$, for any fixed $\sigma > 1$,

$$\begin{aligned}\int_{-\infty}^{\infty}&\Big|\frac{d^k}{ds^k}\Big\{\frac{1}{s}\log\Big(\frac{s-1}{s}\,\zeta(s)\Big)\Big\}\Big|^2 dt\\ &= 2\pi\int_1^\infty x^{-2\sigma-1}(\log x)^{2k}\{\Pi(x) - \Pi_c(x)\}^2\,dx.\end{aligned} \tag{16.31}$$

The main steps in proving Theorem 16.8 are to find a bound for the left side of (16.31) and, with a suitable choice of k, to extract an estimate for $\Pi(x) - \Pi_c(x)$ on the right side of the equation. We begin with the latter task, which we consider a tauberian problem, because we are deriving a pointwise estimate of $\Pi(x) - \Pi_c(x)$ from an integral of this function.

LEMMA 16.9. *Suppose that*

$$\int_{-\infty}^{\infty}\Big|\frac{d^k}{ds^k}\Big\{\frac{1}{s}\log\Big(\frac{s-1}{s}\,\zeta(s)\Big)\Big\}\Big|^2 dt \leq (Mk^{\eta})^{2k} \tag{16.32}$$

holds with some constants $M > 0$ and $\eta \geq 1$, uniformly for $1 < \sigma \leq 2$ and all positive integers k. Then

$$\Pi(x) = \Pi_c(x) + O(x\exp\{-c'(\log x)^{1/\eta}\}) \tag{16.33}$$

holds with some constant $c' > 0$.

PROOF. By (16.31), (16.32), and the monotone convergence theorem, we have

$$\int_1^\infty x^{-3}\{\Pi(x) - \Pi_c(x)\}^2(\log x)^{2k}\,dx \leq (Mk^{\eta})^{2k}. \tag{16.34}$$

We apply a simple tauberian method to the last expression. Let

$$\delta(x) := \Pi(x) - \Pi_c(x), \quad x \geq 1.$$

Suppose first that $\delta(x) > 0$. Since $\Pi \uparrow$ and $(1-u^{-1})/\log u \leq 1$, we see for $y > x$ that

$$\delta(y) = \delta(x) + \int_x^y \Big(d\Pi - \frac{1-u^{-1}}{\log u}\,du\Big) > \delta(x) - (y-x),$$

and so

$$\delta(y) \geq \delta(x)/2 \quad \text{for} \quad x \leq y \leq x + \delta(x)/2 < 3x/2.$$

(The last inequality holds for all sufficiently large x, since $\Pi(x)$ and $\Pi_c(x)$ are each $O(x/\log x)$.) Then, from (16.34),

$$\left(\frac{3x}{2}\right)^{-3}\left(\frac{\delta(x)}{2}\right)^3 (\log\, x)^{2k} \leq \int_x^{x+\delta(x)/2} y^{-3}\delta(y)^2\, (\log\, y)^{2k}\, dy \leq (Mk^\eta)^{2k}$$

and hence

$$\left(\frac{\delta(x)}{x}\right)^3 \leq 27\left(\frac{Mk^\eta}{\log\, x}\right)^{2k}.$$

On the other hand, if $\delta(x) < 0$, consider an interval to the left of x. We have

$$\delta(y) \leq \delta(x)/2 \quad \text{for} \quad x/2 < x + \delta(x)/2 \leq y \leq x,$$

and so

$$\frac{1}{x^3}\left(\frac{|\delta(x)|}{2}\right)^3 \left(\log\, \frac{x}{2}\right)^{2k} \leq \int_{x+\delta(x)/2}^{x} y^{-3}\delta(y)^2\, (\log\, y)^{2k}\, dy \leq (Mk^\eta)^{2k}.$$

Assuming that x is sufficiently large, we find

$$\left(\frac{|\delta(x)|}{x}\right)^3 \leq \left(\frac{3Mk^\eta}{\log\, x}\right)^{2k}.$$

Thus, in either case,

$$\left(\frac{|\delta(x)|}{x}\right)^3 \leq \left(\frac{M'k^\eta}{\log\, x}\right)^{2k}$$

holds with some constant $M' \geq 1$, uniformly for all $k \in \mathbb{N}$.

Then we have

$$\left(|\delta(x)|/x\right)^3 \leq \exp\{-c''(\log\, x)^{1/\eta}\}$$

with $c'' > 0$ by taking

$$k = \left\lfloor \frac{1}{e}\left(\frac{\log\, x}{M'}\right)^{1/\eta} \right\rfloor.$$

This proves (16.33). □

Now we set out to obtain the required bound for the left side of (16.32). Recall the function $f(s) = (s-1)\zeta(s)/s$, $\sigma > 1$. The main contribution to the integrand on the left side of (16.32) comes from terms of the form

$$\frac{d^n}{ds^n} \log \left\{\frac{(s-1)\zeta(s)}{s}\right\} = \left(\frac{f'(s)}{f(s)}\right)^{(n-1)}, \quad 1 \leq n \leq k.$$

We find from formula (16.12) that

$$\left(\frac{f'(s)}{f(s)}\right)^{(n-1)} = \sum_{\ell=1}^{n} (-1)^{\ell-1} P_{\ell,n}\left(\frac{f'(s)}{f(s)}, \ldots, \frac{f^{(n)}(s)}{f(s)}\right).$$

To use this relation, we need upper bounds for $|f|^{-1}$ and for the modulus of derivatives of f; we start with the derivatives.

LEMMA 16.10. *If* (16.28) *is satisfied then*

$$|f^{(\ell)}(\sigma+it)| \le Bc^{-(\ell+1)/\alpha}\Big(\max\Big\{\frac{\ell+1}{\alpha},\ \log(|t|+e)\Big\}\Big)^{(\ell+1)/\alpha} \tag{16.35}$$

with some constant $B>0$, *uniformly for* $1<\sigma\le 2$, $t\in\mathbb{R}$, *and all integers* $\ell\ge 0$.

PROOF. We have

$$f(s)=\Big(1-\frac{1}{s}\Big)\zeta(s)=\int_{1-}^{\infty}x^{-s}\Big(dN(x)-\frac{N(x)\,dx}{x}\Big)=\int_{1-}^{\infty}x^{-s}dN_1(x),$$

where

$$\begin{aligned}N_1(x) &:= \int_{1-}^{x}\Big(dN(u)-\frac{N(u)\,du}{u}\Big)\\ &= N(x)-Ax-\int_{1-}^{x}\frac{(N(u)-Au)\,du}{u}+A \ll xe^{-c(\log x)^{\alpha}}\end{aligned}$$

(the last by (16.28) and L'Hospital's rule).

For any $X>1$, write

$$f(s)=\int_{1-}^{X}x^{-s}\,dN_1(x)-N_1(X)X^{-s}+s\int_{X}^{\infty}x^{-s-1}N_1(x)\,dx.$$

Differentiating ℓ times, we have by Leibniz's rule

$$\begin{aligned}(-1)^{\ell}f^{(\ell)}(s) &= \int_{1-}^{X}x^{-s}\log^{\ell}x\,dN_1(x)-N_1(X)X^{-s}\log^{\ell}X\\ &\quad + s\int_{X}^{\infty}x^{-s-1}\log^{\ell}x\,N_1(x)\,dx\\ &\quad - \ell\int_{X}^{\infty}x^{-s-1}\log^{\ell-1}x\,N_1(x)\,dx\\ &:= I+II+III+IV,\ \text{say.}\end{aligned}$$

Integration by parts shows

$$|I|<\int_{1-}^{X}x^{-1}\log^{\ell}x\Big(dN(x)+\frac{N(x)\,dx}{x}\Big)\ll(\log X)^{\ell+1}$$

uniformly for $\sigma>1$ and all real t. Also, on this set, $II\ll e^{-c(\log X)^{\alpha}}\log^{\ell}X$ and

$$III\ll(|t|+e)\int_{X}^{\infty}x^{-1}e^{-c(\log x)^{\alpha}}\log^{\ell}x\,dx\,;$$

we estimate this integral using an inequality for the incomplete gamma function.

If $x\ge b\ge 1$ then we claim

$$\Gamma(x,b):=\int_{x}^{\infty}y^{b-1}e^{-y}\,dy\le x^{b}e^{-x}. \tag{16.36}$$

The change of integration variable $y=x(1+v)$ gives

$$\Gamma(x,b)=x^{b}e^{-x}\int_{0}^{\infty}(1+v)^{b-1}e^{-xv}\,dv\le x^{b}e^{-x}\int_{0}^{\infty}(1+v)^{b-1}e^{-bv}\,dv.$$

The integral on the right hand side is a decreasing function of $b\ge 1$ and equals 1 at $b=1$. Thus (16.36) holds.

Set $y = c\log^\alpha x$ in the integral estimate for III. We find

$$\int_X^\infty x^{-1}e^{-c(\log x)^\alpha}\log^\ell x\,dx = \frac{1}{\alpha c^{(\ell+1)/\alpha}}\int_{c\log^\alpha X}^\infty y^{(\ell+1)/\alpha-1}e^{-y}dy$$
$$= \frac{\Gamma(c\log^\alpha X,(\ell+1)/\alpha)}{\alpha c^{(\ell+1)/\alpha}} \le \frac{1}{\alpha}\log^{\ell+1} X\exp(-c\log^\alpha X),$$

provided that

$$c\log^\alpha X \ge (\ell+1)/\alpha. \tag{16.37}$$

In this case,

$$III \ll (|t|+e)\log^{\ell+1} X\exp(-c\log^\alpha X).$$

Similarly, for $\ell \ge 1$, again assuming (16.37),

$$IV \ll \ell\int_X^\infty x^{-1}e^{-c(\log x)^\alpha}\log^{\ell-1} x\,dx \ll \log^{\ell+\alpha} X\exp(-c\log^\alpha X).$$

Now II and IV are negligible, so we obtain

$$f^{(\ell)}(s) \ll \log^{\ell+1} X\{1 + (|t|+e)\exp(-c\log^\alpha X)\}.$$

Setting

$$\log X = \max\Big\{\Big(\frac{\ell+1}{c\alpha}\Big)^{1/\alpha}, \Big(\frac{\log(|t|+e)}{c}\Big)^{1/\alpha}\Big\},$$

we find $f^{(\ell)}(\sigma+it) \ll \log^{\ell+1} X$, giving the claimed bound. □

We find bounds for f'/f much as in the last section, but this time we take

$$g(t) := (\log(|t|+e))^{1/\alpha}.$$

We want to apply Lemma 16.1, for which we need

$$|f^{(\ell)}(\sigma+it)| \le Bg(t)^{\ell+1}, \quad \ell = 0,\, 1, \tag{16.24 bis}$$

as well as

$$1 \le g(nt) \le Dg(t), \quad n = 1,2,\dots,7. \tag{16.4 bis}$$

The last inequalities clearly hold. Also, (16.24 bis) follows from (16.35) by taking a larger constant B, because $\alpha \le 1$ and

$$\Big(\max\Big\{\frac{\ell+1}{\alpha}, \log(|t|+e)\Big\}\Big)^{(\ell+1)/\alpha} \le \Big(\frac{2}{\alpha}\Big)^{2/\alpha}\{\log(|t|+e)\}^{(\ell+1)/\alpha}, \quad \ell = 0,1.$$

Now by Lemma 16.1,

$$|f(\sigma+it)|^{-1} \le Cg(t)^\lambda$$

holds with constants $C > 0$ and $\lambda = 5$ for $1 < \sigma \le 2$ and all $t \in \mathbb{R}$.

To evaluate the left hand side of (16.32) we give pointwise bounds for derivatives of $(1/s)\log f(s)$ by using Lemma 16.2. Let $C_\ell := ((\ell+1)/\alpha)^{1/\alpha}$.

LEMMA 16.11. *Assume that the function $f(s) = (s-1)\zeta(s)/s$ satisfies*

$$|f^{(\ell)}(\sigma+it)| \le M_1\big(c^{-1/\alpha}\max\{C_\ell,\, g(t)\}\big)^{\ell+1} \tag{16.35 bis}$$

for $\ell = 0,1,\dots,$ and

$$|f(\sigma+it)|^{-1} \le M_2 g(t)^\lambda \tag{16.38}$$

with positive constants M_1, M_2 and λ, uniformly for $1 < \sigma \le 2$. Then

$$\left|\frac{d^k}{ds^k}\{s^{-1}\log f(s)\}\right| \le \frac{1}{|s|}\big(M_5 k \max\{C_k^2,\, g(t)^2\}\, g(t)^\lambda\big)^k$$

holds with some constant $M_5 > 1$ for $1 < \sigma \le 2$ and all integers $k \ge 1$.

PROOF. By (16.35 bis) and (16.38), we have for each $j \ge 1$,

$$\left|\frac{f^{(j)}(\sigma+it)}{f(\sigma+it)}\right| \le M_3\big(c^{-1/\alpha}\max\{C_j,\, g(t)\}\big)^{j+1} g(t)^\lambda. \tag{16.39}$$

Thus, for $n \ge 1$ and nonnegative integers $k_1, \dots, k_n$ satisfying $k_1 + \cdots + k_n = k$ and $k_1 + 2k_2 + \cdots + nk_n = n$,

$$\begin{aligned}\prod_{j=1}^n \left|\frac{f^{(j)}(\sigma+it)}{f(\sigma+it)}\right|^{k_j} &\le M_3^k \prod_{j=1}^n \Big\{\big(c^{-1/\alpha}\max\{C_n,\, g(t)\}\big)^{j+1} g(t)^\lambda\Big\}^{k_j} \\ &\le \Big(M_4 \max\{C_n^2,\, g(t)^2\} g(t)^\lambda\Big)^n.\end{aligned}$$

(We have used the observations that $C_j \le C_n$ for $1 \le j \le n$ and that $k \le n$.)

Hence, by Lemma 16.2, for $n \ge 1$, all real t, and $1 < \sigma \le 2$,

$$\begin{aligned}\left|\frac{d^n}{ds^n}\log f(s)\right| &= \left|\left(\frac{f'(s)}{f(s)}\right)^{(n-1)}\right| \\ &\le \sum_{k=1}^n \sum_{\mathbf{k}} C_{\mathbf{k},k}\big(M_4 \max\{C_n^2,\, g(t)^2\} g(t)^\lambda\big)^n \\ &\le (n-1)!\big(2M_4 \max\{C_n^2,\, g(t)^2\} g(t)^\lambda\big)^n.\end{aligned} \tag{16.40}$$

On the same set, we have

$$|\log f(s)| \ll |s| \tag{16.41}$$

since, by Theorem 16.3, the integral on the right hand side of (16.30) is uniformly bounded for $1 < \sigma$ and all $t \in \mathbb{R}$.

Now we find derivatives of $s^{-1}\log f(s)$ using Leibniz's rule. For each positive integer k, we have

$$\frac{d^k}{ds^k}\{s^{-1}\log f(s)\} = \sum_{n=0}^k \binom{k}{n}\frac{(-1)^{k-n}(k-n)!}{s^{k-n+1}}\{\log f(s)\}^{(n)}.$$

Using (16.41) and (16.40) with $C_n \le C_k$, we get

$$\begin{aligned}\left|\frac{d^k}{ds^k}\{s^{-1}\log f(s)\}\right| &\le \frac{Bk!}{|s^k|} \\ &\quad + \sum_{n=1}^k \binom{k}{n}\frac{(k-n)!(n-1)!}{|s|^{k-n+1}}\big(2M_4 \max\{C_n^2,\, g(t)^2\} g(t)^\lambda\big)^n \\ &\le \frac{k!}{|s|}\big(2M_4 \max\{C_k^2,\, g(t)^2\} g(t)^\lambda\big)^k \Big(B + \sum_{n=1}^k \frac{1}{n}\Big) \\ &\le \frac{1}{|s|}\big(M_5 k \max\{C_k^2,\, g(t)^2\} g(t)^\lambda\big)^k.\end{aligned}$$

□

PROOF OF THEOREM 16.8. It remains to show that the integral condition (16.32) is satisfied. If we replace $\max\{C_k^2, g(t)^2\}$ by $C_k^2 + g(t)^2$ in the last lemma, we find for $k \geq 1$,

$$\begin{aligned}(16.42)\qquad &\int_{-\infty}^{\infty}\left|\frac{d^k}{ds^k}\left\{\frac{1}{s}\log\left(\frac{s-1}{s}\zeta(s)\right)\right\}\right|^2 dt \\ &\leq (M_5\, k)^{2k}\left\{C_k^{4k}\int_{-\infty}^{\infty}\frac{g(t)^{2\lambda k}}{1+t^2}\,dt + \int_{-\infty}^{\infty}\frac{g(t)^{(2\lambda+4)k}}{1+t^2}\,dt\right\}.\end{aligned}$$

For any $\beta \geq 0$ we have

$$\begin{aligned}(16.43)\qquad \int_0^{\infty}\frac{\log^{\beta}(t+e)\,dt}{1+t^2} &= \int_e^{\infty}\frac{\log^{\beta} w\,dw}{1+(w-e)^2} \leq (e^2+2)\int_e^{\infty}\frac{\log^{\beta} w\,dw}{1+w^2} \\ &\leq (e^2+2)\int_0^{\infty} e^{-v}v^{\beta}dv = (e^2+2)\Gamma(\beta+1).\end{aligned}$$

Applying (16.43) with $\beta = 2\lambda k/\alpha$ and $(2\lambda+4)k/\alpha$ respectively, we find that the right hand side of (16.42) is at most

$$M_5^{2k}k^{2k}\left\{\left(\frac{k+1}{\alpha}\right)^{4k/\alpha}\left(\frac{2\lambda k}{\alpha}\right)^{2\lambda k/\alpha}\left(\frac{2(\lambda+2)k}{\alpha}\right)^{(2\lambda+4)k/\alpha}\right\} \leq (Mk^{1+(\lambda+2)/\alpha})^{2k}.$$

It follows with $\eta = 1 + (\lambda+2)/\alpha$, and in particular, with $\eta = 1 + 7/\alpha$, that

$$\int_{-\infty}^{\infty}\left|\frac{d^k}{ds^k}\left\{\frac{1}{s}\log\left(\frac{s-1}{s}\zeta(s)\right)\right\}\right|^2 dt \leq (Mk^{\eta})^{2k},$$

i.e. (16.32) holds uniformly for $1 < \sigma \leq 2$ and all positive integers k. Thus the conditions of Lemma 16.9 are satisfied, and Theorem 16.8 is established. □

16.5. Notes

§16.3. H. Wegmann [**We66**] has applied the elementary prime number methodology of E. Wirsing to show that if $N(x) = Ax + O(x\log^{-\gamma} x)$, then $\pi(x) = \mathrm{li}(x) + O(x(\log x)^{-k})$ with any $k < \gamma/3$. This is a stronger result than ours, but, with details, would take us too far afield to present.

Nyman [**Ny49**] asserted that the results of Corollary 16.4 were equivalent to the statement that "to every $\epsilon > 0$ and every nonnegative integer n, a constant $A = A(\epsilon, n)$ can be chosen such that

$$(16.44)\qquad |\zeta^{(n)}(s)| < A|t|^{\epsilon} \quad\text{and}\quad |1/\zeta(s)| < A|t|^{\epsilon}$$

hold uniformly in the region $\sigma > 1$, $|t| \geq \epsilon$." It was observed by A. E. Ingham in Math Reviews [MR 0032693] that these conditions do not imply even the PNT. As a simple example, let

$$\mathcal{P} := \{2, 2, 3, 3, 5, 5, \ldots, p, p, \ldots\}$$

be a set of Beurling generalized primes; that is, $\mathcal{P}$ consists of the rational primes p each taken exactly twice. Let $\mathcal{N}$ be the associated set of g-integers. The function $\zeta_{\mathcal{N}}(s)$ is the square of the Riemann zeta function. It is known from classical number theory that the conditions (16.44) are satisfied for $\zeta^2(s)$. On the other hand, $\mathcal{N}$

consists of rational integers n, each repeated $d(n)$ times, where $d(n)$ is the classical divisor function. Here

$$N(x) = x\log x + (2\gamma - 1)x + O(\sqrt{x})$$

and

$$\psi(x) = 2x + O\Big(x\exp\{-c\sqrt{\log x}\,\}\Big).$$

Thus neither of the assertions of Corollary 16.4 holds.

L. Tschakaloff (Chakalov) [**Ts40**] proved the existence of a cosine polynomial p of order 7 with $S(p) = 5.90529\ldots$. An earlier version of his work appears in the Yearbook of the University of Sofia, 19 (1923) (in Bulgarian). This topic is studied more generally by S. Gy. Révész [**Rv07**].

Formula (16.5) is valid with some positive $\lambda < 4.91$ by use of the full force of Tschakaloff's trigonometric inequality.

Formula (16.12) for the n-th derivative of the logarithm of a function is a special case of an inverse Bell polynomial relation or of Faà di Bruno's identity for differentiating composite functions.

§16.4. The main result of this section is based on an article of R. S. Hall [**Ha72**].

CHAPTER 17

Optimality of the dlVP Remainder Term

How good is too good?

SUMMARY. Can one establish a better PNT remainder estimate for a discrete g-number system than that of de la Vallée Poussin? Here we construct an example showing that the answer is in general, No. Also, we show, in the other direction, that there exist discrete g-number systems for which the analogue of the Riemann Hypothesis holds.

17.1. Background

The first proofs of the Prime Number Theorem relied on a global analysis of the Riemann zeta function, mainly, its analytic continuability and Hadamard product representation. In this way, C. J. de la Vallée Poussin proved in 1899 that

$$\zeta(s) \neq 0 \quad \text{for} \quad \sigma > 1 - c/\log t, \ t \geq 2, \tag{17.1}$$

where c is some positive constant. This is the so-called "classical" zero-free region of $\zeta(s)$, from which the PNT was deduced by contour integration, with an remainder term

$$\pi(x) = \mathrm{li}(x) + O(x \exp\{-c\sqrt{\log x}\}), \tag{17.2}$$

where $\mathrm{li}(x) = \int_2^x du/\log u$ and c is some positive constant (not necessarily the one in (17.1)).

Early in the twentieth century, E. Landau showed that the classical zero-free region (17.1) could be derived by means of local analysis. He showed also that essentially the same arguments used in proving the PNT could be applied to establish the Prime Ideal Theorem for a fixed algebraic number field. His reasoning in fact proves that if the number $N(x)$ of integral ideals with norm at most x satisfies

$$N(x) = kx + O(x^\theta) \tag{17.3}$$

with $k > 0$ and $\theta \in (0, 1)$ (a result that had been proved earlier by Weber), then the zeta function of the number field satisfies (17.1) and the number $\pi(x)$ of prime ideals with norm not exceeding x satisfies (17.2).

In this chapter, we revisit questions (2) and (5) of Chapter 1, giving examples that illustrate two cases of extreme behavior of the prime number distributions of (discrete!) g-number systems:

(DLVP) to establish the optimality of (17.1) and (17.2), apart from the numerical value of the constants c; the result is given in Theorem 17.14.

(RH) to give, on the other hand, a g-number system for which the Riemann Hypothesis holds for the associated zeta function and the prime counting function satisfies $\pi(x) = \mathrm{li}(x) + O(x^{1/2})$. This result, which is slightly better than the classical conditional estimate, is given in Theorem 17.11.

In each case, the Beurling g-number system that is constructed satisfies (17.3) with $\theta \in (1/2, 1)$ (optimality is not known for $\theta \leq 1/2$)

Our arguments show the existence of g-number systems having the desired properties, but they are "ineffective" in the sense that they do not provide explicit constructions. In each case, the process proceeds in two main steps. First, we construct an explicit "template" continuous distribution or, equivalently, a template zeta function. This provides the main term in the distribution of a g-prime system that will then be selected probabilistically.

For RH, the template zeta function is $s/(s-1)$; that for DLVP includes $s/(s-1)$ along with a second factor that has infinitely many zeros on $\{\sigma = 1 - 1/\log|t|\}$ and none to the right of this curve. We will show that the Chebyshev function $\psi(x)$ for DLVP satisfies

$$\limsup_{x\to\infty} \frac{\psi(x) - x}{x\exp\{-2\sqrt{\log x}\}} = 2, \qquad \liminf_{x\to\infty} \frac{\psi(x) - x}{x\exp\{-2\sqrt{\log x}\}} = -2.$$

For the second step, random g-primes are created from a sequence of positive real numbers $\{v_k\}$ as the nonzero values of $\{v_k X_k\}$, where $\{X_k\}$ are independent Bernoulli variables, with a "success" probability p_k assigned to each number v_k. The sequence $\{v_k\}$ is chosen to be sufficiently dense in a neighborhood of infinity and $\{p_k\}$ are chosen to mimic the template density. In this case, the expectation of the Fourier transform of a truncation of the random g-prime distribution closely approximates that of the similarly truncated template distribution. Then a familiar inequality of Kolmogorov is used to select g-primes from $\{v_k\}$ so that the Fourier transform of its truncation is suitably close to that of the truncated template distribution, yielding the claimed result in each case.

In the following discussion, c denotes a positive constant which may have different values in several consecutive occurrences. However, where it seems necessary to show clearly the relation between consecutive values of constants, we will instead write $c_1, c_2, \ldots$.

17.2. Discrete random approximation

17.2.1. A general form. Our starting point is an inequality variously ascribed to Bernstein or Kolmogorov [**DMV06, Lo77**]. For completeness, we give a simple proof.

LEMMA 17.1. *Let Y_k, $k = 1, \ldots, K$, be independent random variables such that $|Y_k| \leq 1$ and the expectation $EY_k = 0$. Also, let r_k, $k = 1, \ldots, K$, be real numbers such that $|r_k| \leq 1$. Let $Y = \sum_{k=1}^{K} r_k Y_k$ and set $\sigma^2 = \operatorname{Var} Y$. Then, for any number $\sigma_e^2 \geq \sigma^2$ and $0 \leq v \leq 2\sigma_e^2$, we have*

$$P[Y \geq v] \leq \exp\{-v^2/(4\sigma_e^2)\}.$$

PROOF. Let $\lambda \geq 0$ be a parameter (to be specified later). Then $P[Y \geq v] \leq Ee^{\lambda(Y-v)}$. Since the Y_k are independent,

$$Ee^{\lambda(Y-v)} = e^{-\lambda v} E \prod_{k=1}^{K} e^{\lambda r_k Y_k} = e^{-\lambda v} \prod_{k=1}^{K} Ee^{\lambda r_k Y_k}.$$

It is easily verified that $e^u \le 1 + u + u^2$ for $-1 \le u \le 1$. Thus, for $0 \le \lambda \le 1$,

$$Ee^{\lambda r_k Y_k} \le E(1 + \lambda r_k Y_k + \lambda^2 r_k^2 Y_k^2) = 1 + \lambda^2 r_k^2 EY_k^2 \le \exp\{\lambda^2 r_k^2 EY_k^2\}$$

by the inequality $1 + w \le e^w$, which holds for all real w. Hence we have

$$P[Y \ge v] \le \exp\{-\lambda v + \lambda^2 \sigma^2\}. \tag{17.4}$$

If $0 \le v \le 2\sigma_e^2$, by taking $\lambda = v/(2\sigma_e^2)$, we obtain

$$P[Y \ge v] \le \exp\Big\{-\frac{v^2}{2\sigma_e^2} + \frac{v^2\sigma^2}{4\sigma_e^4}\Big\} \le \exp\Big\{-\frac{v^2}{4\sigma_e^2}\Big\}. \qquad \square$$

We next describe how to make g-numbers using random choices, and then we show that a reasonably simple special case "works" for the two constructions. The argument also will reveal how much latitude there is for our choices.

LEMMA 17.2. *Let*

$$1 \le v_0 < v_1 < \cdots < v_k < \cdots$$

be a sequence such that $v_k \to \infty$ *as* $k \to \infty$. *Also, let* $\{p_k\}_{k=1}^{\infty}$ *be a sequence of real numbers such that* $0 < p_k \le 1$ *for* $k \ge k_0$. *Assume that there exists an increasing function* $F(x)$ *on* $(-\infty, \infty)$ *with support on* $[1, \infty)$ *which satisfies*

$$\sum_{v_k \le x} p_k \ll F(x),$$

$$\sum_{x < v_k \le x+1} p_k \ll \sqrt{F(x)\,(1 + \log x)}\,,$$

$$\log x = o(F(x)),$$

$$\int_1^x v^{-1}\sqrt{F(v)}\,dv \ll \sqrt{F(x)},$$

and

$$F(x+1) \ll F(x).$$

Then there exists a subsequence $\{v_{k_j}\}$, $j = 1, 2, \ldots$, *such that*

$$\sum_{v_{k_j} \le x} v_{k_j}^{it} - \sum_{v_k \le x} v_k^{it} p_k \ll \sqrt{F(x)}\,\{\sqrt{1 + \log x} + \sqrt{\log(t+1)}\,\} \tag{17.5}$$

for $1 \le x < \infty$ *and* $t \ge 0$.

REMARK 17.3. It is convenient to treat quantities of the form v^{it}; our estimates clearly apply to v^{-it} as well, and we shall use the latter form in applications.

REMARK 17.4. Property (17.5) will be shown to hold for almost all subsequences (in the statistical sense) of $\{v_k\}$, by the easy part of the Borel-Cantelli lemma. Later, we shall make an explicit choice of a function F.

PROOF. Without loss of generality, we may assume $k_0 = 1$. Let $X_k(\omega)$, $k = 1, 2, \ldots$, be independent Bernoulli variables on the probability space $([0,1], \mathcal{B}^1, P)$, that is,

$$P[X_k = 1] = p_k, \qquad P[X_k = 0] = 1 - p_k,$$

where P is Lebesgue measure on the Borel sets $\mathcal{B}^1$ on the interval $[0, 1]$.

Let

$$X(x,t) := \sum_{v_k \le x} v_k^{it} X_k.$$

Then the expectation satisfies

$$EX(x,t) = \sum_{v_k \le x} v_k^{it} p_k.$$

Note that, since the variables X_k are independent, the variance satisfies

$$\mathrm{Var}\,(\Re \mathrm{X(x,t)}) = \sum_{\mathrm{v_k \le x}} \cos^2(\mathrm{t \log v_k}) \mathrm{p_k}(1 - \mathrm{p_k}) \le \mathrm{cF(x)}$$

and

$$\mathrm{Var}\,(\Im \mathrm{X(x,t)}) = \sum_{\mathrm{v_k \le x}} \sin^2(\mathrm{t \log v_k}) \mathrm{p_k}(1 - \mathrm{p_k}) \le \mathrm{cF(x)}$$

with some constant $c \ge 1$, uniformly for $x \ge 1$ and $t \ge 0$.

Let

$$\begin{aligned} R^s(x,t) := \Big\{\omega : s\Big(\Re X(x,t) - \Re \sum_{v_k \le x} v_k^{it} p_k\Big) \\ \ge \sqrt{8cF(x)}\,\big(\sqrt{1+\log x} + \sqrt{\log(t+1)}\,\big)\Big\} \end{aligned}$$

and

$$\begin{aligned} I^s(x,t) := \Big\{\omega : s\Big(\Im X(x,t) - \Im \sum_{v_k \le x} v_k^{it} p_k\Big) \\ \ge \sqrt{8cF(x)}\,\big(\sqrt{1+\log x} + \sqrt{\log(t+1)}\,\big)\Big\}, \end{aligned}$$

where s denotes $+$ or $-$. These are the sets on which the desired inequality fails to hold; we shall show them each to be very small.

To satisfy the condition $0 \le v \le 2\sigma_e^2$ of Lemma 17.1 for $t = 0$, note that

$$\sqrt{8cF(x)(1+\log x)} \le 2cF(x)$$

for $x \ge x_0$. The lemma now gives

$$P[R^+(x,0)] \le \exp\Big\{-\frac{8cF(x)(1+\log x)}{4cF(x)}\Big\} \le \exp\{-2\log x\} = x^{-2},$$

and a similar inequality holds for $P[R^-(x,0)]$. Hence,

$$\sum_{m=1}^{\infty} (P[R^+(m,0)] + P[R^-(m,0)]) < \infty.$$

For $t \ne 0$, we satisfy the condition $0 \le v \le 2\sigma_e^2$ of the lemma by noting that

$$\sqrt{8cF(x)}\,\big(\sqrt{1+\log x} + \sqrt{\log(t+1)}\,\big) \le 2cF(x)$$

for $F(x) \ge 3\log(t+1)$, $t \ge t_0$. This time, the lemma gives

$$\begin{aligned} P[R^+(x,t)] &\le \exp\Big\{-\frac{8cF(x)\big(\sqrt{1+\log x} + \sqrt{\log(t+1)}\,\big)^2}{4cF(x)}\Big\} \\ &\le \exp\{-2(\log x + \log(t+1))\} = x^{-2}(t+1)^{-2}. \end{aligned}$$

In particular,

$$P[R^+(m,n)] \le \frac{1}{m^2(n+1)^2}$$

for $n \geq n_0$ and $m \geq m_n$ with $m_n := \min\{m : F(m) \geq 3\log(n+1)\}$. Similar inequalities hold for $P[R^-(m,n)]$, $P[I^+(m,n)]$, and $P[I^-(m,n)]$ as well. Hence

$$\sum_{n=n_0}^{\infty} \sum_{m=m_n}^{\infty} \big(P[R^+(m,n)] + P[R^-(m,n)] + P[I^+(m,n)] + P[I^-(m,n)]\big) < \infty.$$

Therefore, by the first Borel-Cantelli lemma, the subset of ω for which

$$\left| X(m,0) - \sum_{v_k \leq m} p_k \right| \geq \sqrt{8cF(m)\,(1+\log m)} \tag{17.6}$$

holds for infinitely many m or

$$\left| X(m,n) - \sum_{v_k \leq m} v_k^{in} p_k \right| \geq 2\sqrt{8cF(m)}\,\big(\sqrt{1+\log m} + \sqrt{\log(n+1)}\,\big) \tag{17.7}$$

holds for infinitely many pairs (m,n) with $n \geq n_0$ and $m \geq m_n$ has measure zero. Choose any ω_0 in the complement of this subset and fix it. Then there are only finitely many numbers m for which (17.6) holds and only finitely many pairs (m,n) with $n \geq n_0$ and $m \geq m_n$ for which (17.7) is true. It follows that there are m_0 and $n_1 \geq n_0$ such that

$$\left| X(m,0) - \sum_{v_k \leq m} p_k \right| < \sqrt{8cF(m)\,(1+\log m)} \tag{17.8}$$

holds for all $m \geq m_0$ and such that

$$\left| X(m,n) - \sum_{v_k \leq m} v_k^{in} p_k \right| < 2\sqrt{8cF(m)}\,\big(\sqrt{1+\log m} + \sqrt{\log(n+1)}\,\big) \tag{17.9}$$

holds for all $n \geq n_1$ and $m \geq m_n$.

Define

$$k_0 = 1,\ k_j = \min\{k : k > k_{j-1},\ X_k(\omega_0) = 1\},\ j = 1, 2, \ldots,$$

and set

$$G(x,t) := \sum_{v_{k_j} \leq x} v_{k_j}^{it} - \sum_{v_k \leq x} v_k^{it} p_k\,.$$

From the definition of v_{k_j} and (17.8),

$$|G(m,0)| = \left| \sum_{v_{k_j} \leq m} 1 - \sum_{v_k \leq m} p_k \right| \leq \sqrt{8cF(m)\,(1+\log\, m)}$$

for $m \geq m_0$. This inequality is then true for all $m \geq 1$ if $\sqrt{8c}$ is replaced by a sufficiently large constant c_1. Given $x \geq 1$, let $m \leq x < m+1$. Then

$$\begin{aligned} G(x,0) &\leq \left(\sum_{v_{k_j} \leq m+1} 1 - \sum_{v_k \leq m+1} p_k \right) + \sum_{m < v_k \leq m+1} p_k \\ &\leq c_1\sqrt{F(m+1)\,(1+\log(m+1))} + c\sqrt{F(m)\,(1+\log\, m)} + 2 \\ &\leq c_2\sqrt{F(x)\,(1+\log\, x)} \end{aligned}$$

and similarly

$$G(x,0) \geq \left(\sum_{v_{k_j} \leq m} 1 - \sum_{v_k \leq m} p_k \right) - \sum_{m < v_k \leq m+1} p_k \geq -c_2 \sqrt{F(x)\,(1+\log x)}\,.$$

Therefore,

$$|G(x,0)| \leq c_2 \sqrt{F(x)\,(1+\log x)}\,.$$

This proves (17.5) for $1 \leq x < \infty$ and $t = 0$.

Now we consider nonzero t. For the bounded range $n \leq n_1$,

$$G(m,n) = \int_1^m v^{in}\, dG(v,0),$$

and hence, by integration by parts,

$$\begin{aligned} |G(m,n)| &\leq c_2 \sqrt{F(m)\,(1+\log m)} + c_2 n_1 \int_1^m v^{-1} \sqrt{F(v)\,(1+\log v)}\, dv \\ &\leq c_3 \sqrt{F(m)\,(1+\log m)}\,. \end{aligned}$$

Also, for $n \geq n_1$ and $m < m_n$,

$$\begin{aligned} |G(m,n)| &\leq \sum_{v_{k_j} \leq m} 1 + \sum_{v_k \leq m} p_k \leq \left| \sum_{v_{k_j} \leq m} 1 - \sum_{v_k \leq m} p_k \right| + 2 \sum_{v_k \leq m} p_k \\ &\ll \sqrt{F(m)\,(1+\log m)} + F(m) \\ &\ll \sqrt{F(m)} \left(\sqrt{1+\log m} + \sqrt{\log(n+1)}\, \right), \end{aligned} \tag{17.10}$$

since $F(m) \leq 3\log(n+1)$ for $m < m_n$ (by the definition of m_n). Now, for $n \geq n_1$ and $m \geq m_n$, from the definition of v_{k_j} and (17.9),

$$|G(m,n)| \leq 2\sqrt{8cF(m)} \left(\sqrt{1+\log m} + \sqrt{\log(n+1)}\, \right),$$

and hence the inequality between the first and last terms of (17.10) holds also for $m \geq 1$ and $n \geq 0$.

Finally, given $x \geq 1$ and $t > 0$, let $m \leq x < m+1$ and $n \leq t < n+1$. Then

$$|G(x,n)| \leq |G(m,n)| + \sum_{m < v_{k_j} \leq m+1} 1 + \sum_{m < v_k \leq m+1} p_k.$$

The sum of the last two terms on the right hand side equals

$$\begin{aligned} &\left(\sum_{v_{k_j} \leq m+1} 1 - \sum_{v_k \leq m+1} p_k \right) - \left(\sum_{v_{k_j} \leq m} 1 - \sum_{v_k \leq m} p_k \right) + 2 \sum_{m < v_k \leq m+1} p_k \\ &\qquad \ll |G(m+1,0)| + |G(m,0)| + \sqrt{F(m)\,(1+\log m)} + 1. \end{aligned}$$

Hence

$$|G(x,n)| \ll \sqrt{F(x)} \left(\sqrt{1+\log x} + \sqrt{\log(n+1)}\, \right)$$

for $x \geq 1$ and $n \geq 0$. Now write

$$G(x,t) = \int_{v=1}^x v^{it-in} dG(v,n)$$

and integrate by parts. We find that the left hand side of (17.5) satisfies

$$\begin{aligned} G(x,t) &= x^{i(t-n)}G(x,n) - i(t-n)\int_1^x v^{i(t-n)-1}G(v,n)\,dv \\ &\ll \sqrt{F(x)}\,\big(\sqrt{1+\log x}+\sqrt{\log(n+1)}\,\big) \\ &\quad + \int_1^x v^{-1}\sqrt{F(v)}\,\big(\sqrt{1+\log v}+\sqrt{\log(n+1)}\,\big)\,dv \\ &\ll \sqrt{F(x)}\,\big\{\sqrt{1+\log x}+\sqrt{\log(t+1)}\,\big\}. \end{aligned}$$

□

17.2.2. A special form. For our construction, we shall use a special form of Lemma 17.2 for a continuous distribution.

LEMMA 17.5. *Let $f(v)$ be a nonnegative-valued Lebesgue measurable function on $(-\infty,\infty)$ with support $[1,\infty)$ and*

$$1 \le v_0 < v_1 < \cdots < v_k < \cdots$$

be a sequence such that $v_k \to \infty$. Assume that the sequence $\{p_k\}$ defined by

$$p_k := \int_{v_{k-1}}^{v_k} f(v)\,dv$$

satisfies $0 < p_k \le 1$ for $k \ge k_0$. Also, assume that there is a function $F(x)$ satisfying the conditions of Lemma 17.2 and that

$$\sum_{v_k \le x} v_k^{it} p_k - \int_1^x v^{it} f(v)\,dv \ll \sqrt{F(x)}\,\big(\sqrt{1+\log x}+\sqrt{\log(t+1)}\,\big) \tag{17.11}$$

for $F(x) \ge c\log(t+1)$ with some constant $c > 0$. Then there is a subsequence $\{v_{k_j}\}$, $j = 1,2,\ldots$, such that

$$\sum_{v_{k_j} \le x} v_{k_j}^{it} - \int_1^x v^{it} f(v)\,dv \ll \sqrt{F(x)}\,\big(\sqrt{1+\log x}+\sqrt{\log(t+1)}\,\big) \tag{17.12}$$

holds for $1 \le x < \infty$ and all $t \ge 0$.

PROOF. For $(0 <)\, x_0 \le x$ and t satisfying $F(x) < c\log(t+1)$, we have

$$\begin{aligned} \left| \sum_{v_k \le x} v_k^{it} p_k - \int_1^x v^{it} f(v)\,dv \right| &\le 2 \sum_{v_k \le x} p_k + 1 \\ &\ll \sqrt{F(x)}\big(\sqrt{1+\log x}+\sqrt{\log(t+1)}\,\big). \end{aligned}$$

We note that $F(1) > 0$. Hence (17.11) holds for $1 \le x < \infty$ and $t \ge 0$. Then (17.12) follows from Lemma 17.2. □

Let v_k, $k = 0,1,2,\ldots$, be the sequence defined by

$$v_0 = 1 \text{ and } v_k = n + \ell/2^n, \text{ for } k = 2^n + \ell,\ n = 0,1,2,\ldots,\ 1 \le \ell < 2^n. \tag{17.13}$$

REMARK 17.6. It is easily verified that $\{v_k\}$ is closed under addition and multiplication, but a subsequence $\{v_{k_j}\}$ is unlikely to have these properties. It would be of much interest to find a Beurling g-number system having arithmetic structure.

LEMMA 17.7. *Let $f(x)$ be a Lebesgue measurable function on $(-\infty, \infty)$ with support $[1, \infty)$ which satisfies*

$$0 \le f(x) \ll \frac{1 - x^{-1}}{\log x}.$$

Also, let the sequence $\{p_k\}$ be defined as in Lemma 17.5 with $\{v_k\}$ given in (17.13). Then the function

$$F(x) := x/(1 + \log x)$$

and the sequence $\{v_k\}$ defined by (17.13) together satisfy the conditions of Lemmas 17.2 and 17.5. Therefore $\{v_k\}$ has a subsequence $\{v_{k_j}\}$ satisfying

$$\sum_{v_{k_j} \le x} v_{k_j}^{it} - \int_1^x v^{it} f(v)\, dv \ll \sqrt{x}\left(1 + \sqrt{\frac{\log(t+1)}{1 + \log x}}\right) \tag{17.14}$$

for $1 \le x < \infty$ and $t \ge 0$.

REMARK 17.8. The subsequence will provide the Beurling primes having the desired distribution.

PROOF. We first show that $F(x)$ satisfies the conditions of Lemma 17.2. To begin, note that

$$F'(x) = (\log x)/(1 + \log x)^2$$

for $x > 1$, and hence $F(x)$ is increasing. Then we have

$$\sum_{v_k \le x} p_k \le \int_1^x f(v)\, dv \ll \int_1^x \frac{1 - v^{-1}}{\log v}\, dv \ll \frac{x}{1 + \log x}.$$

Similarly,

$$\begin{aligned}\sum_{x < v_k \le x+1} p_k &\le 1 + \int_x^{x+1} f(v)\, dv \ll 1 + \int_x^{x+1} \frac{1 - v^{-1}}{\log v}\, dv \\ &\le 1 + \frac{1}{\log x} \ll \sqrt{F(x)\,(1 + \log x)}.\end{aligned}$$

Next,

$$\frac{\log x}{F(x)} \ll \frac{\log^2 x}{x} = o(1), \quad x \to \infty.$$

Also, by integration by parts,

$$\begin{aligned}\int_1^x v^{-1}\sqrt{F(v)}\, dv &= \int_1^x \frac{dv}{\sqrt{v\,(1 + \log v)}} \\ &= 2(\sqrt{F(x)} - 1) + \int_1^x \frac{dv}{\sqrt{v\,(1 + \log v)^3}} \ll \sqrt{F(x)}.\end{aligned}$$

Finally,

$$F(x+1) = \frac{x+1}{1 + \log(x+1)} \le 2F(x).$$

We next show that the sequence v_k satisfies the conditions of Lemma 17.5 by proving that, for all $t \geq 1$,

$$\sum_{v_k \leq n} v_k^{it} p_k - \int_1^n v^{it} f(v)\, dv \ll \log(t+1). \tag{17.15}$$

We can assume that $n \geq 2\log(t+1)$, for otherwise the result is trivial. Now, the left hand side of (17.15) is at most

$$1 + 2\int_1^{n_0} f(v)\, dv + \sum_{k=2^{n_0}+1}^{2^n} \int_{v_{k-1}}^{v_k} |v_k^{it} - v^{it}| f(v)\, dv.$$

The sum on the right hand side equals

$$\begin{aligned}
&\sum_{\ell=n_0}^{n-1} \sum_{k=1}^{2^\ell} \int_{v_{2^\ell+k-1}}^{v_{2^\ell+k}} |v_{2^\ell+k}^{it} - v^{it}| f(v)\, dv \\
&\qquad \ll \sum_{\ell=n_0}^{n-1} \sum_{k=1}^{2^\ell} \left(\frac{t}{2^\ell \ell}\right) \frac{1}{2^\ell(1+\log \ell)} \ll \frac{t}{2^{n_0} n_0 \log(1+n_0)},
\end{aligned}$$

since

$$|v_{2^\ell+k}^{it} - v^{it}| = |e^{it\log(v_{2^\ell+k}/v)} - 1| \leq t \log \frac{v_{2^\ell+k}}{v_{2^\ell+k-1}} \leq \frac{t}{2^\ell \ell}$$

and

$$f(v) \ll \frac{1}{1+\log(v_{2^\ell+k-1})} \leq \frac{1}{1+\log \ell}$$

for $v_{2^\ell+k-1} \leq v \leq v_{2^\ell+k}$. Hence the left hand side of (17.15) is

$$\ll \frac{n_0}{1+\log n_0} + \frac{t}{2^{n_0} n_0 \log(1+n_0)}.$$

Then (17.15) follows by taking

$$n_0 := 1 + \left\lfloor \frac{\log(t+1)}{\log 2} \right\rfloor,$$

which is less than n.

Now if $F(n) \geq 2\log(t+1)$, then $n \geq 2\log(t+1)$ and

$$\log(t+1) \ll \sqrt{F(n)\log(t+1)}.$$

Therefore, by (17.15),

$$\sum_{v_k \leq x} v_k^{it} p_k - \int_1^x v^{it} f(v)\, dv \ll \sqrt{F(x)}\left(\sqrt{1+\log x} + \sqrt{\log(t+1)}\right)$$

whenever $x = n \in \mathbb{N}$ and $F(n) \geq 2\log(t+1)$. Then, for $F(x) \geq 2\log(t+1)$ and $n < x < n+1$, the modulus of the left hand side is at most

$$\begin{aligned}
&\left|\sum_{v_k \leq n} v_k^{it} p_k - \int_1^n v^{it} f(v)\, dv\right| + 2\sum_{n < v_k \leq n+1} p_k + 1 \\
&\qquad \ll \sqrt{F(n)}\left(\sqrt{1+\log n} + \sqrt{\log(t+1)}\right) \\
&\qquad \leq \sqrt{F(x)}\left(\sqrt{1+\log x} + \sqrt{\log(t+1)}\right).
\end{aligned}$$

This shows that the condition (17.11) of Lemma 17.5 is satisfied, completing the proof of Lemma 17.7. □

17.3. Generalized primes satisfying the Riemann Hypothesis

17.3.1. A Perron-type formula. We first prove a variant of the Perron inversion formula for Mellin transforms. It will be used to derive asymptotics of the counting function $N(x)$ of the g-integers in each of our cases.

LEMMA 17.9. *Let $F \in \mathcal{V}$ and $\widehat{F}(s)$ be the Mellin transform of dF. Assume that $\sigma_a(\widehat{F}) < \infty$. We set*

$$H(x) := \int_1^x F(u)\,du, \qquad x \geq 1.$$

Then, for any $x \geq 1$,

$$H(x) = \lim_{T\to\infty} \frac{1}{2\pi i}\int_{b-iT}^{b+iT} \widehat{F}(s)\frac{x^{s+1}}{s(s+1)}\,ds,$$

where $b > \max\{\sigma_c(\widehat{F}), 0\}$.

PROOF. We have

$$\widehat{H}(s) = \int_1^\infty x^{-s}F(x)\,dx = \frac{\widehat{F}(s-1)}{s-1}$$

for $\sigma > \sigma_c(\widehat{F}) + 1$. Note that $H(x)$ is continuous. Hence by the (usual form of the) Perron inversion formula [**Ap76, Te95, BD04, MV07**],

$$H(x) = \lim_{T\to\infty}\frac{1}{2\pi i}\int_{b_1-iT}^{b_1+iT} x^s\frac{\widehat{F}(s-1)}{(s-1)s}\,ds = \lim_{T\to\infty}\frac{1}{2\pi i}\int_{b-iT}^{b+iT}\widehat{F}(s)\frac{x^{s+1}\,ds}{s(s+1)},$$

where $b = b_1 - 1 > \max\{\sigma_c(\widehat{F}), 0\}$. □

This lemma has the following immediate consequence:

LEMMA 17.10. *If the function $F(x)$ of the last lemma is nondecreasing then*

$$F(x) \leq \lim_{T\to\infty}\frac{1}{2\pi i}\int_{b-iT}^{b+iT}\widehat{F}(s)\frac{(x+1)^{s+1}-x^{s+1}}{s(s+1)}\,ds$$

and

$$F(x) \geq \lim_{T\to\infty}\frac{1}{2\pi i}\int_{b-iT}^{b+iT}\widehat{F}(s)\frac{x^{s+1}-(x-1)^{s+1}}{s(s+1)}\,ds$$

for $x \geq 2$.

PROOF. By Lemma 17.9,

$$H(x+1) - H(x) = \lim_{T\to\infty}\frac{1}{2\pi i}\int_{b-iT}^{b+iT}\widehat{F}(s)\frac{(x+1)^{s+1}-x^{s+1}}{s(s+1)}\,ds\,.$$

Also,

$$H(x+1) - H(x) = \int_x^{x+1} F(u)\,du \geq F(x),$$

since F is increasing. This proves the first inequality of the lemma. The second inequality is proved analogously. □

As an example, take $F(x) = x$ with support $[1, \infty)$. Then $\widehat{F}(s) = s/(s-1)$ and

$$\begin{aligned}\lim_{T\to\infty} \frac{1}{2\pi i} \int_{2-iT}^{2+iT} \widehat{F}(s) \frac{(x+1)^{s+1} - x^{s+1}}{s(s+1)} \, ds \\ = \frac{1}{2\pi i} \int_{\sigma=2} \frac{(x+1)^{s+1} - x^{s+1}}{(s-1)(s+1)} \, ds = \frac{(x+1)^2 - x^2}{x} = x + \frac{1}{2}.\end{aligned}$$

17.3.2. Riemann Hypothesis example. In this subsection we construct a discrete g-number system satisfying the Riemann Hypothesis.

THEOREM 17.11. *There is a system of Beurling primes $\mathcal{P}_R$ such that*
(i) the associated Beurling integers satisfy

$$N_R(x) = k_1 x + O(x^{1/2} \exp\{c(\log x)^{2/3}\})$$

with $k_1 > 0$;
(ii) the associated zeta function $\zeta_R(s)$ is analytic for $\sigma > 1/2$ except for a simple pole at $s = 1$ with residue k_1;
(iii) the function $\zeta_R(s)$ has no zeros on the half plane $\sigma > 1/2$;
(iv) the prime counting function $\pi_R(x)$ satisfies

$$\pi_R(x) = \mathrm{li}(x) + O(x^{1/2}).$$

REMARK 17.12. As we indicate in §17.10, by making a finite number of changes in the g-primes of this theorem, we can produce a system that preserves any specified initial string of g-primes and satisfies $k_1 = 1$ with no changes other than the size of O−constants. (These remarks apply as well for the example in Theorem 17.14.)

To prove Theorem 17.11, we first make a general estimate, which will be used again later.

LEMMA 17.13. *Let $F(x,t)$ be a function defined for $1 \le x < \infty$ and $t \ge 0$ that is locally of bounded variation in x and satisfies $F(1,t) = 0$ as well as*

$$F(x,t) \ll \sqrt{x}\left(1 + \sqrt{\frac{\log(t+1)}{\log x}}\right).$$

Then

$$\int_{v=1}^{\infty} v^{-\sigma} \, dF(v,t) \ll \frac{\sigma}{\sigma - 1/2} + \sigma \sqrt{\frac{\log(t+1)}{\sigma - 1/2}}$$

for all $\sigma > 1/2$ and $t \ge 0$.

PROOF. After integration by parts, the integral on the left hand side equals

$$\sigma \int_1^{\infty} v^{-\sigma-1} F(v,t) \, dv \ll \sigma \int_1^{\infty} v^{-\sigma-1/2} \left(1 + \sqrt{\frac{\log(t+1)}{\log v}}\right) dv.$$

The first part of the right hand side is $\sigma/(\sigma - 1/2)$, and the second part, setting aside the factor $\sigma\sqrt{\log(t+1)}$, is

$$\int_1^{\infty} v^{-\sigma-1/2} \frac{dv}{\sqrt{\log v}} = \frac{1}{\sqrt{\sigma - 1/2}} \int_0^{\infty} e^{-w} w^{1/2} \frac{dw}{w} = \frac{\Gamma(1/2)}{\sqrt{\sigma - 1/2}},$$

where Γ is the Euler gamma function. □

PROOF OF THEOREM 17.11. Consider the template continuous density

$$f_R(v)\,dv := \frac{1-v^{-1}}{\log v}\,dv \quad \text{for} \quad v \geq 1.$$

By Lemma 17.7, there is a sequence $\{p_j\} := \{v_{k_j}\}$, $j = 1, 2, \ldots$, such that

$$\sum_{p_j \leq x} p_j^{-it} - \int_1^x v^{-it} f_R(v)\,dv \ll \sqrt{x}\left(1 + \sqrt{\frac{\log(t+1)}{1+\log x}}\right) \tag{17.16}$$

for $1 \leq x < \infty$ and $t \geq 0$. In particular, when $t = 0$,

$$\sum_{p_j \leq x} 1 - \int_1^x f_R(v)\,dv = O(x^{1/2}), \tag{17.17}$$

and hence

$$\pi_R(x) := \sum_{p_j \leq x} 1 = \mathrm{li}(x) + O(x^{1/2}). \tag{17.18}$$

Recall the general zeta function representation (1.5). By (17.18), our sought-after zeta function

$$\zeta_R(s) = \exp\left\{\int_1^\infty \log\left(1 - x^{-s}\right)^{-1} d\pi_R(x)\right\}$$

is convergent for $\sigma > 1$. Also, by Theorem 1.5, the template zeta function is

$$\exp\left\{\int_1^\infty x^{-s} f_R(x)\,dx\right\} = \frac{s}{s-1}.$$

Then $\zeta_R(s)$ can be written as

$$\zeta_R(s) = \frac{s}{s-1}\exp\{-F_1(s) + F_2(s)\},$$

where

$$F_1(s) = \int_1^\infty \{v^{-s} + \log(1 - v^{-s})\}\,d\pi_R(v)$$

and

$$F_2(s) = \int_1^\infty v^{-s}\{d\pi_R(v) - f_R(v)\,dv\}.$$

Since

$$\log(1 - v^{-s}) = -v^{-s} + O(v^{-2\sigma}) \quad \text{for} \quad v \geq \lambda_1 \ (> 1) \tag{17.19}$$

and (17.18) implies the crude bound $\pi_R(x) \ll x$, it follows that $F_1(s)$ converges uniformly for $\sigma \geq \frac{1}{2} + \epsilon$ with each $\epsilon > 0$. Hence $F_1(s)$ is analytic for $\sigma > 1/2$. By (17.17), $F_2(s)$ has this property too. Therefore $\zeta_R(s)$ has an analytic continuation in the half plane $\sigma > 1/2$ except at $s = 1$, where it has a simple pole with residue

$$k_1 = \exp\{-F_1(1) + F_2(1)\} > 0.$$

Also, from its representation as an exponential, $\zeta_R(s)$ has no zeros on the half plane $\sigma > 1/2$, i.e., the Riemann Hypothesis holds for ζ_R.

It remains to prove (i). Given x, a large *fixed* positive number, let

$$\sigma_1 = \frac{1}{2} + \frac{1}{(\log x)^{1/3}}.$$

By (17.16) and Lemma 17.13,

$$F_2(\sigma+it) \ll (\log x)^{1/3} + (\log x)^{1/6}\sqrt{\log(t+1)}$$

uniformly for $\sigma_1 \le \sigma \le 2$. Also, (17.19) and $\pi_R(x) \ll x$ imply that

$$\begin{aligned} F_1(\sigma+it) &\ll \int_1^\infty v^{-2\sigma}\, d\pi_R(v) = 2\sigma \int_1^\infty v^{-2\sigma-1}\pi_R(v)\, dv \\ &\ll \int_1^\infty v^{-2\sigma}\, dv = \frac{1}{2\sigma-1} \ll (\log x)^{1/3} \end{aligned}$$

for $\sigma_1 \le \sigma \le 2$. Hence ζ_R is analytic on this σ-strip, aside from a pole at $s=1$, and satisfies there uniformly

$$\begin{aligned} (17.20) \qquad & |\zeta_R(\sigma+it)| \\ & \le \left|\frac{\sigma+it}{\sigma-1+it}\right| \exp\left(c\left\{(\log x)^{1/3} + (\log x)^{1/6}\sqrt{\log(t+1)}\,\right\}\right). \end{aligned}$$

Now apply the first inequality of Lemma 17.10 with $b>1$. We see that

$$N_R(x) \le \frac{1}{2\pi i}\int_{b-i\infty}^{b+i\infty} \zeta_R(s)\,\frac{(x+1)^{s+1}-x^{s+1}}{s(s+1)}\, ds.$$

By (17.20), $|\zeta_R(s)| = o(|t|^{1/2})$ uniformly for $\sigma_1 \le \sigma \le 2$ and $|t|>1$, so we can shift the path of integration on the right hand side to the line $\sigma=\sigma_1$ to obtain

$$k_1\cdot(x+1/2) + \frac{1}{2\pi i}\int_{\sigma_1-i\infty}^{\sigma_1+i\infty} \zeta_R(s)\,\frac{(x+1)^{s+1}-x^{s+1}}{s(s+1)}\, ds,$$

where k_1 denotes the residue of $\zeta_R(s)$ at $s=1$. Using the inequalities

$$(17.21) \qquad \frac{(x+1)^{s+1}-x^{s+1}}{s+1} \ll \begin{cases} x^\sigma, & \text{for } |t|\le x, \\ x^{\sigma+1}/|t|, & \text{for } |t|>x, \end{cases}$$

we see that the last integral has order of magnitude

$$\int_0^x |\zeta_R(\sigma_1+it)|\,\frac{x^{\sigma_1}}{t+1}\, dt + \int_x^\infty |\zeta_R(\sigma_1+it)|\,\frac{x^{\sigma_1+1}}{t^2}\, dt =: I + II.$$

By (17.20),

$$I \ll x^{\sigma_1}\exp\left\{c(\log x)^{2/3}\right\}\log x \ll x^{1/2}\exp\left\{c'(\log x)^{2/3}\right\}.$$

To estimate II, use (17.20) again:

$$\begin{aligned} &\int_x^\infty |\zeta_R(\sigma_1+it)|\frac{x^{\sigma_1+1}}{t^2}\, dt \\ &\qquad \ll x^{\sigma_1+1}\exp\left\{c(\log x)^{1/3}\right\}\int_x^\infty \exp\left\{c(\log x)^{1/6}\sqrt{\log(t+1)}\,\right\}\frac{dt}{t^2}. \end{aligned}$$

We have

$$\frac{\int_x^\infty \exp\left\{c(\log x)^{1/6}\sqrt{\log(t+1)}\,\right\}t^{-2}\, dt}{x^{-1}\exp\left\{c(\log x)^{2/3}\right\}} \to 1$$

as $x\to\infty$, by l'Hôpital's rule. This yields

$$\int_x^\infty |\zeta_R(\sigma_1+it)|\frac{x^{\sigma_1+1}}{t^2}\, dt \ll x^{1/2}\exp\left\{c(\log x)^{2/3}\right\}.$$

Together, we obtain

$$N_R(x) \le k_1 \cdot (x + 1/2) + O\big(x^{1/2} \exp\{c(\log x)^{2/3}\}\big).$$

A similar argument, using the second inequality of Lemma 17.10, yields the lower bound

$$N_R(x) \ge k_1 \cdot (x - 1/2) + O\big(x^{1/2} \exp\{c(\log x)^{2/3}\}\big).$$

Thus (i) holds and the proof of Theorem 17.11 is complete. □

17.4. Generalized primes with large oscillation

The rest of this chapter will be devoted to constructing a g-prime system having a large oscillation.

THEOREM 17.14. *There is a system of Beurling primes $\mathcal{P}_B$ such that*

(i) the resulting Beurling integers satisfy

$$N_B(x) = k_2 x + O\big(x^{1/2} \exp\{c(\log x)^{2/3}\}\big)$$

with $k_2 > 0$;

(ii) the associated zeta function $\zeta_B(s)$ is analytic for $\sigma > 1/2$ except for a simple pole at $s = 1$ with residue k_2;

(iii) the function $\zeta_B(s)$ has infinitely many zeros on the curve $\sigma = 1 - 1/\log t$, $t \ge e^2$, and no zeros to its right;

(iv) the Chebyshev function $\psi_B(x)$ associated with $\mathcal{P}_B$ satisfies

$$\limsup_{x\to\infty} \frac{\psi_B(x) - x}{x \exp\{-2\sqrt{\log x}\,\}} = 2 \quad \text{and} \quad \liminf_{x\to\infty} \frac{\psi_B(x) - x}{x \exp\{-2\sqrt{\log x}\,\}} = -2.$$

We note that $\zeta_B(s)$ and $\mathcal{P}_B$ of Theorem 17.14 realize the classical zero-free region (17.1) and the PNT remainder (17.2) of de la Vallée Poussin with explicit $O-$constants. On the other hand, Landau's argument applies to both $\mathcal{P}_B$ and the system $\mathcal{P}_R$ of Theorem 17.11 and produces the same remainder term (17.2). This establishes the optimality of the de la Vallée Poussin remainder for g-primes. It also hints at the difficulty of improving this remainder estimate for rational primes.

In constructing the system $\mathcal{P}_B$ with large oscillation, the function

$$G(z) := 1 - \frac{e^{-z} - e^{-2z}}{z}, \quad G(0) := 0, \tag{17.22}$$

will play a key role. It is easy to see that this function is entire. It will be shown that $G(z)$ has infinitely many other zeros $z_{\pm n} = x_n \pm i y_n$, $n \in \mathbb{N}$ with

$$-c\log\frac{n\pi}{2} \le x_n < -\frac{1}{2}\log\frac{n\pi}{2} \quad \text{and} \quad n\pi < y_n < (n+1)\pi.$$

These zeros, suitably rescaled, translated, and combined in the template zeta function

$$\zeta_C(s) = \frac{s}{s-1}\prod_{k=1}^{\infty} G(\ell_k\{s - \rho_k\})\, G(\ell_k\{s - \bar\rho_k\}),$$

provide the zeros of $\zeta_B(s)$. Also, we shall show that

$$\zeta_C(s) = \exp\Big\{\int_1^\infty v^{-s} f_C(v)\, dv\Big\}$$

for a function $f_C(v)$ satisfying the conditions of Lemma 17.7. The proof of Theorem 17.14 will then follow along the lines of that of Theorem 17.11.

17.5. Properties of $G(z)$

In this section we establish properties of the function $G(z)$ of (17.22) that will be used in the later construction.

LEMMA 17.15. *$G(z)$ is an entire function whose zeros are all simple. It has no zeros on the horizontal lines $\{y = \pm n\pi\}$, $n = 1, 2, \ldots$. It has exactly one zero z_n in the strip $\{n\pi < y < (n+1)\pi\}$ and one zero $z_{-n} = \bar{z}_n$ in the conjugate strip. Also, it has exactly one zero $z_0 = 0$ in the strip $\{-\pi < y < \pi\}$. Furthermore, if $z_n = x_n + iy_n$, then $n\pi < y_n < (n+1)\pi$ and, for some constant $c > 0$,*

$$-c\log(n\pi/2) < x_n < -\frac{1}{2}\log(n\pi/2). \tag{17.23}$$

PROOF. First, we note that $z = 0$ is a removable singular point of $G(z)$. With $G(0) = 0$, the function is entire. Also, it has the Maclaurin series

$$G(z) = \sum_{n=2}^{\infty} \frac{(-1)^n(2^n - 1)}{n!} z^{n-1}.$$

In particular, $z = 0$ is a simple zero. For the rest of this proof, it is convenient to work with the function

$$G_1(z) := zG(z) = z - e^{-z} + e^{-2z},$$

which has the same zeros as $G(z)$, except for a double zero at $z = 0$. Since $G(x)$ is real for x real, by the reflection principle, it suffices to prove the lemma for $y \geq 0$.

Next, observe that

$$G_1(x + in\pi) = \begin{cases} (x + e^{-x} + e^{-2x}) + in\pi, & \text{if } n \text{ is odd,} \\ (x - e^{-x} + e^{-2x}) + in\pi, & \text{if } n \text{ is even.} \end{cases}$$

Hence $G_1(z)$ has no zeros on the lines $y = n\pi$, $n = 1, 2, \ldots$. Also we note that

$$\lim_{x \to \pm\infty} (x \pm e^{-x} + e^{-2x}) = +\infty$$

and, by a little calculus,

$$\min_{-\infty < x < \infty} \{x + e^{-x} + e^{-2x}\} = \log 2 + 3/4, \qquad \min_{-\infty < x < \infty} \{x - e^{-x} + e^{-2x}\} = 0.$$

For n a positive integer and X a large positive number, consider a rectangular contour C on the z-plane with vertices $\pm X + in\pi$ and $\pm X + i(n+1)\pi$. In case of n odd, when $z = x + iy$ travels along C in the positive direction with the initial point $-X + in\pi$, $w = G_1(z)$ travels on the w-plane along the horizontal line $y = n\pi$ from

$$(-X + e^X + e^{2X}) + in\pi \quad \text{to} \quad (\log 2 + 3/4) + in\pi$$

and back to $(X + e^{-X} + e^{-2X}) + in\pi$; then along a "nearly vertical" line segment going from

$$(X + e^{-X} + e^{-2X}) + in\pi \quad \text{to} \quad (X - e^{-X} + e^{-2X}) + i(n+1)\pi$$

(since $|e^{-z}|$ and $|e^{-2z}|$ are each near 0 for $x = X$ large); and continues along the horizontal line $y = (n+1)\pi$ from

$$(X - e^{-X} + e^{-2X}) + i(n+1)\pi \quad \text{to} \quad i(n+1)\pi$$

and back to $(-X - e^X + e^{2X}) + i(n+1)\pi$; and finally around a "near circle" with center $w = 0$ and approximate radius e^{2X} in the positive direction back to $(-X + e^X + e^{2X}) + in\pi$ (since $|z|$ and $|e^{-z}|$ are each $o(e^{2X})$). The argument of

e^{-2z} increases by 2π over the last interval. Hence the increment of the argument of $G_1(z)$ on the contour C is 2π. By the argument principle, $G_1(z)$ has exactly one simple zero z_n in the strip $\{n\pi < y < (n+1)\pi\}$. Similarly, for n even, we can prove the same result.

For the region near the real axis, we consider the rectangular contour C^* with vertices $\pm X \pm i\pi$. In a similar way, we can show that, when z travels along C^* in the positive direction, the increment of the argument of $G_1(z)$ is 4π. In particular, when z travels along the left side of C^* from $-X + i\pi$ to $-X - i\pi$, the argument of e^{-2z} increases by 4π, and hence $G_1(z)$ travels around two "near circles" with center $w = 0$ and approximate radius e^{2X} in the positive direction from

$$(-X + e^X + e^{2X}) + i\pi \quad \text{to} \quad (-X + e^X + e^{2X}) - i\pi.$$

Therefore $G_1(z)$ has exactly two zeros in the strip $\{-\pi < y < \pi\}$, which must coincide with the double zero at $z = 0$.

Now we prove (17.23). If $z_n = x_n + iy_n$ is the zero of G_1 in the strip $\{n\pi < y < (n+1)\pi\}$, then (of course) $n\pi < y_n < (n+1)\pi$; also $x_n < 0$ since

$$|G(x + iy_n)| \geq 1 - \frac{e^{-x} + e^{-2x}}{y_n} > 0$$

for $x \geq 0$ and $y_n > n\pi > 2$. Then we have

$$e^{-x_n} \cos y_n - e^{-2x_n} \cos 2y_n = x_n, \tag{17.24}$$

$$-e^{-x_n} \sin y_n + e^{-2x_n} \sin 2y_n = y_n. \tag{17.25}$$

It follows from (17.25) that

$$2e^{-2x_n} > y_n > n\pi$$

which establishes the upper bound for x_n in (17.23).

It remains to establish the lower bound for x_n. Since $x_n < -(1/2)\log(n\pi/2)$, we have $x_n e^{2x_n} \to 0$ as $n \to \infty$. Writing (17.24) in the form

$$e^{x_n} \cos y_n - \cos 2y_n = x_n e^{2x_n},$$

we see that $\cos 2y_n \to 0$, and hence $\sin 2y_n \to \pm 1$. Next write (17.25) in the form

$$e^{-2x_n}(-e^{x_n} \sin y_n + \sin 2y_n) = y_n.$$

We see that $\sin 2y_n \to +1$, since $y_n > n\pi > 0$. Therefore, for n sufficiently large,

$$\frac{1}{2} e^{-2x_n} < y_n < (n+1)\pi.$$

This in turn implies that

$$x_n \geq -(1/2)\log\{2(n+1)\pi\} > -c\log(n\pi/2)$$

for some constant $c > 0$. □

17.6. Representation of $\log G(z)$ as a Mellin transform

Equation (17.22) yields the simple approximation

$$G(z) = 1 + \theta\big(e^{-x} + e^{-2x}\big)/|z| \tag{17.26}$$

for some θ of modulus at most 1, provided that $z \neq 0$. The function $G(z)$ also has the representation

$$G(z) = 1 - \int_1^2 e^{-uz}\, du, \tag{17.27}$$

and for $\Re z \geq -1$, this gives the bound

$$|G(z)| \leq 1 + \int_1^2 e^u \, du = 1 + e^2 - e < 6. \tag{17.28}$$

LEMMA 17.16. *The function* $\log G(z)$ *is well-defined for* $x > 0$. *It has a representation there as a Mellin transform, namely,*

$$\log G(z) = -\int_1^\infty g(u) u^{-z-1} \, du, \tag{17.29}$$

where

$$g(u) := \sum_{n=1}^\infty \frac{1}{n} \chi^{*n}(u), \tag{17.30}$$

with $\chi(u)$ *the indicator function of the interval* $[e, e^2]$, *and* $*$ *denoting multiplicative convolution of functions with support contained in* $[1, \infty)$ *(see §2.7). Moreover,* $g(u)$ *is nonnegative and has support* $[e, \infty)$. *On the interval* $[e, e^4]$, *it is given by*

$$g(u) = \begin{cases} 1, & \text{for } e \leq u \leq e^2, \\ \frac{1}{2} \log u - 1, & \text{for } e^2 < u \leq e^3, \\ \frac{1}{6} \log^2 u - \frac{3}{2} \log u + \frac{7}{2}, & \text{for } e^3 < u \leq e^4. \end{cases} \tag{17.31}$$

Finally, for each $m \geq 1$, *the function* $g(u)$ *is a polynomial in* $\log u$ *of degree at most* $m - 1$ *on the interval* $e^m < u < e^{m+1}$.

PROOF. By Lemma 17.15, $G(z)$ is zero-free on the half plane $x > 0$. Hence $\log G(z)$ is well-defined there. For $x > 0$, the modulus of the integral in (17.27) is less than 1. Also, we can write the integral as a Mellin transform

$$\int_e^{e^2} u^{-z-1} du = \int_1^\infty \chi(u) u^{-z-1} \, du \, .$$

Using the series for $\log(1 - w)$ and (17.27), we get

$$\log G(z) = -\sum_{n=1}^\infty \frac{1}{n} \left(\int_1^\infty \chi(u) u^{-z-1} \, du \right)^n \tag{17.32}$$

We note that (2.10) gives

$$\left(\int_1^\infty \chi(u) u^{-z-1} \, du \right)^n = \int_1^\infty \chi^{*n}(u) u^{-z-1} \, du.$$

Then (17.29) follows, since the series in (17.32) converges absolutely for $x > 0$.

To show (17.31), we first note that $\chi^{*n}(x) = 0$ for $n \geq \log x$, $n \geq 2$. In particular, $\chi^{*n}(x) = 0$ on the interval $[e, e^2]$ for $n \geq 2$, on the interval $(e^2, e^3]$ for $n \geq 3$, and on $(e^3, e^4]$ for $n \geq 4$. Then, on $(e^2, e^3]$, by definition,

$$\chi^{*2}(x) = \int_1^x \chi(x/u) \chi(u) u^{-1} \, du = \int_e^{x/e} u^{-1} \, du = \log x - 2$$

and on $(e^3, e^4]$,

$$\chi^{*2}(x) = \int_{x/e^2}^{e^2} u^{-1} \, du = 4 - \log x.$$

For the last expression in (17.31), we have on $(e^3, e^4]$,

$$\begin{aligned}\chi^{*3}(x) &= \int_e^{e^2} \chi^{*2}(x/u)u^{-1}\,du = \int_e^{x/e^2} \left(\log\frac{x}{u} - 2\right) u^{-1}\,du \\ &= \frac{1}{2}\log^2 x - 3\log x + \frac{9}{2}\,.\end{aligned}$$

Now (17.31) follows from (17.30).

It remains to show the polynomial behavior of $g(x)$ on each interval $(e^m, e^{m+1}]$ for $m \geq 4$. Since $\chi^{*j}(x) = 0$ for $x \leq e^j$ for all $j \geq 2$, only terms $\chi^{*j}(x)$ with $j \leq m$ can contribute to $g(x)$ on this interval. Thus it suffices to show inductively that $\chi^{*m}(x)$ is a polynomial of degree at most $m-1$ (or identically 0) on each interval $e^j < x \leq e^{j+1}$ with $j \geq m$. We have

$$\chi^{*m}(x) = \int_1^x \chi^{*(m-1)}(u)\chi(x/u)u^{-1}\,du = \int_{x/e^2}^{x/e} \chi^{*(m-1)}(u)u^{-1}\,du\,.$$

The condition $e^j < x \leq e^{j+1}$ implies that $x/e^2 \leq e^{j-1} < x/e$, so we write

$$\chi^{*m}(x) = \int_{x/e^2}^{e^{j-1}} \chi^{*(m-1)}(u)u^{-1}\,du + \int_{e^{j-1}}^{x/e} \chi^{*(m-1)}(u)u^{-1}\,du\,.$$

On each interval, $\chi^{*(m-1)}(u)$ is a polynomial in $\log u$ of degree at most $m-2$, and integration produces a sum of polynomials, each of degree at most $m-1$. □

LEMMA 17.17. *For $u > e^2$, the function $g(u)$ defined in Lemma 17.16 satisfies*

$$g(u)\log u = \sum_{n=-\infty}^{\infty} u^{z_n}, \tag{17.33}$$

where z_0 and $\{z_{\pm n}\}$ are the zeros of $G(z)$.

PROOF. Differentiating (17.29) gives

$$\frac{G'(z)}{G(z)} = \int_1^\infty u^{-z-1}g(u)\log u\,du$$

for $x > 0$. By the definition of multiplicative convolution, $\chi^{*n}(u)$, $n \geq 2$, are all continuous and their derivatives are piecewise continuous. Hence the function $g(u)\log u$ is continuous for $u > e^2$ and locally of bounded variation. Integration by parts shows that $zG'(z)/G(z)$ is the Mellin transform of $d\{g(u)\log u\}$. Therefore, by the Perron inversion formula,

$$g(u)\log u = \lim_{T\to\infty} \frac{1}{2\pi i}\int_{c-iT}^{c+iT} \frac{G'(z)}{G(z)}u^z\,dz$$

for fixed $c \in (0, 1)$. We note that $G'(z)/G(z)$ is meromorphic since $G(z)$ is entire.

The integral on the right hand side can be evaluated by contour integration. Let J be a positive integer. By the left hand inequality in (17.23), all zeros z_m of $G(z)$ with $|m| < J$ are interior to the rectangle with vertices $-J \pm iJ\pi$ and $c \pm iJ\pi$. Contour integration around the rectangle in the positive direction yields

$$\frac{1}{2\pi i}\int_{c-iJ\pi}^{c+iJ\pi} \frac{G'(z)}{G(z)}u^z\,dz = \sum_{|m|<J} u^{z_m} - \frac{1}{2\pi i}\left(I_T + I_B + I_L\right), \tag{17.34}$$

where I_T, I_B, and I_L denote the integrals along the top, bottom, and left side of the rectangle, respectively. The integrand in (17.34) is

$$\frac{G'(z)}{G(z)}\, u^z = \frac{e^{-z} - 2e^{-2z} + e^{-z}z^{-1} - e^{-2z}z^{-1}}{z - e^{-z} + e^{-2z}}\, u^z.$$

On the left side of the rectangle, $x = -J$ and the integrand is $O\bigl(u^{-J}\bigr)$. Hence

$$|I_L| \ll Ju^{-J} \ll Je^{-2J}$$

for $u > e^2$. On the top side, $y = J\pi$ and so the imaginary part of the denominator of $G'(z)/G(z)$ is $J\pi$. Thus the integrand is $O\bigl(e^{-2x}e^{x \log u}/J\bigr)$, and

$$|I_T| \ll J^{-1} \int_{-\infty}^{c} \exp\{x(\log u - 2)\}dx \ll J^{-1}$$

for any fixed $u > e^2$. Since $|I_B| = |I_T|$, we obtain

$$|I_T + I_B + I_L| \ll J^{-1} + Je^{-2J} \to 0$$

as $J \to \infty$. Letting $J \to \infty$ in (17.34), we obtain (17.33). □

We can deduce useful estimates for $g(u)\log u$ from (17.33).

LEMMA 17.18. *For $u > e^2$,*

$$\bigl|g(u)\log u - 1\bigr| < 2u^{-(1/2)\log(\pi/2)}\,\zeta\Bigl(\frac{\log u}{2}\Bigr), \tag{17.35}$$

where $\zeta(s)$ is the Riemann zeta function. In particular, for $u \ge e^4$,

$$g(u)\log u < 7/3.$$

And for $u \ge e^5$, we have

$$(g(u)\log u)' \ll u^{-1-(1/2)\log(\pi/2)}. \tag{17.36}$$

PROOF. By (17.33) and (17.23) and the fact that $u^{x_0} = 1$, we have

$$\Bigl|g(u)\log u - 1\Bigr| \le 2\sum_{n=1}^{\infty} u^{x_n} < 2\sum_{n=1}^{\infty} \exp\Bigl\{-\frac{\log u}{2}\log\Bigl(\frac{n\pi}{2}\Bigr)\Bigr\}.$$

This proves (17.35). For $u \ge e^4$, the right hand side of the last inequality is at most

$$2(2/\pi)^2\zeta(2) = 4/3.$$

Next, differentiating (17.33) gives

$$(g(u)\log u)' = \sum_{n=-\infty}^{\infty} z_n u^{z_n - 1}.$$

This series is absolutely summable, since

$$\sum_{n=1}^{\infty} |z_n| u^{x_n - 1} \ll u^{-1-(1/2)\log(\pi/2)} \sum_{n=1}^{\infty} n^{1-(1/2)\log u}$$

by the inequalities of Lemma 17.15. For $u \ge e^5$, the last series is bounded by $\zeta(3/2)$, and we obtain the estimate (17.36). □

17.7. A template zeta function

Let

$$1 < \ell_1 < \ell_2 < \cdots < \ell_k < \cdots$$

be a sequence such that $\sum_k 1/\ell_k$ converges. Let

$$\gamma_k = e^{\ell_k}, \qquad \beta_k = 1 - \frac{1}{\ell_k}, \qquad \text{and} \qquad \rho_k = \beta_k + i\gamma_k.$$

Consider

$$\prod_{k=1}^{\infty} G(\ell_k(s-\rho_k))\, G(\ell_k(s-\bar{\rho}_k)), \tag{17.37}$$

where $G(z)$ is the function defined in (17.22). Given $T > 0$, let

$$S = S(T) := \{s = \sigma + it : \sigma \geq 1/2,\ |t| \leq T\}.$$

We note that for $k \geq k_0 = k_0(T)$ and all $s \in S$,

$$|s - \rho_k| \geq \frac{1}{2}\gamma_k \quad \text{and} \quad \Re\{\ell_k(s-\rho_k)\} \geq -\frac{\ell_k}{2} + 1.$$

By (17.26),

$$G(\ell_k(s-\rho_k)) = 1 + \theta_1 \frac{2e^{\ell_k - 2}}{\ell_k(\frac{1}{2}\gamma_k)} = 1 + \frac{4\theta_1}{e^2 \ell_k} \tag{17.38}$$

with $|\theta_1| < 1$. A similar equality holds for $G(\ell_k(s-\bar{\rho}_k))$. Hence $G(\ell_k(s-\rho_k))$ and $G(\ell_k(s-\bar{\rho}_k))$ have no zeroes in S for $k \geq k_0$. Also, the product in (17.37) converges absolutely and uniformly on S.

In particular, we take

$$\ell_k = 4^k, \ k = 1, 2, \ldots, \tag{ℓ_k}$$

and hence

$$\gamma_k = e^{4^k} \quad \text{and} \quad \beta_k = 1 - 4^{-k}. \tag{γ_k, β_k}$$

We now define our template zeta function on $\{s : \sigma \geq 1/2\}$ by

$$\zeta_C(s) := \frac{s}{s-1} \prod_{k=1}^{\infty} G(\ell_k(s-\rho_k))\, G(\ell_k(s-\bar{\rho}_k))\,. \tag{17.39}$$

Then $(s-1)\zeta_C(s)$ is continuous on $\sigma \geq 1/2$ and analytic for $\sigma > 1/2$.

LEMMA 17.19. *The function $\zeta_C(s)$ has zeros*

$$\rho_k = 1 - \frac{1}{\log \gamma_k} + i\gamma_k \ \textit{and conjugates}\ \bar{\rho}_k,\ k = 1, 2, \ldots,$$

on the curve $\sigma = 1 - 1/\log|t|$, $|t| \geq e^2$. All other zeros lie to the left of this curve.

PROOF. On its domain of definition, $\{s : \sigma \geq 1/2\}$, $\zeta_C(s)$ has zeros

$$\rho_k + 4^{-k} z_{\pm n}, \quad \bar{\rho}_k + 4^{-k} z_{\pm n}, \quad k = 1, 2, \ldots,\ n = 0, 1, 2, \ldots,$$

with k, n satisfying the condition

$$\Re(\rho_k + 4^{-k} z_{\pm n}) = 1 + 4^{-k}(x_n - 1) \geq 1/2, \tag{17.40}$$

where $z_{\pm n}$ are zeros of the function $G(z)$. We note that $z_0 = 0$ and for $k \geq 1$,

$$1 + 4^{-k}(x_0 - 1) = 1 - 4^{-k} > 1/2.$$

Hence all the zeros of the form

$$\rho_k = \rho_k + 4^{-k} z_0 = 1 - \frac{1}{\log \gamma_k} + i\gamma_k, \quad k = 1, 2, \ldots$$

and their conjugates $\bar{\rho}_k$ are on the curve $\sigma = 1 - 1/\log|t|$, $|t| \geq e^2$.

It remains to show that all zeros $\rho_k + 4^{-k} z_{\pm n}$ with $n \geq 1$ are to the left of the curve; the conjugate zeros,

$$\rho_k + 4^{-k} z_{\pm n} = (\rho_k + 4^{-k} z_{\mp n})^{-},$$

will clearly satisfy the same condition. Numbers $\rho_k + 4^{-k} z_{\pm n}$ for which (17.40) does not hold lie outside the domain of $\zeta_C(s)$, so we henceforth restrict to k and n values for which (17.40) holds.

By Lemma 17.15, $x_n < 0$ and $y_n > 0$ and hence

$$\Re(\rho_k + 4^{-k} z_n) < 1 - \frac{1}{\log \gamma_k} < 1 - \frac{1}{\log\left(\Im\{\rho_k + 4^{-k} z_n\}\right)}.$$

Thus all zeros $\rho_k + 4^{-k} z_n$ are to the left of the curve $\sigma = 1 - 1/\log t$. For the remaining zeros we have

$$\Re(\rho_k + 4^{-k} z_{-n}) = 1 + 4^{-k}(x_n - 1), \quad \Im(\rho_k + 4^{-k} z_{-n}) = \gamma_k - 4^{-k} y_n.$$

By (17.40) and Lemma 17.15,

$$\frac{1}{2} 4^k - 1 \geq -x_n > \frac{1}{2} \log \frac{n\pi}{2},$$

and hence $e^{4^k} 2e^{-2} > n\pi$. We now have

$$\begin{aligned}
\gamma_k - \exp\left\{\frac{4^k}{1 + \frac{1}{2}\log\frac{n\pi}{2}}\right\} &= e^{4^k}\left(1 - \exp\left\{-4^k + \frac{4^k}{1 + \frac{1}{2}\log(n\pi/2)}\right\}\right) \\
&= e^{4^k}\left(1 - \exp\left\{\frac{-4^k \log(n\pi/2)}{2 + \log(n\pi/2)}\right\}\right) \\
&\geq e^{4^k}(1 - e^{-1/2}) > e^{4^k} 2e^{-2},
\end{aligned}$$

since

$$\frac{4^k \log(n\pi/2)}{2 + \log(n\pi/2)} \geq \frac{4 \log(\pi/2)}{2 + \log(\pi/2)} > \frac{1}{2}$$

and $1 - e^{-1/2} > 2e^{-2}$. Hence

$$\gamma_k - \exp\left\{\frac{4^k}{1 + \frac{1}{2}\log\frac{n\pi}{2}}\right\} > n\pi > 4^{-k}(n+1)\pi.$$

Hence, by Lemma 17.15,

$$\gamma_k - 4^{-k} y_n > \exp\left\{\frac{4^k}{1 - x_n}\right\},$$

and so

$$4^{-k}(x_n - 1) < -\frac{1}{\log(\gamma_k - 4^{-k} y_n)}.$$

This shows that

$$\Re(\rho_k + 4^{-k} z_{-n}) < 1 - \frac{1}{\log(\Im\{\rho_k + 4^{-k} z_{-n}\})},$$

i.e., all the zeros $\rho_k + 4^{-k} z_{-n}$ of $\zeta_C(s)$ are to the left of the curve too. □

The following estimate will enable us to show that the template continuous "prime counting" measure is nonnegative and satisfies the strong Chebyshev-type condition (17.45).

LEMMA 17.20. *For* $v \geq e^4$,

$$\sum_{k\geq 1} g(v^{4^{-k}})\, 4^{-k} v^{-4^{-k}} \leq \frac{c_1}{\log v} \tag{17.41}$$

with a positive constant $c_1 < (1-e^{-4})/2$.

PROOF. The sum in (17.41) extends over k for which $v^{4^{-k}} \geq e$, since $g = 0$ for other values of k. We split the sum into two parts S_1 and S_2, over the ranges $e \leq v^{4^{-k}} < e^4$ and $e^4 \leq v^{4^{-k}}$ respectively.

There is exactly one positive integer k which satisfies $e \leq v^{4^{-k}} < e^4$, namely $k = \lfloor (\log\log v)/\log 4 \rfloor$. Hence

$$S_1 = \sum_{\substack{k \\ e\leq v^{4^{-k}}<e^4}} g(v^{4^{-k}}) 4^{-k} v^{-4^{-k}} \leq \frac{1}{\log v} \max_{e\leq u<e^4} \left\{ g(u)(\log u)u^{-1} \right\}.$$

From (17.31), it is easily seen that the maximum value of $g(u)u^{-1}\log u$ on the interval $[e, e^2]$ is e^{-1}. Then, on $[e^2, e^3]$, the derivative of the function is

$$u^{-2}\{-(\log u)^2/2 + 2\log u - 1\},$$

which is positive. Hence $g(u)u^{-1}\log u$ is increasing and has the maximum value $(3/2)e^{-3}$ at the point $u = e^3$. Also, on $[e^3, e^4]$, the derivative is

$$-u^{-2}(\log^3 u - 12\log^2 u + 39\log u - 21)/6,$$

which is negative. Hence the maximum on $[e^3, e^4]$ is $(3/2)e^{-3}$ at $u = e^3$ again. Therefore

$$S_1 \leq e^{-1}/\log v.$$

Now we consider S_2. By Lemma 17.18,

$$S_2 < \frac{7}{3\log v} \sum_{\substack{k \\ v^{4^{-k}}\geq e^4}} v^{-4^{-k}}. \tag{17.42}$$

If $e^4 \leq v < e^{16}$, then S_2 is an empty sum. For $v > e^{16}$, define the positive integer K by $v^{4^{-K}} \geq e^4 > v^{4^{-K-1}}$. Then the sum in (17.42) is at most

$$\sum_{k=1}^{K} e^{-4^{K+1-k}} = e^{-4} + \sum_{n=2}^{K} e^{-4^n} < e^{-4} + \sum_{n=2}^{\infty} e^{-4^n}.$$

Hence

$$S_2 < \frac{1}{\log v}\left\{ \frac{7}{3e^4} + \epsilon \right\}$$

where

$$\epsilon = \frac{7}{3}\sum_{n=2}^{\infty} e^{-4^n} < 3\cdot 10^{-7}.$$

Together, the inequalities for S_1 and S_2 give (17.41) with

$$c_1 = \frac{1}{e} + \frac{7}{3e^4} + \epsilon = 0.410616\ldots < (1-e^{-4})/2. \qquad \square$$

In (1.5), we represented $\log \zeta(s)$ as the Mellin transform of the weighted g-prime counting measure, or equivalently,

$$\zeta(s) = \exp\left\{ \int_1^\infty v^{-s}\, d\Pi(v) \right\}.$$

Now we show that our template zeta function, defined in (17.39), has such a form, with an absolutely continuous "prime counting" measure, and that the density function of this measure has the expected properties.

LEMMA 17.21. *For $\sigma > 1$,*

$$\zeta_C(s) = \exp\left\{ \int_1^\infty v^{-s} f_C(v)\, dv \right\}, \tag{17.43}$$

with

$$f_C(v) := \frac{1 - v^{-1}}{\log v} - 2\sum_{k\ge 1} \frac{g(v^{1/\ell_k})}{\ell_k} v^{-1/\ell_k} \cos(\gamma_k \log v) \tag{17.44}$$

for $v \ge 1$. We have $f_C(v) > 0$ for $v > 1$. Furthermore there is a constant $c \in (0, 1)$ such that

$$(1-c)\frac{1 - v^{-1}}{\log v} \le f_C(v) \le (1+c)\frac{1 - v^{-1}}{\log v} \tag{17.45}$$

holds for $v \ge e^4$.

PROOF. Recall from (17.39) that our template zeta function is represented as $s/(s-1)$ times a product of G factors involving zeros $\{\rho_k\}$ and their conjugates. For $\sigma > 1$, $\Re\{\ell_k(s - \rho_k)\} > 1$, and so, by Lemma 17.16,

$$G(\ell_k(s-\rho_k)) = \exp\left\{ -\int_1^\infty \frac{g(v^{4^{-k}})}{4^k} v^{-4^{-k}+i\gamma_k} v^{-s}\, dv \right\}.$$

Lemma 17.20 implies that

$$\sum_{k\ge 1} g(v^{4^{-k}})\, 4^{-k} v^{-4^{-k}} \ll 1$$

for $v \ge e^4$, and hence

$$\sum_{k=1}^\infty \int_1^\infty \frac{g(v^{4^{-k}})}{4^k} v^{-4^{-k}} \cos(\gamma_k \log v)\, v^{-s}\, dv$$

converges absolutely for $\sigma > 1$. Thus we can write

$$\begin{aligned}
&\prod_{k=1}^\infty G(\ell_k(s-\rho_k))\, G(\ell_k(s-\bar\rho_k)) \\
&\quad = \prod_{k=1}^\infty \exp\left\{ -2\int_1^\infty \frac{g(v^{4^{-k}})}{4^k} v^{-4^{-k}} \cos(\gamma_k \log v)\, v^{-s}\, dv \right\} \\
&\quad = \exp \int_1^\infty \left\{ -2\sum_{k=1}^\infty \frac{g(v^{4^{-k}})}{4^k} v^{-4^{-k}} \cos(\gamma_k \log v)\, v^{-s}\, dv \right\}.
\end{aligned}$$

We showed in Theorem 1.5 that

$$\frac{s}{s-1} = \exp\left\{ \int_1^\infty v^{-s} \frac{1 - v^{-1}}{\log v}\, dv \right\}.$$

Combining the preceding formulas, we obtain (17.43).

For $v < e^4$, we have $v^{1/\ell_k} < e$ and so $g(v^{1/\ell_k})$ in (17.44) is 0. Thus

$$f_C(v) = \frac{1 - v^{-1}}{\log v} > 0.$$

On the other hand, for $v \geq e^4$, (17.45) with $c = 2c_1/(1-e^{-4})$ follows from (17.41), since $2c_1 < 1 - e^{-4}$. □

17.8. Asymptotics of $N_B(x)$

The last lemma shows that $f_C(v)$ satisfies the conditions of Lemma 17.7. Thus there exists a sequence $\{p_j\} := \{v_{k_j}\}$, $j = 1, 2, \ldots$, such that

$$\sum_{p_j \leq x} p_j^{-it} - \int_1^x v^{-it} f_C(v)\, dv \ll \sqrt{x}\left(1 + \sqrt{\frac{\log(t+1)}{1 + \log x}}\right) \tag{17.46}$$

for $1 \leq x < \infty$ and $t \geq 0$. In particular,

$$\pi_B(x) = \sum_{p_j \leq x} 1 = \int_1^x f_C(v)\, dv + O(x^{1/2}). \tag{17.47}$$

We proceed as in the proof of Theorem 17.11. Let

$$F_1(s) = \int_1^\infty \{v^{-s} + \log(1 - v^{-s})\}\, d\pi_B(v),$$

$$F_2(s) = \int_1^\infty v^{-s}\{d\pi_B(v) - f_C(v)\, dv\}.$$

By (17.47) and (17.45), $F_1(s)$ and $F_2(s)$ converge uniformly for $\sigma \geq \frac{1}{2} + \epsilon$ for each $\epsilon > 0$ and hence are analytic for $\sigma > 1/2$. The sought-after zeta function $\zeta_B(s)$ is convergent for $\sigma > 1$ and can be written as

$$\zeta_B(s) = \zeta_C(s) \exp\{-F_1(s) + F_2(s)\}.$$

It follows that $\zeta_B(s)$ has an analytic continuation to the same half plane as $\zeta_C(s)$, and it has there the same set of zeros and poles. In particular, $\zeta_B(s)$ has infinitely many zeros on the curve $\sigma = 1 - 1/\log|t|$, $|t| \geq e^2$, and no zeros to its right. Also, it has a simple pole at $s = 1$ with residue $k_2 > 0$. (This number will, as usual, be the density of the g-integer system.)

To establish the asymptotics of $N_B(x)$, we need estimates of $\zeta_B(s)$ both close to and far from zeros.

LEMMA 17.22. *Given $x \geq x_0\,(> 1)$, let*

$$\sigma_1 = 1/2 + (\log x)^{-1/3}.$$

The following estimates hold uniformly for $\sigma_1 \leq \sigma \leq 2$, $t \geq 0$, and $|s - 1| \geq c > 0$.

(i) if $|t - \gamma_k| > \frac{1}{2}\gamma_k$ for all $k \in \mathbb{N}$, then

$$\zeta_B(\sigma + it) \ll \exp\Big\{c\Big((\log x)^{1/3} + (\log x)^{1/6}\sqrt{\log(t+1)}\Big)\Big\};$$

(ii) if $|t-\gamma_{k_0}| \le \frac{1}{2}\gamma_{k_0}$ *for some* $k_0 \in \mathbb{N}$, *then*

$$\zeta_B(\sigma+it) \ll \exp\left\{c\Big((\log x)^{1/3} + (\log x)^{1/6}\sqrt{\log(t+1)}\Big)\right\} \times \left(1 + \frac{t}{4^{k_0}|\sigma+it-\rho_{k_0}|}\right).$$

REMARK 17.23. Our argument actually rules out the apparent blow-up of the last estimate at $s=\rho_{k_0}$, but we do not need the sharper bound.

PROOF. Analogously with the proof of Theorem 17.11 (there with formulas (17.17) and (17.18) and Lemma 17.13), we have

$$|\zeta_B(\sigma+it)| \le |\zeta_C(\sigma+it)| \exp\left\{c\big((\log x)^{1/3} + (\log x)^{1/6}\sqrt{\log(t+1)}\,\big)\right\} \tag{17.48}$$

uniformly for $\sigma_1 \le \sigma \le 2$ and $t \ge 0$. Now we examine $\zeta_C(\sigma+it)$, omitting the factor $s/(s-1)$, since we excluded a neighborhood of $s=1$.

By (17.38), if $|t-\gamma_k| > \gamma_k/2$, then

$$G(4^k(s-\rho_k)) = 1 + \frac{4\theta_1}{e^2 4^k},$$

and $G(4^k(s-\bar\rho_k))$ satisfies the same equality. Thus, if $|t-\gamma_k| > \frac{1}{2}\gamma_k$ holds for all $k \in \mathbb{N}$, then the remaining factors of $\zeta_C(\sigma+it)$ satisfy

$$\left|\prod_{k=1}^{\infty} G(4^k(s-\rho_k))G(4^k(s-\bar\rho_k))\right| \le \prod_{k=1}^{\infty}\left(1+\frac{4}{e^2 4^k}\right)^2 \ll 1,$$

uniformly for $\sigma \ge 1/2$ and all such $t \ge 0$. This inequality together with (17.48) proves (i).

Next, we prove (ii). For any given $t \ge 0$, there is at most one k such that $|t-\gamma_k| \le \gamma_k/2$, i.e., $\gamma_k/2 \le t \le 3\gamma_k/2$; indeed, the big gap between γ_k and γ_{k+1} insures that $3\gamma_k/2 < \gamma_{k+1}/2$. If there is some k_0 such that $|t-\gamma_{k_0}| \le \frac{1}{2}\gamma_{k_0}$, then $|t-\gamma_k| > \frac{1}{2}\gamma_k$ for all $k \ne k_0$. Hence

$$\prod_{\substack{k=1\\ k\ne k_0}}^{\infty} G(4^k(s-\rho_k))G(4^k(s-\bar\rho_k)) \ll 1$$

uniformly for $\sigma \ge 1/2$ and for t satisfying $|t-\gamma_{k_0}| \le \frac{1}{2}\gamma_{k_0}$ with some k_0. Then

$$\frac{s-1}{s}\,\zeta_C(\sigma+it) \ll |G(4^{k_0}(\sigma+it-\rho_{k_0}))|, \tag{17.49}$$

since

$$G(4^{k_0}(s-\bar\rho_{k_0})) = 1 + \frac{4\theta_1}{e^2 4^{k_0}} \ll 1.$$

To estimate the right hand side of (17.49), suppose first that

$$1 - 2\cdot 4^{-k_0} =: \sigma_2 \le \sigma \le 2.$$

In this case,

$$Z := \Re\{4^{k_0}(\sigma+it-\rho_{k_0})\} = 4^{k_0}(\sigma-1)+1 \ge -1,$$

and by (17.28), $G(4^{k_0}(\sigma+it-\rho_{k_0})) \ll 1$.

Now suppose $\sigma_1 \le \sigma < \sigma_2$. In this case $-Z > 1 > 0$ and so

$$-Z < -2Z \le -2\cdot\{4^{k_0}(\sigma_1-1)+1\} =: -2Z_1.$$

Applying (17.26), the right hand side of (17.49) is bounded by

$$1+\frac{2\exp\{-2Z_1\}}{4^{k_0}|\sigma+it-\rho_{k_0}|}\,.$$

We have

$$\begin{aligned}\exp\{-2Z_1\} &= \exp\big\{-2\cdot 4^{k_0}(\sigma_1-1)-2\big\}\\ &\le \exp\big\{-2\cdot 4^{k_0}\big(-1/2+(\log x)^{-1/3}\big)\big\}\\ &\ll \gamma_{k_0}\exp\left\{\frac{-2\log\gamma_{k_0}}{(\log x)^{1/3}}\right\}\le \gamma_{k_0}\ll t\,.\end{aligned}$$

It follows that

$$G(4^{k_0}(\sigma+it-\rho_{k_0}))\ll 1+\frac{t}{4^{k_0}|\sigma+it-\rho_{k_0}|}$$

for $\sigma_1\le\sigma<\sigma_2$ and t satisfying $|t-\gamma_{k_0}|\le\frac12\gamma_{k_0}$. Combining this estimate with (17.49) and (17.48) establishes (*ii*), which completes the proof of the lemma. □

Proceeding as in the proof of Theorem 17.11, write

$$N_B(x)\le\lim_{T\to\infty}\frac{1}{2\pi i}\int_{b-iT}^{b+iT}\zeta_B(s)\frac{(x+1)^{s+1}-x^{s+1}}{s(s+1)}\,ds\,.$$

We apply contour integration to evaluate the last integral. Let $T_n=2\gamma_n$. Consider the contour integral taken along the rectangle with vertices $b\pm iT_n$ and $\sigma_1\pm iT_n$. Then $|\pm T_n-\gamma_k|>\frac12\gamma_k$ for all $k\in\mathbb{N}$. By (*i*) of the last lemma, along the top of the rectangle we have

$$\begin{aligned}&\left|\int_{\sigma_1+iT_n}^{b+iT_n}\zeta_B(s)\frac{(x+1)^{s+1}-x^{s+1}}{s(s+1)}\,ds\right|\\ &\qquad\ll\frac{(x+1)^{b+1}}{T_n^2}\exp\Big(c\big\{(\log x)^{1/3}+(\log x)^{1/6}\sqrt{\log(T_n+1)}\,\big\}\Big)=o(1)\end{aligned}$$

as $n\to\infty$. A similar inequality holds for the integral along the bottom side of the rectangle.

Using the residue theorem, we shift the integration path to the line $\sigma=\sigma_1$. Then break the integral at $t=x$ and apply (17.21) to the two parts to obtain

$$\begin{aligned}N_B(x)&\le k_2\cdot(x+1/2)+\lim_{n\to\infty}\frac{1}{2\pi i}\int_{\sigma_1-iT_n}^{\sigma_1+iT_n}\zeta_B(s)\frac{(x+1)^{s+1}-x^{s+1}}{s(s+1)}\,ds\\ &=k_2\cdot(x+1/2)+O(I_1)+O(I_2),\end{aligned}$$

with k_2 the residue of $\zeta_B(s)$ at $s=1$,

$$I_1:=\int_0^x|\zeta_B(\sigma_1+it)|\frac{x^{\sigma_1}}{t+1}\,dt\quad\text{and}\quad I_2:=\int_x^\infty|\zeta_B(\sigma_1+it)|\frac{x^{\sigma_1+1}}{t^2}\,dt\,.$$

To estimate these integrals, we next show that

$$\int_T^{2T}|\zeta_B(\sigma_1+it)|\,dt\ll T\exp\big\{c(\log x)^{2/3}\big\},\quad T\le x,\tag{17.50}$$

and

$$\int_T^{2T}|\zeta_B(\sigma_1+it)|\,dt\ll T\exp\big\{c(\log x)^{1/6}\sqrt{\log T}\,\big\},\quad T>x.\tag{17.51}$$

In (17.50), suppose first that there exists no $\gamma_k \in [2T/3, 4T]$. In this case, $|t-\gamma_k| > \gamma_k/2$ holds for each point $t \in [T, 2T]$. Indeed,

$$\text{if } \gamma_k > 4T, \text{ then } \gamma_k - t \geq \gamma_k - 2T > \gamma_k/2,$$
$$\text{if } \gamma_k < 2T/3, \text{ then } t - \gamma_k \geq T - \gamma_k > \gamma_k/2.$$

Now Case (i) of Lemma 17.22 applies, and we have $\zeta_B(\sigma_1+it) \ll \exp\{c(\log x)^{2/3}\}$ for $T \leq t \leq 2T$, which establishes (17.50).

On the other hand, suppose $\gamma_K \in [2T/3, 4T]$ for some K. To avoid multiple cases, we extend the integral in (17.50) to the interval $[\gamma_K/4, 3\gamma_K]$. It is easy to see that $\gamma_K/4 \leq T$ and $2T \leq 3\gamma_K$. Since estimate (ii) of Lemma 17.22 is at least as large as that of (i), it is valid to use (ii) for all t in the integration range. We find

$$\zeta_B(\sigma_1+it) \ll \exp\left\{c(\log x)^{2/3}\right\}\left\{1+\frac{t}{4^K|\sigma_1+it-\rho_K|}\right\}.$$

Now the denominator of the last expression is

$$4^K\Big|-1/2+(\log x)^{-1/3}+4^{-K}+i(t-\gamma_K)\Big| \gg 4^K(1+|t-\gamma_K|),$$

and so

$$\int_{\gamma_K/4}^{3\gamma_K}|\zeta_B(\sigma_1+it)|\,dt \ll \exp\left\{c(\log x)^{2/3}\right\}\int_{\gamma_K/4}^{3\gamma_K}\left\{1+\frac{t}{4^K(1+|t-\gamma_K|)}\right\}dt.$$

The last integral is bounded by

$$3\gamma_K + 2\cdot 4^{-K}\cdot 3\gamma_K\int_0^{2\gamma_K}\frac{dv}{1+v} \ll \gamma_K \ll T,$$

and so (17.50) is true in this case too.

The same argument, with the obvious modification in applying Lemma 17.22, yields (17.51).

We return to the estimates of $N_B(x)$. By (17.50), we have

$$\begin{aligned}
I_1 &= \int_0^x |\zeta_B(\sigma_1+it)|\,\frac{x^{\sigma_1}}{t+1}\,dt \\
&\ll x^{1/2}\exp\left\{c(\log x)^{2/3}\right\}\sum_{0\leq k\leq\frac{\log x}{\log 2}}\frac{2^{k+1}}{x}\int_{x/2^{k+1}}^{x/2^k}|\zeta_B(\sigma_1+it)|\,dt \\
&\ll x^{1/2}\exp\left\{c(\log x)^{2/3}\right\}.
\end{aligned}$$

Also, from (17.51), we have

$$\begin{aligned}
I_2 &= \int_x^\infty |\zeta_B(\sigma_1+it)|\frac{x^{\sigma_1+1}}{t^2}\,dt \\
&\ll x^{\sigma_1+1}\sum_{k=0}^\infty 2^{-2k}x^{-2}\int_{2^kx}^{2^{k+1}x}|\zeta_B(\sigma_1+it)|\,dt \\
&\ll x^{\sigma_1}\sum_{k=0}^\infty 2^{-k}\exp\left\{c(\log x)^{1/6}(\log\{2^kx\})^{1/2}\right\}.
\end{aligned}$$

Substituting for σ_1, the right hand side becomes

$$\ll x^{1/2}\exp\{c(\log x)^{2/3}\}\sum_{k=0}^\infty\frac{1}{2^k}\exp\left\{c(\log x)^{1/6}(\log\{2^kx\})^{1/2}\right\}.$$

We split the last sum into two parts, over the ranges $2^k \le x$ and $2^k > x$ respectively. On the first range $(\log\{2^k x\})^{1/2} \ll (\log x)^{1/2}$, and so the contribution of the first part is $\ll \exp\{c(\log x)^{2/3}\}$. The second part is

$$\ll \sum_{k>\frac{\log x}{\log 2}} 2^{-k}\exp\{c(\log\{2^k\})^{2/3}\} \le \sum_{k=1}^{\infty} 2^{-k/2} \ll 1,$$

since for $k > \frac{\log x}{\log 2} \ge 8c^3/\log 2$,

$$2^{-k}\exp\{c(\log\{2^k\})^{2/3}\} \le 2^{-k/2}.$$

Putting together the estimates, we find that

$$\int_x^\infty |\zeta_B(\sigma_1+it)|\frac{x^{\sigma_1+1}}{t^2}\,dt \ll x^{1/2}\exp\{c(\log x)^{2/3}\},$$

and hence

$$N_B(x) \le k_2\cdot(x+1/2) + O\big(x^{1/2}\exp\{c(\log x)^{2/3}\}\big).$$

A similar argument shows that

$$N_B(x) \ge k_2\cdot(x-1/2) + O\big(x^{1/2}\exp\{c(\log x)^{2/3}\}\big),$$

and together, the two inequalities for $N_B(x)$ establish (i) of Theorem 17.14.

17.9. Asymptotics of $\psi_B(x)$

To complete the proof of Theorem 17.14, we analyze the asymptotics of $\psi_B(x)$ with a direct calculation. This will yield the sharp $\Omega_\pm$-constants of (iv). Recall from §1.2 that the Chebyshev function is defined by

$$\psi_B(x) := \sum_{{p_i}^{\alpha_i}\le x} \log p_i = \int_1^x \log v\, d\Pi_B(v),$$

where, as usual,

$$\Pi_B(x) := \pi_B(x) + \frac{1}{2}\pi_B(x^{1/2}) + \frac{1}{3}\pi_B(x^{1/3}) + \dots.$$

Lemma 1.1 implies that

$$\Pi_B(x) = \pi_B(x) + O(x^{1/2}).$$

(A slightly better remainder term can be shown, but that is not needed here.)

The preceding formula and integration by parts give

$$\psi_B(x) = \int_1^x \log v\, d\pi_B(v) + O\big(x^{1/2}\log x\big).$$

Define the template psi function by

$$\psi_C(x) := \int_1^x \log\, v\, f_C(v)\,dv.$$

Then, by (17.47) and more integration by parts,

$$\begin{aligned}\psi_B(x) - \psi_C(x) &= \int_1^x \log\, v\{d\pi_B(v) - f_C(v)\,dv\} + O\big(x^{1/2}\log\, x\big)\\ &= O\big(x^{1/2}\log\, x\big).\end{aligned}$$

Given $x > e^4$, define $K \in \mathbb{N}$ by $\gamma_K \le x < \gamma_{K+1}$. (As usual, $\gamma_k = e^{4^k}$.) By (17.44),

$$\psi_C(x) = x - 2F(x) + O(\log x),$$

where

$$F(x) := \sum_{k=1}^{K} \int_{\gamma_k}^{x} (\log v)\, 4^{-k} g(v^{4^{-k}})\, v^{-4^{-k}} \cos(\gamma_k \log v)\, dv =: \sum_{k=1}^{K} I_k(x),$$

say. To show the oscillation of ψ_B, it suffices to prove that

$$\limsup_{x\to\infty} \frac{F(x)}{x \exp\{-2\sqrt{\log x}\,\}} = 1 \tag{17.52}$$

and

$$\liminf_{x\to\infty} \frac{F(x)}{x \exp\{-2\sqrt{\log x}\,\}} = -1. \tag{17.53}$$

We shall show that the behavior of $F(x)$ is dominated by the contribution of a single term $I_{k_0}(x)$ or $I_{k_0+1}(x)$ with $k_0 = K/2$ or $k_0 = (K-1)/2$.

For $k \le K-2$, change the variable and write

$$I_k(x) := 4^k \int_{e}^{x^{4^{-k}}} g(u)\, (\log u)\, u^{4^k - 2} \cos(4^k \gamma_k \log u)\, du. \tag{17.54}$$

We split the integration interval into two subintervals $[e, e^5]$ and $[e^5, x^{4^{-k}}]$ and let $I_{k,1}$ and $I_{k,2}(x)$ denote the respective partial integrals. We estimate $I_{k,1}$ trivially, using the information that $g(u)\log u$ and the cosine are bounded. We find that

$$I_{k,1} \ll (e^5)^{4^k} \le x^{5/16}.$$

To estimate $I_{k,2}(x)$, we exploit the bounds we have found for $g(u)\log u$ and its derivative. Integrating by parts and applying Lemma 17.18, we obtain

$$\begin{aligned} \gamma_k I_{k,2}(x) &= x^{1-4^{-k}} \sin(\gamma_k \log x) + O\big(x^{1-4^{-k}(1+\frac{1}{2}\log\{\pi/2\})}\big) \\ &\quad - \int_{e^5}^{x^{4^{-k}}} \sin(4^k \gamma_k \log u) u^{4^k - 1} (g(u) \log u)'\, du \\ &\quad - \int_{e^5}^{x^{4^{-k}}} \sin(4^k \gamma_k \log u) g(u) (\log u)(4^k - 1) u^{4^k - 2}\, du. \end{aligned} \tag{17.55}$$

(The integrated term at the lower limit has order $\exp(5 \cdot 4^k)$; when divided by γ_k, this is $O(x^{1/4})$ for $k \le K-2$.) By (17.36), the first integral in (17.55) is

$$\ll \int_{e^5}^{x^{4^{-k}}} u^{4^k - 2 - \frac{1}{2}\log(\pi/2)}\, du \ll 4^{-k} x^{1-4^{-k}(1+\frac{1}{2}\log\{\pi/2\})}.$$

To estimate the second integral in (17.55), subtract and add 1 in the integrand. Apply Lemma 17.18 (again) to the part with $g(u)\log u - 1$; we find

$$\begin{aligned} &(4^k - 1) \int_{e^5}^{x^{4^{-k}}} \sin(4^k \gamma_k \log u)(g(u) \log u - 1) u^{4^k - 2}\, du \\ &\qquad \ll 4^k \int_{e^5}^{x^{4^{-k}}} u^{4^k - 2 - \frac{1}{2}\log(\pi/2)}\, du \ll x^{1-4^{-k}(1+\frac{1}{2}\log\{\pi/2\})}. \end{aligned}$$

For the remaining portion of the integral, integration by parts (or application of the second mean value theorem) yields

$$(4^k-1)\int_{e^5}^{x^{4^{-k}}} \sin(4^k\gamma_k \log u)\, u^{4^k-2}\, du \ll \frac{x^{1-4^{-k}}}{\gamma_k}.$$

Together, we obtain

$$(17.56)\quad I_k(x) = \frac{x^{1-4^{-k}}\sin(\gamma_k \log x)}{\gamma_k} + O\Big(\frac{x^{1-4^{-k}(1+\frac{1}{2}\log(\pi/2))}}{\gamma_k} + \frac{x^{1-4^{-k}}}{\gamma_k^2} + x^{\frac{5}{16}}\Big)$$

for $k \le K-2$. We estimate the maximum of each of the error terms as k varies. With $4^k = \alpha$, the first error term in (17.56) is

$$x \exp\Big\{-\frac{(\log x)(1+\frac{1}{2}\log(\pi/2))}{\alpha} - \alpha\Big\}.$$

By calculus, the maximum of this expression for $\alpha > 0$ is

$$x\exp\big\{-2\sqrt{1+(1/2)\log(\pi/2)}\,\sqrt{\log x}\big\} < x\exp\big\{-2.21\sqrt{\log x}\,\big\}.$$

A similar calculation shows the second error estimate is at most

$$x\exp\big\{-2\sqrt{2}\,\sqrt{\log x}\,\big\}.$$

Then, the third error term in (17.56) is independent of k and is smaller than the estimate of the other two. We use estimate of the first error term in (17.56) for $1 \le k \le K-2$ and find that the sum of the I_k errors has order

$$x\exp\big\{-2.21\sqrt{\log x}\,\big\}\log\log x \ll x\exp\big\{-2.2\sqrt{\log x}\,\big\}.$$

Thus

$$\sum_{k=1}^{K-2} I_k(x) = \sum_{k=1}^{K-2}\frac{x^{1-4^{-k}}\sin(\gamma_k\log x)}{\gamma_k} + O\big(x\exp\big\{-2.2\sqrt{\log x}\,\big\}\big).$$

It remains to estimate $I_K(x)$ and $I_{K-1}(x)$. We note from the definition of K, viz. $\gamma_K \le x < \gamma_{K+1}$, that $x^{4^{-k}}$, the upper limit of the integration range in (17.54), can extend up to $\exp\big(4^{K+1-k}\big)$, that is, up to e^4 for $I_K(x)$ and up to e^{16} for $I_{K-1}(x)$. The key fact we shall use (Lemma 17.16) is that $g(u)$ is a polynomial in $\log u$ of degree at most $m-1$ on each interval $(e^m, e^{m+1}]$.

Combining formula (17.54) with the representation of $g(u)$ as a polynomial, we can express $I_k(x)$ with $k = K$ or $K-1$ as a linear combination of terms

$$I_k(x,a) := \int_{e^a}^{\mu} 4^k(\log u)^j\, u^{4^k-1}\cos(4^k\gamma_k\log u)\, du/u,$$

with $1 \le a < 4^{-k}\log x$ and $\mu = \min(e^{a+1}, x^{4^{-k}}) \le e^{16}$. Integrating by parts yields

$$\begin{aligned} I_k(x,a) &= \gamma_k^{-1}\sin(4^k\gamma_k\log u)u^{4^k-1}(\log u)^j\Big|_{e^a}^{\mu} \\ &\quad - \gamma_k^{-1}\int_{e^a}^{\mu}\sin(4^k\gamma_k\log u)\{u^{4^k-1}(\log u)^j\}'\,du \\ &\ll \gamma_k^{-1}\mu^{4^k}(\log\mu)^j + \gamma_k^{-1}\int_{e^a}^{\mu}|\{u^{4^k-1}(\log u)^j\}'|\,du. \end{aligned}$$

Now $u^{4^k-1}(\log u)^j$ is clearly increasing for $\mu^{4^k} \le x$, $\log\mu \le 16$, and

$$\gamma_k^{-1} = \begin{cases} e^{-4^K} & \le x^{-1/4}, \quad k = K, \\ e^{-4^{K-1}} & \le x^{-1/16}, \quad k = K-1. \end{cases}$$

Thus we have $I_K(x) \ll x^{3/4}$ and $I_{K-1}(x) \ll x^{15/16}$, and each of these quantities is $\ll x\exp\{-2.2\sqrt{\log x}\}$.

Thus we arrive at the representation

$$F(x) = \sum_{k=1}^{K-2} \frac{x^{1-4^{-k}} \sin(\gamma_k \log x)}{\gamma_k} + O\big(x\exp\{-2.2\sqrt{\log x}\}\big). \tag{17.57}$$

We are now close to establishing the desired estimates (17.52) and (17.53).

To analyze the last formula, we first look where the maximum of

$$x^{-4^{-k}}/\gamma_k = \exp\{-4^{-k}\log x - 4^k\}$$

occurs. Setting $\alpha := 4^k$, we see that $\exp\{-(\log x)/\alpha - \alpha\}$ achieves its maximum when $\alpha = \sqrt{\log x}$ (regarding α as a real variable). Given x sufficiently large, let

$$4^{k_0} \le \sqrt{\log x} < 4^{k_0+1}.$$

The main contribution to the sum in (17.57) comes from the terms having indices k_0 or k_0+1. Indeed, we see that

$$\sum_{\substack{1\le k\le K-2\\ k\ne k_0, k_0+1}} \frac{x^{1-4^{-k}}}{\gamma_k} = O(x\exp\{-4\sqrt{\log x}\})$$

by evaluating the sum at the next-largest terms, $k = k_0 - 1$ and $k = k_0 + 2$, and then multiplying by $K - 4 \ll \log\log x$.

The last sum is smaller than the error term in (17.57), so we drop all summands from that formula except those with indices k_0 and k_0+1. We now have

$$F(x) = \sum_{k=k_0,\, k_0+1} \frac{x^{1-4^{-k}} \sin(\gamma_k \log x)}{\gamma_k} + O\big(x\exp\{-2.2\sqrt{\log x}\}\big). \tag{17.58}$$

In choosing k_0, we noted that $x^{1-4^{-k}}/e^{4^k}$ has its maximum at $4^k = \sqrt{\log x}$. (In this instance, k is not an integer, in general.) The value at the maximum is $x\exp(-2\sqrt{\log x})$. Next, we show that at most one of the terms with indices k_0, k_0+1 can achieve this maximum. Suppose first that $4^{k_0} \le \sqrt{\log x} < 2\cdot 4^{k_0}$. In this case,

$$\frac{x^{1-4^{-k_0-1}}}{\gamma_{k_0+1}} \le x\exp\big\{-(2.25)\sqrt{\log x}\big\}.$$

On the other hand, if $2\cdot 4^{k_0} \le \sqrt{\log x} < 4^{k_0+1}$, again we have

$$\frac{x^{1-4^{-k_0}}}{\gamma_{k_0}} \le x\exp\big\{-(2.25)\sqrt{\log x}\big\}.$$

It follows that

$$\limsup_{x\to\infty} \frac{|F(x)|}{x\exp\{-2\sqrt{\log x}\}} \le 1.$$

It remains to show that $F(x)/(x\exp\{-2\sqrt{\log x}\})$ approaches ± 1 infinitely often. By (17.58) and the analysis that follows, for any given (large) k and for x sufficiently near $x_1 := \exp\left(4^{2k}\right)$ we have

$$F(x) \sim x^{1-4^{-k}} e^{-4^k} \sin(\gamma_k \log x) \sim x\exp\{-2\sqrt{\log x}\}\sin(\gamma_k \log x).$$

We conclude the argument by making a small perturbation of x_1 whose only significant effect is to make the value of sine be ± 1.

Set $x_2 = x_1 e^{\epsilon}$, where ϵ is a small number to be determined. We have

$$x_2^{1-4^{-k}} e^{-4^k} = x_1^{1-4^{-k}} e^{-4^k} e^{\epsilon(1-4^{-k})} \to x_1 \exp\{-2\sqrt{\log x_1}\}$$

as $\epsilon \to 0$. For the argument of sine, suppose

$$e^{4^k} \log x_1 = 2\pi m + \pi\theta$$

for some positive integer m and $|\theta| \le 1$. Then we have

$$\gamma_k \log x_2 = e^{4^k}(\log x_1 + \epsilon) = 2\pi m + \pi\theta + e^{4^k}\epsilon = 2\pi m \pm \pi/2$$

by taking

$$\epsilon = e^{-4^k}(\pm\pi/2 - \theta\pi).$$

The latter is a number that goes to zero as $k \to \infty$, and thus $F(x_2(k))$ satisfies (17.52) (resp. (17.53)). This completes the proof of Theorem 17.14.

17.10. Normalization and hybrid

We can construct other g-prime systems, based on $\mathcal{P}_R$ and $\mathcal{P}_B$ of Theorems 17.11 and 17.14, that have the same properties as the primes in $\mathbb{N}$ (assuming the truth of the Riemann hypothesis for (*iv*) and (*v*)). For instance, given any positive integer n_X, there is a g-prime system $\mathcal{P}_X$ such that

(*i*) $\mathcal{P}_X$ consists of numbers in $\{v_k\}$ given in (17.13) and $\mathcal{P}_X \cap [1, n_X]$ is identical with the primes in $\mathbb{N}$ of size at most n_X;

(*ii*) the counting function $N_X(x)$ of the resulting g-integers satisfies

$$N_X(x) = x + O(x^{1/2}\exp\{c(\log x)^{2/3}\});$$

(*iii*) the associated zeta function $\zeta_X(s)$ is analytic for $\sigma > 1/2$ except for a simple pole at $s = 1$ with residue 1;

(*iv*) the function $\zeta_X(s)$ has no zeros on the half plane $\sigma > 1/2$;

(*v*) the g-prime counting function $\pi_X(x)$ satisfies

$$\pi_X(x) = \mathrm{li}(x) + O(x^{1/2}).$$

The construction can be done by successively inserting some rational primes into the g-prime system $\mathcal{P}_R$ of Theorem 17.11 and deleting some other g-primes from it. Specifically, we can first insert all rational primes up to n_X that are not already in $\mathcal{P}_R$ and then delete all g-primes up to n_X that are not rational primes. In this way, the new system satisfies all the properties (*i*) – (*v*) except that the residue k of $\zeta_X(s)$ at $s = 1$, i.e., the density of the resulting g-integers, may not be 1. The desired estimates can be shown for (*ii*) by using the inclusion-exclusion principle, since only a finite number of g-primes are involved. We may suppose that $k < 1$ in the new system, for otherwise we delete a finite number of additional g-primes. Finally, there is a finite or infinite sequence $\{w_n\}$ of numbers $w_n \in \mathbb{N}$ such that

$$w_n > n_X, \quad \prod(1 - w_n^{-1}) = k, \quad \sum w_n^{-1/2} < \infty, \quad \sum_{w_n \le x} 1 = O(\log x).$$

We enlarge $\mathcal{P}_X$ to contain the collection $\{w_n\}$. Then the resulting g-number system has all the expected properties.

17.11. Notes

§17.1. The material of this chapter is based on the articles [**DMV06**] and [**Zh07**].

§17.2.1. A reference for the Bernstein-Kolmogorov Lemma is §19.1A(i) in [**Lo77**]. If we have $v > 2\sigma_e^2$, then the choice $\lambda = 1$ in (17.4) in the proof yields the further estimate

$$P[Y \ge v] \le \exp\{-v + \sigma^2\} \le \exp\{-v/2\}.$$

Besides the sequence $\{v_k\}$ defined by (17.13), any other sequence that is suitably dense (in some sense) in a neighborhood of ∞ also will satisfy the conditions of Lemmas 17.2 and 17.5. For example, the sequences $\{v_k\}$ defined by

$$v_k := \sqrt{\log(k + k_0)}\,, \quad k = 0, 1, 2, \ldots,$$

and

$$v_k := \log(k + k_0) \log\log(k + k_0), \quad k = 0, 1, 2, \ldots,$$

satisfy the conditions of Lemmas 17.2 and 17.5 as well.

§17.2.2. In addition to our choice of the pair f and F, it is easy to show that if

$$f(v) \ll \frac{v^\alpha}{(1 + \log v)^\beta} \quad \text{on} \quad [1, \infty),$$

where α and β are constants with $-1 < \alpha < 1$ and $\beta \ge 0$, then

$$F(x) = \frac{x^{\alpha+1}}{(1 + \log x)^\beta}$$

too satisfies the conditions of Lemmas 17.2 and 17.5.

§17.3.2. The system $\mathcal{P}_R$ of generalized primes in Theorem 17.11 can be taken as subsequences of the sequence $\{v_k\}$ given in (17.13) as well as the sequences mentioned above.

CHAPTER 18

The Dickman and Buchstab Functions

Real numbers want to be sieved too

SUMMARY. The Dickman and Buchstab functions appear in approximations to the counting functions of rational integers without large prime factors, resp. those having only large prime factors. We use the framework of Beurling numbers to suggest how these functions arise and give several related formulas.

18.1. Introduction

For $x \geq 1,\ y > 1$, the function $\psi(x, y)$ denotes the number of positive rational integers not exceeding x all of whose prime factors are at most y. (This ψ is not related to Chebyshev's function.) The integers counted here are called "y-smooth" or simply "smooth." To see the connection of $\psi(x, y)$ to primes and their powers, note that the generating function of $\psi(x, y)$ is a truncation of the product form of the Riemann zeta function: for $\Re s > 1$,

$$\zeta(s, y) := \int_{x=1-}^{\infty} x^{-s} d\psi(x, y) = \sum_{\substack{n \geq 1 \\ p|n \Rightarrow p \leq y}} n^{-s} = \prod_{p \leq y} \left(1 - p^{-s}\right)^{-1}.$$

Thus

$$\log \zeta(s, y) = -\sum_{p \leq y} \log \left\{1 - p^{-s}\right\} = \sum_{p \leq y} \sum_{\alpha=1}^{\infty} \frac{1}{\alpha} p^{-\alpha s} = \int_{x=1}^{\infty} x^{-s} d\Pi(x, y),$$

where

$$\Pi(x, y) := \sum_{\substack{p^{\alpha} \leq x \\ p \leq y}} \frac{1}{\alpha}.$$

It follows by the argument that established (3.4) that

$$\int_{1-}^{\infty} x^{-s} d\psi(x, y) = \exp \int_{x=1}^{\infty} x^{-s} d\Pi(x, y) = \int_{x=1-}^{\infty} x^{-s} \exp^{*} d\Pi(x, y),$$

whence, by the identity theorem for Mellin transforms,

$$\psi(x, y) = \int_{t=1-}^{x} \exp^{*} d\Pi(t, y). \tag{18.1}$$

On the other hand, the function $\phi(x, y)$ denotes the number of positive rational integers not exceeding x without any prime factors in $(1, y]$. Note that if $y \geq x \geq 1$, then 1 is the only number in $[1, x]$ having no prime factors in the range $(1, y]$, and in this case, $\phi(x, y) = 1$.

For given y, the Mellin transform of $d\phi(x,y)$ is

$$\int_{1-}^{\infty} x^{-s} d\phi(x,y) = \sum_{\substack{n\geq 1 \\ p|n \Rightarrow p>y}} n^{-s} = \prod_{p>y} \left(1-p^{-s}\right)^{-1}, \quad \Re s > 1.$$

Also,

$$-\sum_{p>y} \log\left\{1-p^{-s}\right\} = \sum_{p>y} \sum_{\alpha=1}^{\infty} \frac{1}{\alpha} p^{-\alpha s} = \int_{x=1}^{\infty} x^{-s} d\Upsilon(x,y)$$

with

$$\Upsilon(x,y) := \sum_{\substack{p^\alpha \leq x \\ p>y}} \frac{1}{\alpha},$$

whence

$$\phi(x,y) = \int_{t=1-}^{x} \exp^* d\Upsilon(t,y). \tag{18.2}$$

We have, for each $y > 1$,

$$\Pi(x,y) + \Upsilon(x,y) = \Pi(x).$$

The preceding definitions and properties of exp from Chapter 3 give

$$\int_{1-}^{x} d\psi(\cdot,y) * d\phi(\cdot,y) = \int_{1-}^{x} \exp^*\{d\Pi(\cdot,y) + d\Upsilon(\cdot,y)\}$$
$$= \int_{1-}^{x} \exp^* d\Pi = \lfloor x \rfloor := N(x), \quad x \geq 1.$$

Also, (3.7) gives the relation

$$\int_{1-}^{x} \exp^* d\Pi_c = \int_{1-}^{x} \exp^*\Big(\frac{1-t^{-1}}{\log t} dt\Big) = \int_{1-}^{x} (\delta_1 + dt) = x := N_c(x).$$

We have $N(x) \approx N_c(x)$, and by the classical PNT, $\Pi(x) \approx \Pi_c(x)\,(\approx \mathrm{li}(x))$. In this chapter, we introduce continuous analogues of $\psi(x,y)$ and $\phi(x,y)$ in terms of exponentials involving $d\Pi_c$, and we show that the resulting functions are precisely those occurring in asymptotic formulas of Dickman–de Bruijn and of Buchstab.

18.2. The $\psi(x,y)$ function

By well-known theorems of Dickman and de Bruijn, $\psi(x,y)$ is asymptotic to $x\,\rho(\log x/\log y)$ as $x,\ y \to \infty$ in certain ranges. Here ρ denotes the Dickman function, defined by $\rho(u) = 1$ for $0 \leq u \leq 1$ and for $u \geq 1$ by

$$u\,\rho(u) := \int_{u-1}^{u} \rho(t)\,dt\,.$$

Further information on $\psi(x,y)$ and the ρ function can be found in Chapter III.5 of [**Te95**], §7.1 of [**MV07**], and [**Mo93**]. We make one observation on $\rho(u)$: it vanishes more swiftly at infinity than the reciprocal of Euler's gamma function.

A continuous analogue of $\Pi(x,y)$ can be defined for $x \geq 1,\ y > 1$ by

$$\Pi_c(x,y) := \int_1^x \Big\{\chi_{[1,y]}(t) \frac{(1-t^{-1})\,dt}{\log t} - \chi_{(y,\infty)}(t) \frac{dt}{t\log t}\Big\}, \tag{18.3}$$

with χ_E the indicator function of a set E, and a continuous analogue of $\psi(x,y)$ by

$$\psi_c(x,y) := \int_{1-}^{x} \exp^*\Big\{\chi_{[1,\,y]}(t)\frac{(1-t^{-1})\,dt}{\log t} - \chi_{(y,\infty)}(t)\frac{dt}{t\log t}\Big\}, \tag{18.4}$$

i.e. $\psi_c(x,y) := \int_{1-}^{x} \exp^* d\Pi_c(\cdot,y)$.

Now we give an exact representation of $\psi_c(x,y)$ in terms of the rho function, which resembles the Dickman asymptotic formula.

THEOREM 18.1. *We have* $\psi_c(x,y) = x\,\rho(\log x/\log y)$ *for all* $x \geq 1$, $y > 1$.

Before proving the theorem, let's make a small calculation to show the result plausible. We start with another expression for $\psi_c(x,y)$ created by adding and subtracting a term $\chi_{(y,\infty)}(t)\,dt/\log t$. We find

$$\begin{aligned}\psi_c(x,y) &:= \int_{1-}^{x} \exp^*\Big\{\frac{(1-t^{-1})\,dt}{\log t} - \chi_{(y,\infty)}(t)\frac{dt}{\log t}\Big\}\\ &= \int_{1-}^{x} (\delta + dt) * \exp^*\Big\{-\chi_{(y,\infty)}(t)\frac{dt}{\log t}\Big\}\\ &= \int_{1-}^{x} \frac{x}{t}\exp^*\Big\{-\chi_{(y,\infty)}(t)\frac{dt}{\log t}\Big\},\end{aligned} \tag{18.5}$$

and by the homomorphic property of exp (Lemma 3.8),

$$\psi_c(x,y) = x\int_{1-}^{x} \exp^*\Big\{-\chi_{(y,\infty)}(t)\frac{dt}{t\log t}\Big\}.$$

We expand the last formula as

$$\frac{\psi_c(x,y)}{x} = 1 - \int_1^x \chi_{(y,\infty)}(t)\frac{dt}{t\log t} + \frac{1}{2!}\int_1^x \Big\{\chi_{(y,\infty)}(t)\frac{dt}{t\log t}\Big\}^{*2} \mp \cdots. \tag{18.6}$$

If $1 \leq x \leq y$, then 1 is the only term on the right side of (18.6), and we find

$$\frac{\psi_c(x,y)}{x} = 1 = \rho\Big(\frac{\log x}{\log y}\Big),$$

since $\rho(u) = 1$ for $0 \leq u \leq 1$. Next, suppose $y < x \leq y^2$. In this case,

$$\frac{\psi_c(x,y)}{x} = 1 - \int_y^x \frac{dt}{t\log t} = 1 - \log\frac{\log x}{\log y} = \rho\Big(\frac{\log x}{\log y}\Big),$$

since, by a small calculation, $\rho(u) = 1 - \log u$ for $1 \leq u \leq 2$. Calculations for larger x are more complicated, so we now turn to the

PROOF OF THEOREM 18.1. We establish the formula using Mellin transforms. For $s \in \mathbb{C}$ with $\Re s > 1$ we have, on the one hand,

$$\begin{aligned}\int_1^\infty x^{-s}\psi_c(x,y)\,\frac{dx}{x} &= \frac{1}{s}\int_{x=1-}^{\infty} x^{-s}d\psi_c(x,y)\\ &= \frac{1}{s}\int_{x=1-}^{\infty} x^{-s}\exp^*\Big\{\big(\chi_{[1,\,y]}(x) - x^{-1}\big)\,\frac{dx}{\log x}\Big\}\\ &= \frac{1}{s}\exp\Big\{\int_{x=1}^{\infty} x^{-s}\big(\chi_{[1,\,y]}(x) - x^{-1}\big)\,\frac{dx}{\log x}\Big\}\\ &= \frac{1}{s}\exp\Big\{\int_{x=1}^{\infty} x^{-s}(1 - x^{-1})\,\frac{dx}{\log x} - \int_y^\infty x^{-s}\,\frac{dx}{\log x}\Big\}.\end{aligned}$$

Set

$$E_1(z) := \int_z^\infty e^{-t}\, dt/t,$$

with the integration along the horizontal line from $z = x + iy$ to $+\infty + iy$. We see that $E_1(z)$ is analytic on the open half-plane $\Re z > 0$ and $E_1{}'(z) = -e^{-z}/z$. Hence

$$\int_1^\infty x^{-s}\psi_c(x,y)\frac{dx}{x} = \frac{1}{s-1}\exp(-E_1(s-1)\log y),$$

where we have used formula (1.8).

On the other hand,

$$\int_1^\infty x^{-s}\rho\Big(\frac{\log x}{\log y}\Big)\, dx = \log y \int_0^\infty e^{-(s-1)u\log y}\rho(u)\, du\,.$$

The last integral is a Laplace transform, and we identify it by forming its differential equation and solving. For $\Re z > 0$, let

$$F(z) := \int_0^\infty e^{-zu}\rho(u)\, du\,.$$

It is convenient here to define $\rho(u) = 0$ for $u < 0$, for then we have

$$\{u\,\rho(u)\}' = \rho(u) - \rho(u-1), \quad u > 0,\ u \neq 1.$$

If we differentiate F and integrate by parts the resulting expression (noting that $u\,\rho(u)$ is continuous at $u = 1$, the point of nondifferentiability), we find

$$\begin{aligned} F'(z) &= -\int_0^\infty e^{-zu}u\,\rho(u)\, du = -\frac{1}{z}\int_0^\infty e^{-zu}\{\rho(u)-\rho(u-1)\}\, du \\ &= -z^{-1}(1-e^{-z})\, F(z). \end{aligned}$$

Let

$$f(z) := \log z + E_1(z), \quad \Re z > 0.$$

Then $f'(z) = (1 - e^{-z})/z$ and

$$\{e^{f(z)}F(z)\}' = 0.$$

Therefore

$$F(z) = ce^{-f(z)} = cz^{-1}e^{-E_1(z)}, \quad \Re z > 0.$$

To evaluate the constant c, note that $\rho(u) = 1$ for $0 \le u < 1$, and hence

$$F(z) = \int_0^\infty e^{-zu}du + \int_1^\infty e^{-zu}\{\rho(u)-1\}\, du = \frac{1}{z} + O(e^{-z}), \quad z \to +\infty.$$

Therefore $zF(z) \to 1$, and also $E_1(z) \to 0$ as $z \to +\infty$. Thus $c = 1$ and we have $F(z) = z^{-1}\exp\{-E_1(z)\}$. For $\Re s > 1$, it follows that

$$\begin{aligned} \int_1^\infty x^{-s}\rho\Big(\frac{\log x}{\log y}\Big)\, dx &= F\{(s-1)\log y\}\log y \\ &= \frac{1}{s-1}\exp(-E_1\{(s-1)\log y\}) = \int_1^\infty x^{-s}\psi_c(x,y)\,\frac{dx}{x}\,. \end{aligned}$$

By the Mellin identity theorem, the claimed formula for $\psi_c(x,y)$ holds. □

As an immediate consequence of the theorem and formula (18.6) we have

COROLLARY 18.2. *Let* $x \geq 1,\ y > 1$. *Then*

$$\rho\Big(\frac{\log x}{\log y}\Big) = \frac{\psi_c(x,y)}{x} = \int_{1-}^{x} \exp^*\Big\{ - \chi_{(y,\infty)}(t)\frac{dt}{t\log t}\Big\}.$$

We shall consider $\psi_c(x,y)$ further in §18.4 below.

18.3. The $\phi(x, y)$ function

Buchstab's theorem provides an asymptotic estimate of the number of rational integers up to x whose prime factors all exceed y, namely,

$$\phi(x,y) \sim \frac{x}{\log y}\,\omega\Big(\frac{\log x}{\log y}\Big)$$

as $x,\ y \to \infty$ in certain ranges. ω denotes Buchstab's function, which is defined as

$$\omega(u) = \begin{cases} 0, & u < 1,\\ 1/u, & 1 \leq u \leq 2,\end{cases}$$

and for $u > 2$ as the continuous solution of the difference differential equation

$$\{u\,\omega(u)\}' = \omega(u-1). \tag{18.7}$$

Further study of $\phi(x,y)$ and the omega function can be found in Chapter III.6 of [**Te95**] and in §7.2 of [**MV07**]. We remark that $\omega(u)$ tends rapidly toward $e^{-\gamma}$ ($\gamma =$ Euler's constant) as $u \to \infty$.

Define a continuous analogue of $\phi(x,y)$ for $x \geq 1,\ y > 1$ by

$$\phi_c(x,y) := \int_{1-}^{x} \exp^*\Big\{\chi_{(y,\infty)}(t)\,\frac{dt}{\log t}\Big\} \tag{18.8}$$

If we combine the last formula with (18.5) and (3.7), we find, for any numbers $x \geq y > 1$,

$$\int_{1-}^{x} d\psi_c(\cdot,y) * d\phi_c(\cdot,y) = \int_{1-}^{x}(\delta_1 + dt) = x.$$

Thus $\psi_c(x,y)$ and $\phi_c(x,y)$ fit together in a similar way as their rational integer counterparts $\psi(x,y)$ and $\phi(x,y)$, as was noted in §18.1.

Now we show that $\phi_c(x,y)$ has an exact representation in terms of Buchstab's function. Comparing the present formula with that for $\psi_c(x,y)$, here we have an integral and a constant term.

THEOREM 18.3. *For* $x \geq 1,\ y > 1$,

$$\phi_c(x,y) = 1 + \frac{1}{\log y}\int_{y}^{x} \omega\Big(\frac{\log t}{\log y}\Big)\,dt. \tag{18.9}$$

PROOF. Again, we verify the claim using Mellin transforms. We have

$$\int_{x=1-}^{\infty} x^{-s} d\phi_c(x,y) = \int_{x=1-}^{\infty} x^{-s}\exp^*\Big\{\chi_{(y,\,\infty)}(x)\,\frac{dx}{\log x}\Big\} = \exp\int_{y}^{\infty} x^{-s}\frac{dx}{\log x}$$

$$= \exp\int_{(s-1)\log y}^{\infty} e^{-t}\,\frac{dt}{t} = \exp(E_1\{(s-1)\log y\}).$$

Also, making the substitution $u = \log x/\log y$, we get

$$\int_{x=1-}^{\infty} x^{-s}\Big\{\delta_1 + \omega\Big(\frac{\log x}{\log y}\Big)\frac{dx}{\log y}\Big\} = 1 + \int_{1}^{\infty}\exp\{-(s-1)u\,\log y\}\,\omega(u)\,du.$$

The last integral is a Laplace transform, and we identify it by forming its differential equation and solving. For $\Re z > 0$, let

$$F(z) := \int_1^\infty e^{-zu}\omega(u)\,du.$$

We differentiate F and integrate by parts the resulting expression using the difference differential equation (18.7). That relation in fact holds for $u > 1$, aside from $u = 2$. At this point, $u\,\omega(u)$ is continuous, and thus the integration by parts is valid. We get

$$F'(z) = -\int_1^\infty e^{-zu} u\,\omega(u)\,du = -\frac{e^{-z}}{z} - \frac{e^{-z}}{z}\int_1^\infty e^{-zv}\omega(v)\,dv = -\frac{e^{-z}}{z}\{1 + F(z)\}.$$

Then

$$\left(e^{-E_1(z)}\{F(z)+1\}\right)' = 0,$$

i.e., $F(z)+1 = c\exp\{E_1(z)\}$. Hence $F(z)+1 \to c$ as $z \to +\infty$. On the other hand, $F(z) \to 0$ as $z \to +\infty$. Thus $c = 1$ and $F(z)+1 = \exp\{E_1(z)\}$. Therefore

$$\int_{x=1-}^\infty x^{-s}\Big\{\delta_1 + \omega\Big(\frac{\log x}{\log y}\Big)\frac{dx}{\log x}\Big\} = 1 + F\{(s-1)\log y\} = \exp\Big(E_1\{(s-1)\log y\}\Big).$$

We see that the two Mellin integrals are equal, so

$$d_t\phi_c(t,y) = \delta_1 + \omega\Big(\frac{\log t}{\log y}\Big)\frac{dt}{\log y}. \tag{18.10}$$

Formula (18.9) holds, as we see by integrating the last equation from $1-$ to x and noting that the right side becomes

$$1 + \frac{1}{\log y}\int_{1-}^x \omega\Big(\frac{\log t}{\log y}\Big)\,dt = 1 + \frac{1}{\log y}\int_y^x \omega\Big(\frac{\log t}{\log y}\Big)\,dt. \qquad \square$$

18.4. A Beurling version of $\psi(x, y)$

The reader will notice that $\Pi_c(x,y)$ of §18.2 does not qualify as a g-prime counting function, for it is decreasing in x for $x > y$. (Its density function also has support extending beyond the smoothness bound y, but this is not a fatal flaw, since the classical counterpart $\Pi(x,y)$ also counts the contribution of higher prime powers lying in the range $(y, x]$.)

As a replacement for $\Pi_c(x,y)$ we introduce

$$\Pi_c^+(x,y) := \int_1^x \chi_{[1,\,y]}(t)\,\frac{(1-t^{-1})\,dt}{\log t}$$

and its associated Beurling g-number counting function

$$\psi_c^+(x,y) := \int_{1-}^x \exp\Big\{\chi_{[1,\,y]}(t)\,\frac{(1-t^{-1})\,dt}{\log t}\Big\}.$$

We now represent $\psi_c^+(x,y)$ in terms of the rho and omega functions.

THEOREM 18.4. *For $x \ge y > 1$, set $u = \log x/\log y$. We have*

$$\psi_c^+(x,y) = x\rho(u) + x\int_1^u \rho(u-v)\,y^{-v}\omega(v)\,dv.$$

PROOF. For given y, we have by (18.3)

$$\chi_{[1,\,y]}(t)\,\frac{(1-t^{-1})\,dt}{\log t} = d\Pi_c(t,y) + \chi_{(y,\,\infty)}(t)\,\frac{dt}{t\,\log t}\,.$$

By results from the last two sections,

$$\begin{aligned}\psi_c^+(x,y) &= \int_{1-}^{x} \exp^*\Big\{d\Pi_c(t,y) + \chi_{(y,\,\infty)}\frac{dt}{t\,\log t}\Big\}\\ &= \int_{1-}^{x} d\psi_c(\cdot,y) * T^{-1}d\phi_c(\cdot,y) = \int_{1-}^{x} \psi_c\Big(\frac{x}{t},y\Big)\,t^{-1}d\phi_c(t,y)\\ &= \int_{1-}^{x} \frac{x}{t}\,\rho\Big(\frac{\log x - \log t}{\log y}\Big)\Big\{\delta_1 + \omega\Big(\frac{\log t}{\log y}\Big)\frac{dt}{t\,\log y}\Big\}.\end{aligned}$$

If we substitute $v = \log t/\log y$ and note that $\omega(v) = 0$ for $v < 1$, we obtain the claimed formula for $\psi_c^+(x,y)$. □

18.5. G-numbers with primes from an interval

A series of papers by J. Friedlander [**Fr76**] and E. Saias [**Sa92, Sa95**] studied the counting function $\theta(x,y,z)$ of rational integers in $[1,\,x]$ having all their prime factors in a range $(z,\,y]$. Here we define a continuous version of this function by setting

$$\theta_c(x,y,z) := \int_{1}^{x} \exp^*\Big\{\chi_{(z,\,y]}(t)\,\frac{dt}{\log t}\Big\},$$

and we describe $\theta_c(x,y,z)$ in terms of the Dickman and Buchstab functions.

THEOREM 18.5. *Suppose $x \ge y > z > 1$. We have*

$$\theta_c(x,y,z) = \int_{1-}^{x} \Big\{\delta_1 + \rho'\Big(\frac{\log t}{\log y}\Big)\frac{dt}{\log y}\Big\} * \Big\{\delta_1 + \omega\Big(\frac{\log t}{\log z}\Big)\frac{dt}{\log z}\Big\}. \tag{18.11}$$

PROOF. From the identity

$$\chi_{(z,\,y]}(t) = (\chi_{[1,\,y]}(t) - t^{-1}) - (1-t^{-1}) + \chi_{(z,\,\infty)}(t)$$

we find

$$d_t\theta_c(t,y,z) = \exp^*\Big(\{(\chi_{[1,\,y]}(t) - t^{-1}) - (1-t^{-1}) + \chi_{(\dot z,\,\infty)}(t)\}\,\frac{dt}{\log t}\Big).$$

Using the relations

$$\begin{aligned}\exp^*\Big((\chi_{[1,\,y]}(t) - t^{-1})\,\frac{dt}{\log t}\Big) &= d\psi_c(\cdot,y),\\ \exp^*\Big(-(1-t^{-1})\,\frac{dt}{\log t}\Big) &= \delta_1 - t^{-1}dt,\\ \exp^*\Big(\chi_{(z,\,\infty)}(t)\,\frac{dt}{\log t}\Big) &= d\phi_c(\cdot,z)\end{aligned}$$

and the homomorphic property of exp, we find

$$d_t\theta_c(t,y,z) = d\psi_c(\cdot,y) * (\delta_1 - t^{-1}dt) * d\phi_c(\cdot,z).$$

By (18.10)

$$d_t\phi_c(t,z) = \delta_1 + \omega\Big(\frac{\log t}{\log z}\Big)\frac{dt}{\log z}\,.$$

Using Theorem 18.1 and the fact that $\psi_c(x,y)$ has a jump of 1 at $x=1$, we get

$$d\psi_c(t,y) = \delta_1 + d\Big\{t\rho\Big(\frac{\log t}{\log y}\Big)\Big\} = \delta_1 + \rho\Big(\frac{\log t}{\log y}\Big)\,dt + \rho'\Big(\frac{\log t}{\log y}\Big)\frac{dt}{\log y}.$$

To treat the convolution factor $\delta_1 - t^{-1}dt$, we note that, for any measure $dF \in d\mathcal{V}$, we have

$$\int_{1-}^{x} dF * t^{-1}dt = \int_{1-}^{x} F(x/t)\,dt/t = \int_{1-}^{x} F(v)\,dv/v,$$

whence $dF * t^{-1}dt = F(t)\,dt/t$. Thus

$$d\psi_c(\cdot,y) * t^{-1}dt = t\rho\Big(\frac{\log t}{\log y}\Big)\frac{dt}{t}$$

and hence

$$d\psi_c(\cdot,y) * (\delta_1 - t^{-1}dt) = \delta_1 + \rho'\Big(\frac{\log t}{\log y}\Big)\frac{dt}{\log y}.$$

If we combine the last relation with that for $d_t\phi_c(t,z)$, we obtain the claimed formula for $\theta_c(x,y,z)$. □

Here is another form of the last theorem.

THEOREM 18.6. *For* $x \geq y > z > 1$ *and*

$$u = \log x/\log y, \quad v = \log x/\log z, \quad \sigma = \log z/\log y,$$

we have

$$d_x\theta_c(x,y,z) = \delta_1 + \Big\{\frac{\rho'(u)}{\log y} + \frac{\rho(u-\sigma)}{\log z} + \frac{1}{\log z}\int_1^v \rho(u-t\sigma)\,\omega'(t)\,dt\Big\}\,dx.$$

PROOF. The key is to rewrite the convolution in (18.11). Let

$$F(x) := \int_1^x \rho'\Big(\frac{\log t}{\log y}\Big)\,dt * \omega\Big(\frac{\log t}{\log z}\Big)\,dt = \int_1^x \Big\{\int_1^{x/t} \rho'\Big(\frac{\log s}{\log y}\Big)\,ds\Big\}\,\omega\Big(\frac{\log t}{\log z}\Big)\,dt.$$

We differentiate the double integral, noting that the derivative of the outer integral is 0 because the inner integral vanishes as $t \to x$. We find

$$F'(x) = \int_1^x \rho'\Big(\frac{\log x/t}{\log y}\Big)\,\omega\Big(\frac{\log t}{\log z}\Big)\frac{dt}{t} = \log z\int_1^v \rho'(u-s\sigma)\,\omega(s)\,ds.$$

Now divide the last integral by $\log y\,\log z$ and integrate it by parts. We get

$$\frac{dF(x)}{\log y\,\log z} = \frac{1}{\log z}\Big\{\rho(u-\sigma) - \omega(v) + \int_1^v \rho(u-s\sigma)\,\omega'(s)\,ds\Big\}\,dx,$$

and if we insert this expression in place of the convolution in (18.11) we obtain the claimed formula for $d_x\theta_c(x,y,z)$. □

REMARK. In [**Sa92**], Saias gives the formula

$$\frac{\theta(x,y,z)}{x} \sim \frac{\rho'(u)}{\log y} + \frac{\rho(u-\sigma)}{\log z} + \frac{1}{\log z}\int_1^v \rho(u-t\sigma)\,\omega'(t)\,dt.$$

Although the left side of this formula differs from that of the last theorem, the results are not inconsistent; rather they are instances of a relation

$$\frac{1}{x}\int_1^x g(t)\,dt \sim g(x),$$

which holds under quite mild conditions, e.g. $xg(x) \to \infty$ and $xg'(x) = o(g(x))$.

REMARK. If we want to create a continuous g-number system having prime density that comes from several intervals (disjoint or not) we can convolve together factors $d_x\theta_c(x, y_i, z_i)$ associated with intervals $(z_i, y_i]$, $i = 1, 2, \ldots$.

18.6. Other relations

We conclude with brief mention of continuous analogues of some other Dickman/Buchstab relations.

BUCHSTAB IDENTITY. If $1 < y < z \leq x$, then

$$\psi(x, y) = \psi(x, z) - \sum_{y<p\leq z} \psi\Big(\frac{x}{p}, p\Big).$$

Continuous analogue:

$$\psi_c(x, y) = \psi_c(x, z) - \int_y^z \psi_c\Big(\frac{x}{t}, t\Big) \frac{dt}{\log t}.$$

HILDEBRAND IDENTITY. If $1 < y \leq x$, then

$$\int_1^x \log t \, d\psi(t, y) = \sum_{p\leq y,\ p^m\leq x} \psi\Big(\frac{x}{p^m}, y\Big) \log p.$$

Continuous analogue:

$$\int_1^x \log t \, d\psi_c(t, y) = \int_1^x \psi_c\Big(\frac{x}{t}, y\Big) (\chi_{[1,\,y]}(t) - t^{-1})\, dt \tag{18.12}$$

or, more simply,

$$\psi_c(x, y) \log x = \int_1^y \psi_c\Big(\frac{x}{t}, y\Big)\, dt.$$

In each case, proof of the continuous analogue is an easy verification using Theorem 18.1 and a form of the differential equation for ρ. Notice that (18.12) is the specialization of the Chebyshev identity (3.3) to the continuous "integer" counting function $\psi_c(x, y)$.

A CHEBYSHEV IDENTITY FOR $\phi_c(x, y)$. We have a version of the Chebyshev identity for the continuous analogue of $\phi(x, y)$ as well; it is

$$\int_1^x \log t \, d\phi_c(t, y) = \int_1^x \phi_c\Big(\frac{x}{t}, y\Big) \chi_{(y,\,\infty)}(t)\, dt. \tag{18.13}$$

A direct proof of the formula can be established by using Theorem 18.3 and showing the cumulative function of each side is

$$x\, u\, \omega(u) - y - y\int_0^{u-1} \omega(v)\, y^v\, dv, \quad x \geq y > 1,$$

with $u := \log x/\log y$. However, this approach is complicated by the discontinuity of ω and the need for manipulations. Instead, we prove the relation more simply using Mellin transforms.

Let $F_y(x) := \int_1^x \chi_{(y,\,\infty)}(t)\, dt$. We have, for $\Re s > 1$,

$$\widehat{F_y}(s) = \int_{x=1}^\infty x^{-s} \chi_{(y,\,\infty)}(x)\, dx = \frac{y^{1-s}}{s-1}$$

and, from the proof of Theorem 18.3,

$$\widehat{\phi_c}(s) = \int_{x=1-}^\infty x^{-s}\, d\phi_c(x, y) = \exp(E_1\{(s-1)\log y\}),$$

with $E_1(z) := \int_z^\infty e^{-t}\,dt/t$. We find

$$\begin{aligned} -\widehat{\phi_c}'(s) &= \int_{x=1}^{\infty} x^{-s} \log x \, d\phi_c(x,y) \\ &= \exp(E_1\{(s-1)\log y\}) \, \frac{e^{-(s-1)\log y}}{s-1} = \widehat{\phi_c}(s)\,\widehat{F_y}(s). \end{aligned}$$

By Lemma 2.6 and the Mellin identity theorem, (18.13) holds.

18.7. Notes

§18.2. There is a family of functions j_κ with $\kappa \geq 1$ that plays an important role in sieve theory. Up to a multiplicative constant, j_1' is the function ρ. The κth convolution power of $\exp d\Pi_c$ is expressible in terms of j_κ'. These functions are discussed in Chapter 14 of [**DH08**].

Bibliography

[AS64] M. Abramowitz and I. A. Stegun, *Handbook of Mathematical Functions*, National Bureau of Standards Applied Mathematics Series, 55, 1964. xiv+1046 pp. Reprinted by Dover Publications, New York, 1965.

[AMH15] F. Al-Maamori and T. Hilberdink, *An example in Beurling's theory of generalised primes*, Acta Arith. 168 (2015), 383–395. MR3352439.

[Ap74] T. M. Apostol, *Mathematical analysis, 2nd ed.*, Addison-Wesley, Reading, MA, 1974. xvii+492 pp. MR49:9123.

[Ap76] T. M. Apostol, *Introduction to analytic number theory*, Undergrad. Texts in Math., Springer-Verlag, New York-Heidelberg, 1976. xii+338 pp. MR55:7892.

[Bzd99] M. Balazard, *La version de Diamond de la méthode de l'hyperbole de Dirichlet*, Enseign. Math. (2) 45 (1999), 353–270. MR2001a:11167.

[BS00] R. G. Bartle and D. R. Sherbert, *Introduction to real analysis*, Third ed. John Wiley, New York, 2000. xii+404 pp. MR1135107.

[BD69] P. T. Bateman and H. G. Diamond, *Asymptotic distribution of Beurling's generalized prime numbers*, in *Studies in number theory,* W. J. LeVeque, ed., Math. Assoc. Amer., 1969, 152–210. MR39:4105.

[BD04] P. T. Bateman and H. G. Diamond, *Analytic number theory. An introductory course*, World Scientific, Singapore, 2004. xiv +360 pp. Reprinted, with minor changes, in series Monographs in Number Theory, vol. 1, 2009. MR2005h:11208.

[Be37] A. Beurling, *Analyse de la loi asymptotique de la distribution des nombres premiers généralisés. I*, Acta Math. 68 (1937), 255–291.

[BGT87] N. H. Bingham, C. M. Goldie, and J. L. Teugels, *Regular variation*, Encyclopedia of Mathematics and its Applications, 27. Cambridge U. Press, Cambridge, 1987. xx+491 pp. MR88i:26004.

[Ch68] K. Chandrasekharan, *Introduction to analytic number theory,* Springer–Verlag, Berlin, 1968. viii+140 pp. MR40:2593.

[Ch70] K. Chandrasekharan, *Arithmetical functions,* Springer–Verlag, Berlin, 1970. xi+231 pp. MR43:3223.

[Di69] H. G. Diamond, *The prime number theorem for Beurling's generalized numbers*, J. Number Theory, 1 (1969), 200–207. MR39:4106.

[Di70a] ———, *Asymptotic distribution of Beurling's generalized integers*, Illinois J. Math. 14 (1970), 12–28. MR40:5555.

[Di70b] ———, *A set of generalized numbers showing Beurling's theorem to be sharp*, Illinois J. Math., 14 (1970), 29–34. MR40:5556.

[Di73a] ———, *Chebyshev estimates for Beurling generalized prime numbers*, Proc. Amer. Math. Soc. 39 (1973), 503–508. MR47:3332.

[Di73b] ———, *Chebyshev type estimates in prime number theory*, Sém. de Théorie des Nombres, 1973-1974 (Univ. Bordeaux I, Talence), Exp. No. 24, 11 pp., 1974. MR52:13690.

[Di77] ———, *When do Beurling generalized integers have a density?*, J. Reine Angew. Math. 295 (1977), 22–39. MR56:8518.

[DH08] H. G. Diamond and H. Halberstam, *A higher-dimensional sieve method. With an appendix ("Procedures for computing sieve functions") by William F. Galway.* Cambridge Tracts in Mathematics, 177. Cambridge University Press, Cambridge, 2008. MR2458547 (2009h:11151).

[DMV06] H. G. Diamond, H. L. Montgomery, and U. M. A. Vorhauer, *Beurling primes with large oscillation*, Math. Ann. 334 (2006), 1–36. MR2006j:11131.

[DZ12] H. G. Diamond and W.-B. Zhang, *A PNT equivalence for Beurling numbers*, Funct. Approx. Comment. Math. 46 (2012), 225–234. MR2931668.

[DZ13a] ———, *Chebyshev Bounds for Beurling Numbers*, Acta Arith. 160 (2013), 143–157. MR3105332.

[DZ13b] ———, *Optimality of Chebyshev Bounds for Beurling Generalized Numbers*, Acta Arith. 160 (2013), 259–275. MR3106097.

[Da00] H. Davenport, *Multiplicative number theory, 3rd ed.* Revised and with a preface by Hugh L. Montgomery, Grad. Texts in Math., 74, Springer-Verlag, New York, 2000. xiv+177 pp. MR2001f:11001.

[Fr76] J. B. Friedlander, *Integers free from large and small primes*, Proc. London Math. Soc. (3) 33 (1976), 565–576. MR54:5139.

[Ha72] R. S. Hall, *The prime number theorem for generalized primes*, J. Number Theory 4 (1972), 313–320. MR0308069.

[Ha73] ———, *Beurling generalized prime number systems in which the Chebyshev inequalities fail*, Proc. Amer. Math. Soc. 40 (1973), 79–82. MR47:6634.

[Ha49] G. H. Hardy, *Divergent series,* Oxford University Press, Oxford, 1949. xvi+396 pp. MR11,25a.

[Hi12] T. Hilberdink, *Generalised prime systems with periodic integer counting function*, Acta Arith., 152 (2012), 217–241. MR2885785.

[HL06] T. Hilberdink and M. Lapidus, *Beurling zeta functions, generalised primes, and fractal membranes*, Acta Appl. Math. 94 (2006), 21–48. MR2271675 (2007k:11174).

[In32] A. E. Ingham, *The distribution of prime numbers,* Cambridge Tracts in Math. and Math. Physics, 30, Cambridge Univ. Press, Cambridge, 1932.

[Iv85] A. Ivić, *The Riemann zeta-function*, Wiley–Interscience, New York, 1985. xvi+517 pp. MR87d:11062.

[Kn68] Y. Katznelson, *An introduction to harmonic analysis, 2nd corrected ed.*, Dover Publications, Inc., New York, 1976. xiv+264 pp. MR10976.

[Ka96] J. P. Kahane, *Une formule de Fourier sur les nombres premiers*, Gazette des Mathématiciens 67 (1996), 3–9. MR97b:11113.

[Ka96b] ———, *Une formule de Fourier pour les nombres premiers*, Application aux nombres premiers généralisés de Beurling. (French) [A Fourier formula for primes. Application to Beurling generalized primes] Harmonic analysis from the Pichorides viewpoint (Anogia, 1995, Myriam Déchamps, ed.), 41–49, Publ. Math. Orsay, 96-01, Univ. Paris XI, Orsay, 1996. MR98c:11103.

[Ka97] ———, *A Fourier formula for prime numbers*, Harmonic analysis and number theory (Montreal, PQ, 1996), 89–102, CMS Conf. Proc., 21, Amer. Math. Soc., Providence, RI, 1997. MR98f:11099.

[Ka97b] ———, *Sur les nombres premiers généralisés de Beurling. Preuve d'une conjecture de Bateman et Diamond*, Journal de Théorie des Nombres de Bordeaux 9 (1997), 251–266. MR99f:11127.

[Ka98] ———, *Le rôle des algebres A de Wiener, A^∞ de Beurling et H^1 de Sobolev dans la théorie des nombres premiers généralisés de Beurling*, Ann. Inst. Fourier (Grenoble) 48 (1998), 611–648. MR99k:11152.

[Ko04] J. Korevaar, *Tauberian theory. A century of developments*, Grund. der Math. Wiss. 329, Springer-Verlag, Berlin, 2004. xvi+483 pp. MR2006e:11139.

[Ko05] ———, *Distributional Wiener-Ikehara theorem and twin primes*, Indag. Math. (N.S.) 16 (2005), no. 1, 37–49. MR2006d:11105.

[La03] E. Landau, *Neuer Beweis des Primzahlsatzes und Beweis des Primidealsatzes*, Math. Annalen 56 (1903), 645–670. Reprinted in *Collected Works Vol. 1*, Thales Verlag, Essen, 1987, 327–352.

[La09] ———, *Handbuch der Lehre von der Verteilung der Primzahlen. 2 Bände*, 2d ed. With an appendix by Paul T. Bateman. Chelsea Publishing Co., New York, 1953. xviii+pp. 1–564; ix+pp. 565–1001. MR0068565.

[Lo77] M. Loève, *Probability theory. I*, 4th ed. Graduate Texts in Mathematics 45, Springer-Verlag, New York, 1977. xvii+425 pp. MR58:31324a.

[Mo93] P. Moree, *Psixyology and Diophantine equations*, Ph.D. Dissertation, Rijksuniversiteit, Leiden, 1993. x+196 pp. MR96e:11114.

[MV07] H. L. Montgomery and R. C. Vaughan, *Multiplicative number theory. I. Classical theory*, Cambridge Studies in Advanced Mathematics, 97, Cambridge University Press, Cambridge, 2007. xviii+552 pp. MR2009b:11001.

[Na00] W. Narkiewicz, *The development of prime number theory. From Euclid to Hardy and Littlewood*, Monographs in Math. Springer-Verlag, Berlin, 2000. xii+448 pp. MR2001c:11098.

[Ny49] B. Nyman, *A general prime number theorem*, Acta Math. 81 (1949), 299–307. MR0032693.

[Ol11] R. Olofsson, *Properties of the Beurling generalized primes*, J. Number Theory 131 (2011), 45–58. MR2729208.

[Pl13] P. Pollack, *On Mertens' theorem for Beurling primes*, Canad. Math. Bull. 56 (2013), no. 4, 829–843. MR3121692.

[Po99] Ch. J. de la Vallée Poussin, *Sur la fonction $\zeta(s)$ de Riemann et le nombre des nombres premiers inférieurs à une limite donnée*, Mémoires couronnés et autres mémoires publiés par l'Académie Royale des Sciences, des Lettres et des Beaux-Arts de Belgique 59 (1899–1900), no. 1, 74 pp.

[Rv07] S. Gy. Révész, *On some extremal problems of Landau*, Serdica Math. J. 33 (2007), 125–162. MR2313797.

[Ro88] H. L. Royden, *Real analysis, 3rd ed.*, Macmillan, New York, 1988. xx+444 pp. MR90g:00004.

[Ru87] W. Rudin, *Real and complex analysis, 3rd ed.*, McGraw-Hill, New York, 1987. xiv+416 pp. MR88k:00002.

[Sa92] E. Saias, *Entiers sans grand ni petit facteur premier.I*, Acta Arith. 61 (1992), 347–374. MR93d:11096.

[Sa95] E. Saias, *Entiers sans grand ni petit facteur premier.III*, Acta Arith. 71 (1995), 351–379. MR96g:11113.

[SPV12] J. C. Schlage-Puchta and J. Vindas, *The prime number theorem for Beurling's generalized numbers. New cases*, Acta Arith. 153 (2012), 299–324. MR2912720.

[So36] S. L. Sobolev (Soboleff), *Sur quelques évaluations concernant les familles de fonctions etc.* .... C. R. Acad. Sc. URSS (Doklady)I(X), N.7 (84), 1936, p.279 et III(XII), N.1 98, 1936, p. 107.

[Te95] G. Tenenbaum, *Introduction to analytic and probabilistic number theory*, Cambridge Studies in Adv. Math., 46, Cambridge Univ. Press, Cambridge, 1995. MR97e:11005b.

[Ti39] E. C. Titchmarsh, *Theory of functions, 2nd ed.*, Oxford Univ. Press, Oxford, 1939.

[Ts40] L. Tschakaloff (Chakalov), *Trigonometrische Polynome mit einer Minimumeigenschaft*, Ann. Scuola Norm. Super. Pisa (2) 9, (1940). MR0001365.

[Vn12] J. Vindas, *Chebyshev estimates for Beurling generalized prime numbers. I.*, J. Number Theory 132 (2012), 2371–2376. MR2944760.

[Vn13] ______, *Chebyshev upper estimates for Beurling's generalized prime numbers*, Bull. Belg. Math. Soc. Simon Stevin 20 (2013), 175–180. MR3082752.

[We66] Wegmann, H., Beiträge zur Zahlentheorie auf freien Halbgruppen. II, J. Reine Angew. Math. 221 (1966), 150–159, MR32:4098. (The result of Wegmann and the reference to Wirsing are not correctly stated in the review.)

[Zh86] W.-B. Zhang, *Asymptotic distribution of Beurling's generalized prime numbers and integers*, Ph.D. Thesis, University of Illinois, Urbana, 1986.

[Zh87a] ______, *Chebyshev type estimates for Beurling generalized prime numbers*, Proc. Amer. Math. Soc. 101 (1987), 205–212. MR88m:11081.

[Zh87b] ______, *A generalization of Halsz's theorem to Beurling's generalized integers and its application*, Illinois J. Math. 31 (1987), 645–664. MR89a:11102.

[Zh88] ______, *Density and O-density of Beurling generalized integers*, J. Number Theory 30 (1988), 120–139. MR90a:11111.

[Zh07] ______, *Beurling primes with RH and Beurling primes with large oscillation*, Math. Ann. 337 (2007), 671–704. MR2007k:11148.

[Zh14] ______, *Wiener-Ikehara theorems and the Beurling generalized primes*, Monatsh. Math. 174 (2014), 627–652. MR3233115.

[Zh15] ______, *Extensions of Beurling's prime number theorem*, Int. J. Number Theory 11 (2015), 1589–1616. MR3398753.

Index